CLASSICS IN RADIO ASTRONOMY

STUDIES IN THE HISTORY
OF MODERN SCIENCE

VOLUME 10

Karl Guthe Jansky (1905–1950) and his "merry-go-round" antenna at Bell Telephone Laboratories in Holmdel, New Jersey, U.S.A. (1933).

CLASSICS IN RADIO ASTRONOMY

Selection and commentary by

WOODRUFF TURNER SULLIVAN, III
Department of Astronomy, University of Washington, Seattle

D. REIDEL PUBLISHING COMPANY

DORDRECHT : HOLLAND / BOSTON : U.S.A.

LONDON : ENGLAND

Library of Congress Cataloging in Publication Data
Main entry under title:

Classics in radio astronomy.

(Studies in the history of modern science; v. 10)
Bibliography: p.
Includes indexes
1. Radio astronomy. I. Sullivan, Woodruff Turner. II. Series
QB475.C55 522'.682 81-23476

ISBN-13: 978-94-009-7754-9 e-ISBN-13: 978-94-009-7752-5
DOI:10.1007/978-94-009-7752-5

Published by D. Reidel Publishing Company,
P.O. Box 17, 3300 AA Dordrecht, Holland.

Sold and distributed in the U.S.A. and Canada
by Kluwer Boston Inc.,
190 Old Derby Street, Hingham, MA 02043, U.S.A.

In all other countries, sold and distributed
by Kluwer Academic Publishers Group,
P.O. Box 322, 3300 AH Dordrecht, Holland.

D. Reidel Publishing Company is a member of the Kluwer Group.

TABLE OF CONTENTS

ACKNOWLEDGMENTS

Permission to reprint articles is gratefully acknowledged as follows:

Institute of Electrical and Electronic Engineers, New York (Papers 1, 2, 3, 4, and 7); University of Chicago Press (Papers 5, 6, 8, 22, and 29); American Institute of Physics, New York (Papers 9, 11, and 27); The Royal Society, London (Papers 12 and 19); Macmillan Journals Ltd., London (Papers 16, 18, 20, 21, 23, 24, 25, 26, 28, 30, 31, 32, 33, 36, and 37); Johann Ambrosius Barth Publishers, Leipzig (Paper 13); Gauthier-Villars, Paris (Papers 14 and 15); Franklin Institute, Philadelphia (Paper 17), and the Foundation for Basic Research on Matter, Utrecht (Paper 34). In addition I thank David Jansky for the use of the Frontispiece photograph and the Royal Astronomical Society of Canada for Figure 1. Finally, my deepest gratitude is, of course, to the authors (and translators) of the articles themselves.

The old dream of wireless communication through space has now been realized in an entirely different manner than many had expected. The cosmic shortwaves bring us neither the stockmarket nor jazz from distant worlds. With soft noises they rather tell the physicist of the endless love play between electrons and protons.

A. UNSÖLD (1946)

PREFACE

Radio techniques were the first to lead astronomy away from the quiescent and limited Universe revealed by traditional observations at optical wavelengths. In the earliest days of radio astronomy, a handful of radiophysicists and engineers made one startling discovery after another as they opened up the radio sky. With this collection of classic papers and the extensive introductory material, the reader can experience these exciting discoveries, as well as understand the developing techniques and follow the motivations which prompted the various lines of inquiry. For instance he or she will follow in detail the several attempts to detect radio waves from the sun at the turn of the century; the unravelling by Jansky of a "steady hiss type static"; the incredible story of Reber who built a 9 meter dish in his backyard in 1937 and then mapped the Milky Way; the vital discoveries by Hey and colleagues of radio bursts from the Sun and of a discrete source in the constellation of Cygnus; the development of receivers and interferometry in the post-war years by the groups led by Ryle in Cambridge and Pawsey in Sydney; the first measurements and exciting identifications of Taurus A (the Crab Nebula), Centaurus A, Virgo A, Cassiopeia A, and Cygnus A, the last opening the field of radio cosmology; the early development of synchroton theory; and the prediction and discovery seven years later of the 21 cm line of neutral hydrogen.

My aim in this volume has simply been to make these classics more widely accessible and more appreciated in their scientific and historical context. Webster defines *classic* as "of the highest quality or rank; having recognized and permanent value; of enduring interest and appeal". The papers included in this volume have indeed been primarily selected on precisely these criteria as applied to journal articles in the field of radio astronomy over the period 1896 to 1954. Practitioners as well as students of astronomy, physics, radio science and engineering, and history of science should find the collection of value. Even today's radio astronomer, immersed in the latest issue of *Nature* or *Astrophysical Journal Letters*, often has little knowledge of how his own science has developed. While some may occasionally pay homage to these classics in their own modern publications, I suspect that very few have ever actually sat down and *read* the papers to which they refer. In fact, too often a classic paper might be operationally defined as one whose contribution is generally judged to be so important, so historic and so seminal, that one somehow knows the contents without the bother of reading it for himself!

The collection is devoted solely to those papers whose main thrust is concerned with observations of natural radio emission from extraterrestrial sources. Papers dealing with radar astronomy have not been included; this in particular eliminates both lunar and meteor radar studies. With this exception, the choice of papers runs the entire gamut of observational and theoretical radio astronomical investigations up until the early 1950s. This cutoff date is rather arbitrary; in fact, an earlier version of this book (prepared in 1975) included 54 papers and extended until 1958, but as a result of the present-day economics of publishing, this version was never able to see the light of day. Note also that different research fields within radio astronomy are represented by papers from somewhat differing periods; this has been done in order to achieve a reasonable balance and to emphasize the pioneering papers. For example, the last included paper concerning solar observations was published in 1947, that for spectral line observations and theory in 1951, and that for radio sources in 1954.

The choice of papers was long and carefully considered. Papers which are primarily summaries of research (as often found in the proceedings of meetings) or reviews are not eligible for inclusion. The most important criterion was that the candidate paper should be historically and scientifically significant, i.e., that it represent a quantum leap forward in the development of radio astronomy. Ideally, papers should represent more than a simple report of a detection and should pay attention to astronomical interpretation of the radio observations. Finally, I have given weight not only to the paper as it stands alone, but also to how well it fits into the developmental sequence of its field.

Each section and paper is preceded by introductory remarks in which I have summarized the importance of the following papers to both history and science, explaining how they came about and where they fit into their contemporary science and into ours. Taken as a whole, this text effectively traces the highlights of the development of the field, but it should be understood that it by no means represents a complete history. In a similar vein the References, Name Index, and Subject Index have been created solely for the purpose of their use in connection with this volume; they are not intended to be complete in any sense.

I have felt it important, for the sake of accuracy and faithfulness to the originals, to reprint all papers in their entirety and not to re-draw figures; in fact, most articles are presented as photographic reprints of the original journal pages, giving the reader a better historical sense. Six papers appear in full translation for the first time. The only modifications are new notes inserted where it is thought that they lend to the reader's understanding without substantially detracting from the flow of the original paper.

An additional criterion – unfortunate, but necessary – was the length of the candidate papers. The total collection was required to be of a reasonable length and yet in some sense remain complete. When the choice was between two papers of equal merit, the shorter has been included. The inclusion of each of the few longer articles was deemed compelling. An incomplete list of papers which would have been included if they had been shorter includes (see References): Reber (1942), Ryle and Vonberg (1948), Stanley and Slee (1950), Ryle, Smith and Elsmore (1950), Mills (1952a) and Shklovsky (1952a).

While I have attempted to apply my criteria as uniformly as possible, it is realized that choices cannot help but be somewhat subjective. My guess is that another person making a similar selection of 37 papers would probably agree on ~70% of the selection and would want a different "other 30%". It is nevertheless not obvious *which* 70% would be common to the two selections. It can thus be appreciated that absence of a person's name or of a particular article proves little except the necessary subjectivity and tight constraints inherent in the selection process for such a volume. It should also be emphasized that it is relatively easy for a young whippersnapper in the 1980s to sit back and correct mistakes, criticize, and, in general, kibitz from the vantage point of "20-20 hindsight". Where I have done this, it has always been my goal to shed light on the historical development as I see it; in no way do I intend to detract from the foresight, creativity, ingenuity, and plain hard work of the scientists and engineers involved.

This volume can be profitably studied along side of Lang and Gingerich's recent *Source Book* (1979) of reprints concerning twentieth-century astronomy of all kinds. In fact, ten of the selections here are also found in that volume, but usually in abridged form and with less extensive commentary. That volume is recommended to provide a broader context for the development of radio astronomy, as well as touching on some of the further highlights of the field, extending as far as the pulsar era in 1968.

An endeavor of this sort is never possible without help from many others. Pat Kimpton and Helga Byhre have been indispensable in the preparation of the manuscript at my home base. I have also received considerable assistance from the Kapteyn Institute in Groningen, the Institute of Astronomy at Cambridge University, and the Radiophysics Division of CSIRO near Sydney. Barbara Sullivan has aided my research in many ways, both in substance and in spirit. Gerald Holton and Robert S. Cohen have been very helpful in bringing this volume into existence. Advice concerning the selection has been generously given by David O. Edge, James S. Hey, Frank J. Kerr, and Gart Westerhout. I have been aided with various aspects of the translations by Karl-Heinz Böhm, Tjeerd de Graaf, Thomas L. Hankins, Joke Huizinga, and Robert A. Schommer. Furthermore, detailed criticisms of the manuscript by Edge, Kerr and Schommer have improved its accuracy and style.

I also want to thank the many people around the world who have over the past decade discussed the development of radio astronomy with me. Most of the material which I have gathered from them will be incorporated into a future monograph on the worldwide history and development of radio and radar astronomy. Finally, I urge contact from anyone who wishes either to correct or to add to what I have written, or to inform concerning any aspect of the history of the field.

February 1981

W. T. SULLIVAN III
Department of Astronomy
University of Washington
Seattle, Washington 98195
U.S.A.

KEY

The book is divided into five Parts by topic; within each Part papers are, in general, chronologically arranged except in a few instances where it was thought more logical otherwise. Parts and sections are preceded by introductory remarks. Editor's notes and amendments within the original articles are indicated by a ★ in the text and a bold number in the outside margin; these notes are then gathered together immediately following each paper. Editorial remarks within the translated articles are enclosed in brackets [ ].

The papers in this volume and all articles referred to in the editor's remarks and notes are listed (by year) in the References section. The Name Index includes names mentioned in the editor's remarks, as well as the authors of the papers in this volume. On the other hand, the Subject Index is more extensive and refers to the actual contents of the original papers as well as to the editor's remarks.

INTRODUCTION

Shortly after Heinrich Hertz's laboratory experiments in 1887 conclusively demonstrated the existence of the radio waves predicted by James Clerk Maxwell's theory of electromagnetic radiation, unsuccessful attempts to detect radio waves emitted from the sun were made in England, Germany, and France. Yet notwithstanding the tremendous growth of radio technology, there seem to have been no further searches for extraterrestrial radio radiation during the period 1902–1932. The first discovery (in 1932) in fact was accidental; it was made by an American radio engineer named Karl Jansky who was studying the nature of the interference occurring on the recently inaugurated trans-Atlantic radio telephone service. Despite his recognition of the importance of his discovery and many attempts to interest radio engineers, astronomers, and physicists, Jansky's work was not appreciated. The radio engineers and astronomers simply did not know enough of each other's discipline to realize the potential of this discovery. In that situation it might well have fallen to the physicists to forge the link, but perhaps they were too preoccupied with the many fundamental discoveries being made in their own bailwick at that time. 1932 has, in fact, been called the *annus mirabilis* of nuclear and particle physics: witness Harold Urey and deuterium, James Chadwick and the neutron, Carl Anderson and the positron, and John Cockcroft and E. T. S. Walton and the first disintegration of nuclei, all discoveries taking place in the same year as Jansky's recording of a "steady hiss type static of unknown origin".

In the ensuing decade, only one man, Grote Reber, recognized the importance of this extraterrestrial radiation and carried out extensive investigations as best he could with limited resources. Not until the accelerated research of World War II did Jansky's discovery become more than an idle curiosity. The intense wartime development of the techniques of radar and of radio communication led to a situation afterwards in which many physicists and engineers, highly trained in radio techniques and looking for suitable fields of research, finally began to open up the domain of extraterrestrial radio emission.

Radio technology opened up approximately ten new octaves of the electromagnetic spectrum to astronomical scrutiny. It soon became apparent

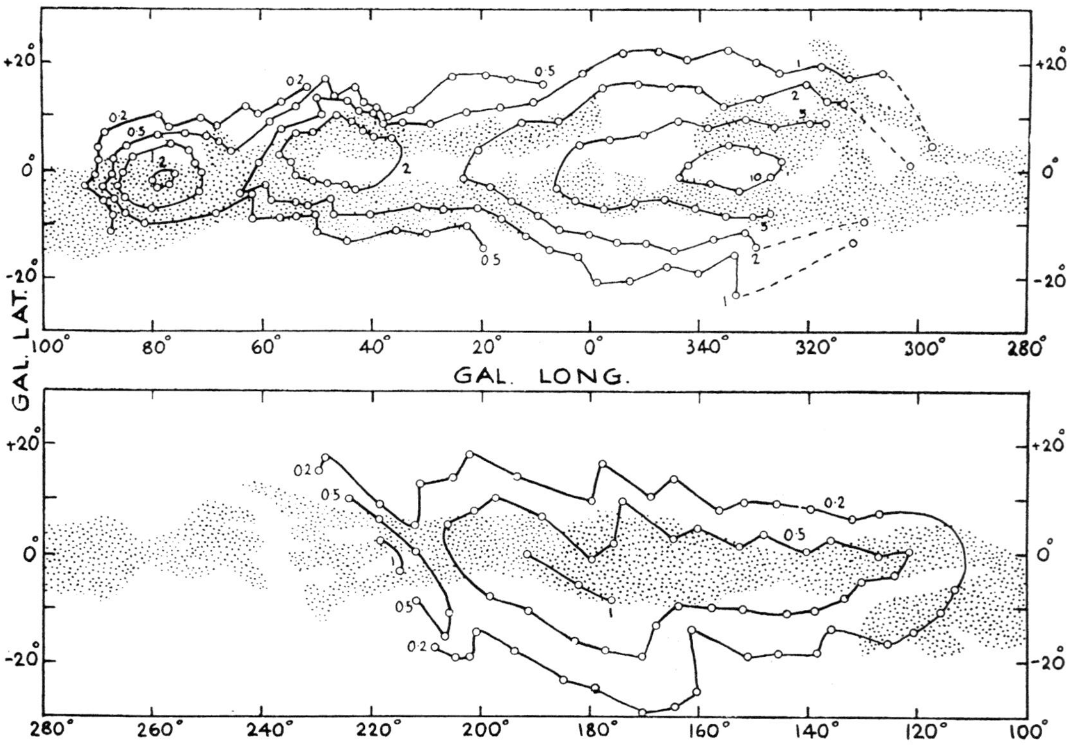

Fig. 1. Radio brightness contours at a wavelength of 1.9 m from Reber (1944; Paper 6) superimposed on the outlines of the Milky Way as given in Norton's Star Atlas (Northcott and Williamson, 1948).

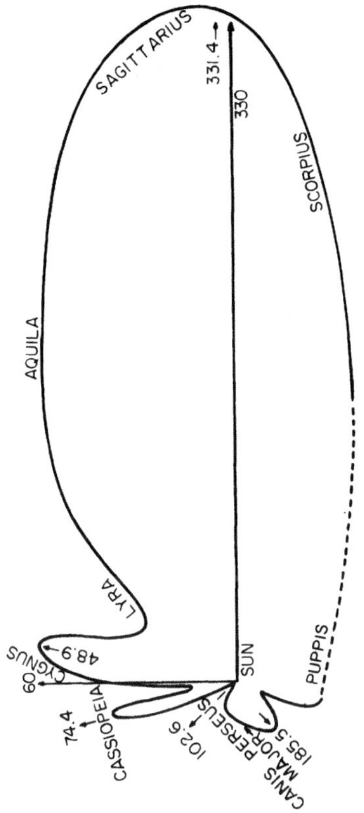

Fig. 2. A representation of Reber's 1.9-m survey as given by him to an engineering audience (Reber, 1948b). The strength of radio emission as a function of longitude in the galactic plane is indicated as a "galactic power pattern". Galactic longitudes and constellations are also given.

that the information gathered from radio observations of both the usual astronomical objects as well as entire new classes of objects called "radio sources" was extremely useful. Yet despite the excitement abounding in radio astronomy, it was not until the late 1950s that the discipline was fully accepted by the astronomical establishment. A couple exceptions only prove the rule – for example, the fact that the first thorough review of radio astronomy was co-authored by an optical astronomer (Reber and Greenstein 1947). Otto Struve (1897–1963) was another astronomer who recognized the importance of these radio data, but even he laid his prejudices bare in a 1949 review when he referred on the one hand to "astronomers" and on the other to "radio technicians". A graphic example of the communication problem between the two camps in the late 1940s is given by two representations of Reber's 1944 survey of the galactic plane at 160 MHz (Paper 6). In Figure 1, Northcott and Williamson (1948) appealed to the optical astronomers by sketching in the features of the optical Milky Way for comparison with Reber's map, now plotted in galactic coordinates. Meanwhile, Reber (1948b) was transforming precisely the same survey data into an equivalent "antenna pattern of the galaxy" (Figure 2) for the edification of radio engineers!

The papers in this collection bear testimony that startling observational discoveries have not been infrequent in the history of radio astronomy. Another way to state this is that, with the notable exception of the field of spectroscopy, the radio observers have, in general, been several steps ahead of the theorists. This situation arose as the engineers and radiophysicists methodically improved their techniques and opened new vistas with ever increasing sensitivity and better angular resolution. Greater sensitivity was aided by low-noise receivers and larger collecting areas while improved resolution was achieved with the techniques of interferometry and the quest for shorter wavelengths. These trends apply equally to both the era covered by this volume as well as subsequent years. The available technology has always strongly influenced the direction of the observations, which in turn blaze the path for our understanding of the astronomy. An idea of the incredible progress in radio astronomy since the early 1950s can be gained by noting that the papers included in this collection almost all concern observations taken under the following constraints: wavelength $\geqslant 21$ cm, detectable flux $\gtrsim 20$ Jy,[1] and resolution $\gtrsim 1$ arcmin.

NOTE

[1] Jy = Jansky $\equiv 10^{-26}$ W m^{-2} Hz^{-1} .

PART I

GALACTIC BACKGROUND RADIATION

We begin with papers concerned with the background continuum radiation associated with the Milky Way. This was first to be detected simply because it usually yields by far the most intense signal when observing at decameter wavelengths with an antenna of low angular resolution. We say "usually" because it turns out that Jansky could easily have detected a strong solar radio burst such as are common during periods of maximum activity in the 11-year solar cycle. But 1931–32 was very near solar minimum and thus, as has happened more than once, the solar cycle was influential in the course taken by radio astronomy. The low level of solar activity also reduced ionospheric effects at that time and thus greatly aided Jansky's ability to distinguish extraterrestrial radiation from other effects.

Jansky's pioneering work at 20.5 MHz (Papers 1–3) was complete enough to establish, right from the start, almost all of the basic empirical features of the background radiation, namely (a) the concentration to the plane of the Milky Way, (b) the maximum in the direction of the galactic center, (c) the relatively high values at high galactic latitudes, and (d) the fact that its intensity was far above that expected to arise from thermal radiation of interstellar material. Reber's later observations at higher frequencies (Papers 4–7) added the most puzzling item of all, namely the sharply non-thermal spectrum. It took almost two decades to resolve this through the concept of cosmic electrons spiraling in a general galactic magnetic field and emitting so-called synchrotron radiation (Papers 9 and 10).

Several other early surveys of background radiation are not included in this selection, but deserve mention. These include post-war studies by: (1) Moxon (1946) who reported observations of the Milky Way taken at 40, 90, and 200 MHz, both on shore and on Royal Navy ships, some of which covered the Southern Hemisphere, (2) Sander (1946) who studied the background intensity at 60 MHz as a function of position, and (3) Herbstreit and Johler (1948) who measured the background at 25 and 110 MHz. In addition Hey, Parsons, and Phillips (1946, 1948) mapped the entire northern sky at 64 MHz using a modified British radar antenna and receiver, finding results similar to those reported by Reber in Paper 6. Also of interest are the detections, but without further study, reported at 9.5–18.6 MHz by Friis and Feldman (1937) (working in the same group as Jansky at Bell Laboratories), and by Fränz (1942) who, working in wartime Germany, measured at 30 MHz a similar brightness to Jansky's for the galactic center region. A further study, albeit unpublished, was that in 1936 by G. W. Potapenko (1895–1979) and D. F. Folland of the California Institute of Technology. Using simple apparatus in the Mojave Desert and elsewhere, they were able to confirm the basic results of Jansky's work.

On the theoretical side we have had to be more selective. In 1935, R. M. Langer, also at Cal Tech, gave a paper at the Berkeley meeting of the American Physical Society in which he attempted to account for Jansky's observations by means of low frequency radiation, supposedly emitted when electrons recombined with many-times ionized dust particles. The first detailed theoretical paper devoted to radio astronomical measurements, however, was that by Whipple and Greenstein (1937). Although they were the only astronomers at that time who had the insight to recognize Jansky's observations as being of importance to an understanding of the interstellar medium, they took an unfruitful tack in their proposed origin for the "interstellar radio disturbances". Whipple and Greenstein proposed that the radio waves might be due to the black-body radiation emitted by dust grains at a temperature of $\lesssim 30$ K, but they themselves then showed that the resultant intensity "fails by a factor of 10^4 to account for the observations". The only other theoretical papers before 1945 were by H. C. van de Hulst of Leiden and by A. Unsöld of Kiel, both written in 1944 but unable to be published until after the cessation of hostilities. The remarkable story of van de Hulst's work is told in the introduction to Paper 34. Unsöld's paper (1946) was largely a review of Jansky's and Reber's observations and a further detailed confirmation of the free-free theory worked out by Henyey and Keenan (1940) in Paper 8. In addition he made a few interesting comments about the possibility of putting limits on the density of any intergalactic gas, as well as using shortwaves to investigate a diverse group of objects including the sun, the galactic center region, and the recently-proposed Strömgren spheres of ionized hydrogen.

After the war many workers attacked the theoretical problem, usually using models of free-free thermal radiation, which tended to give the wrong spectrum, or supposing the observed radiation to be due to the integrated effects of many hypothetical "radio stars", which tended to yield the wrong spatial distribution. The modern era began in 1950 with Kiepenheuer's and

Alfvén and Herlofson's suggestion of synchroton radiation from cosmic electrons (Papers 9 and 27). This was soon followed by extensive work in the Soviet Union, led by Ginzburg (Paper 10) and Shklovsky, which demonstrated the beauty and utility of this mechanism which is still today a basic tenet of the interpretation of a wide variety of astronomical phenomena.

SECTION A

THE BEGINNING OF RADIO ASTRONOMY

The following three papers represent virtually all of the work relevant to radio astronomy which the pioneer Karl Guthe Jansky (1905–50) was able to pursue. After obtaining his Bachelor of Science in Physics at the University of Wisconsin in 1927, he began working at the Bell Telephone Laboratories in northern New Jersey. His first assignment was to study the nature of the static affecting the recently inaugurated trans-Atlantic short-wave radio telephone circuits. The nature of the static at "long waves" (frequencies less than 1 MHz) was by that time well studied, but the technology of the short-wave region (2–20 MHz) was just opening up. Thus, Jansky spent three years carefully designing and building a sensitive 20.5 MHz receiver and rotable antenna which he called "the merry-go-round". Finally, in the summer of 1931, he took his first measurements on the "atmospherics", i.e., radio noise effects created by electrical discharge in the atmosphere.

Paper 1 presents the results of his first half-year of observations. He is able to divide the detected static into three groups, the first two of which are associated with nearby and distant thunderstorms. It is the third group, however, which is the most puzzling and which he describes in his abstract as "a steady hiss type static of unknown origin". Later in the paper he states that "it can hardly be distinguished from the hiss caused by set noise", thus at once pointing out the outstanding audio characteristic of this emission, which was destined to become known as "extraterrestrial radio *noise*". Now it happened that the Sun, in its annual movement along the ecliptic, was in exactly the same region of the sky as the galactic center during the months of December and January when Jansky first recognized this hiss-type static. Thus is he led astray in his brief comments in the first paper on the possible origins of the static, suggesting that it might come directly from the Sun, but more likely from the subsolar point in the atmosphere. He naturally tended to think along these lines for this would then make the static of Group 3 another type of atmospheric, albeit a strange one.

A convincing argument can be made that Paper 2 should really be considered as Jansky's first paper in radio astronomy, but Paper 1 has also been included in order to give a better feeling for the milieu in which radio astronomy was born. In Paper 2, Jansky has now traced the changes in solar arrival time of the hiss-type static over a full year. He conclusively establishes that it originates in a source (the first use of this term) "fixed in space" and not associated with the Sun. He obviously has learned a great deal about astronomy since Paper 1 (in particular, from a close friend, A. M. Skellett, who was both a radio engineer and an astronomer) as he lucidly describes the fundamentals of astronomical coordinates and time-keeping to his fellow engineer-readers. He determines the position of the source to be $\alpha = 18^h00^m \pm 30^m$, $\delta = -10^\circ \pm 30^\circ$, a position which is nicely consistent with the position of the galactic center, namely $\alpha = 17^h40^m$, $\delta = -29^\circ$. The main source of confusion in this paper comes about because Jansky does not recognize that, although the direction of arrival of the *strongest* signal does indeed correspond to one small region of the sky, other emission from all along the plane of the Milky Way is also being detected. This effect is the origin of his problems when in Figure 9 he attempts to fit a single value of declination to his observations. Furthermore, Jansky can only resolve the paradox of a single source at southern declinations which is yet somehow visible for the entire day by invoking ionospheric refraction, but this would have had to have been sufficient to cause the source to be detectable, even when it is 50° below the horizon!

Jansky has not only learned some classical astronomy, but it appears that he has also boned up on modern astrophysics. For example, in Paper 2 he now mentions the possibility of "primary rays" (cosmic rays?) causing secondary radiations in the Earth's atmosphere. In addition, he discusses whether or not there might be *many* cosmic sources, as opposed to a single source. Finally, he suggests two phenomena possibly associated with the apparent direction of arrival of the waves: the center of the galaxy and the solar apex.

In his last astronomical paper two years later, Jansky takes a closer look at the same set of data for the year 1932. Here he establishes that, in fact, "these radiations are received any time the antenna system is directed towards some part of the Milky Way System, the greatest response being obtained when the antenna points towards the center of the system". A new interpretation put forward is the possibility of thermal radiation arising either in an interstellar plasma or in the stars themselves. He gives no quantitative analysis, but already he recognizes the paradox inherent in any explanation for the intense galactic background emission in terms of unresolved stars in the face of the fact that the nearest star, the Sun itself, is not detectable.

Not included in this selection are the following papers by Jansky:

(1) *Nature* (1933b), a one-column note with results intermediate to those of Papers 1 and 2,

(2) *Popular Astronomy* (1933c), a paper aimed towards astronomers, both amateur and professional, with results closer to those of Paper 3 than Paper 2, although submitted only three months after Paper 2. In particular, he shows (a) why $\delta = -20^\circ$ is a better estimate of the source declination than the value $\delta = -10^\circ$ given in Paper 2, and (b) that a disk-like distribution of sources similar to the Milky Way would give a curve similar to that observed.

(3) *Proc. I.R.E.* (1937), wherein he shows that at the frequencies 9.3, 18.0, and 21.4 MHz "interstellar noise always sets the lower noise limit" for a sensitive receiver when man-made interference is not present. Useful astronomical data include *absolute* measurements of galactic radiation power (obtained from drift scans with various antennas) and an attempt to measure the spectrum of the background radiation. These measurements were considerably hindered by the large daytime absorption caused by the dense ionosphere at that time of solar maximum.

After this time, Jansky moved on to other research in radio engineering and the first important chapter in the history of radio astronomy came to a close.

It is of interest to mention here the capabilities of Jansky's receiver and first antenna. We estimate his effective half-power beam width to have been 24° in azimuth by 36° in elevation, with maximum gain at an elevation of 24°. Tying in his 1937 absolute measurements with the complete sky coverage of relative intensity in 1932, one can determine that the maximum brightness temperature T_b[1] which he observed at 20.5 MHz was ~100 000 K (to within 30%), a value agreeing with more recent observations. The data contained in Figures 1 and 2 of Paper 3 can then be incorporated into a contour map of T_b in galactic coordinates (Sullivan, 1978). This map is shown in Figure 3; its outstanding features have already been summarized in the introduction to this section. The detectability of strong decameter sources other than the galactic center region can also be calculated. One finds that Cas A and Cyg A, with respective values T_b ~30 000 K and ~15 000 K, should be definitely influencing the shapes of Jansky's contours in Figure 3 (at longitudes in the galactic plane of 112° and 76°), but that Jupiter and the

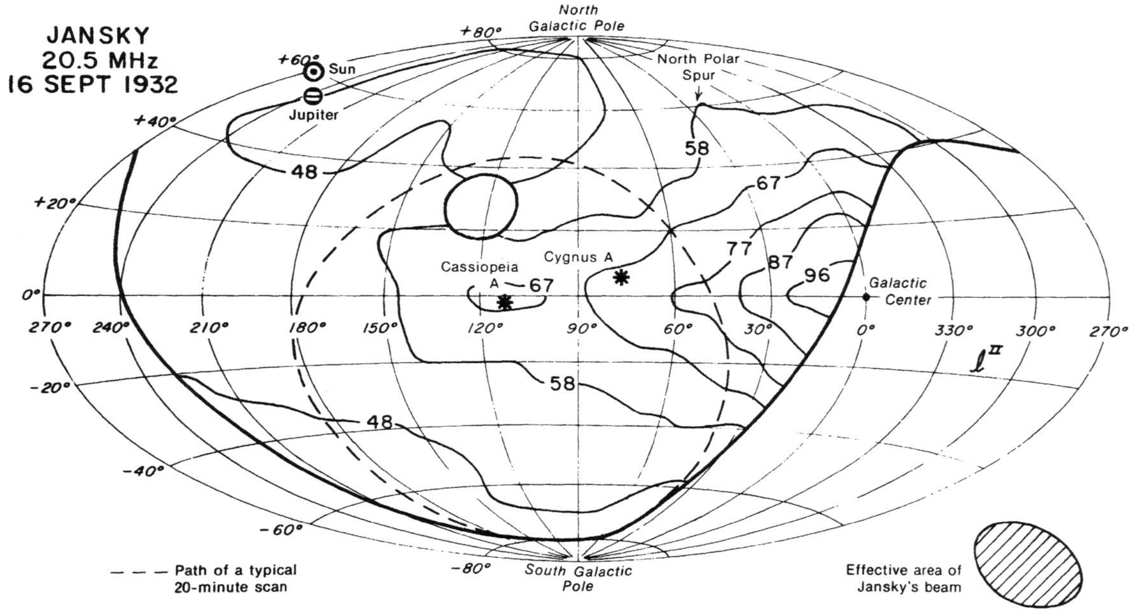

Fig. 3. Radio brightness temperature (in Kelvins) derived (Sullivan, 1978) from the drift scans at a wavelength of 14.6 m given by Jansky in 1935 (Paper 3). Coordinates are *modern* galactic, i.e., $l^{II} = 0°$ corresponds to the galactic center. The approximate size of Jansky's primary beam is also indicated.

Sun, while in a quiescent state, had little influence when they were in the beam. On the other hand, if either Jupiter or the Sun had exhibited strong bursts when in the beam, these would have been easily detectable, although perhaps not recognizable for what they were.

NOTE

[1] *Brightness temperature* is a convenient way to represent a specific intensity at radio wavelengths. The two quantities are defined to be lineary proportional through the Rayleigh–Jeans relation, whether or not the radiation is, in fact, thermal in character.

DIRECTIONAL STUDIES OF ATMOSPHERICS AT HIGH FREQUENCIES*

By

KARL G. JANSKY

(Bell Telephone Laboratories, New York City)

Summary—*A system for recording the direction of arrival and intensity of static on short waves is described. The system consists of a rotating directional antenna array, a double detection receiver and an energy operated automatic recorder. The operation of the system is such that the output of the receiver is kept constant regardless of the intensity of the static.*

Data obtained with this system show the presence of three separate groups of static: Group 1, static from local thunderstorms; Group 2, static from distant thunderstorms, and Group 3, a steady hiss type static of unknown origin.

Curves are given showing the direction of arrival and intensity of static of the first group plotted against time of day and for several different thunderstorms.

Static of the second group was found to correspond to that on long waves in the direction of arrival and is heard only when the long wave static is very strong. The static of this group comes most of the time from directions lying between southeast and southwest as does the long wave static.

Curves are given showing the direction of arrival of static of group three plotted against time of day. The direction varies gradually throughout the day going almost completely around the compass in 24 hours. The evidence indicates that the source of this static is somehow associated with the sun.

INTRODUCTION

FOR some time various investigators have made records of one type or another of the direction of arrival of static on the long wavelengths. Watson Watt has made a comprehensive study of the direction of arrival of static in England. Others working under him have used apparatus similar to his in Australia and Africa. Captain Bureau has done considerable work on the study of static in France. In this country, L. W. Austin with E. B. Judson working with him has worked on the long-wave static problem. Harper and Dean, also of this country, have made a study of the direction of arrival of long-wave static in Maine. Very little work, however, has been done on the direction of arrival of short and very short-wave static with the exception of the series of observations made by Mr. Potter as described in his paper on short-wave noise.[1]

* Decimal classification: R114. Original manuscript received by the Institute, May 26, 1932. Presented at the meeting of the American Section of the U.R.S.I. at Washington, D. C., April 29, 1932.

[1] R. K. Potter, "High-frequency atmospheric noise," PROC. I.R.E., vol. 19, p. 1731; October, (1931).

Reprinted, with permission, from the 'Proceedings of the Institute of Radio Engineers' **20**, 1920–1932 (1932).

Description of Apparatus

Since the middle of August, 1931, records have been taken at
Holmdel, N. J., of the direction of arrival and the intensity of static
on 14.6 meters.* Fig. 1 shows a schematic diagram of the recording sys- 1
tem.* It consists of a rotating antenna array, a short-wave measuring 2
set, and a Leeds and Northrup temperature recorder revamped to re-
cord field strengths.[2]

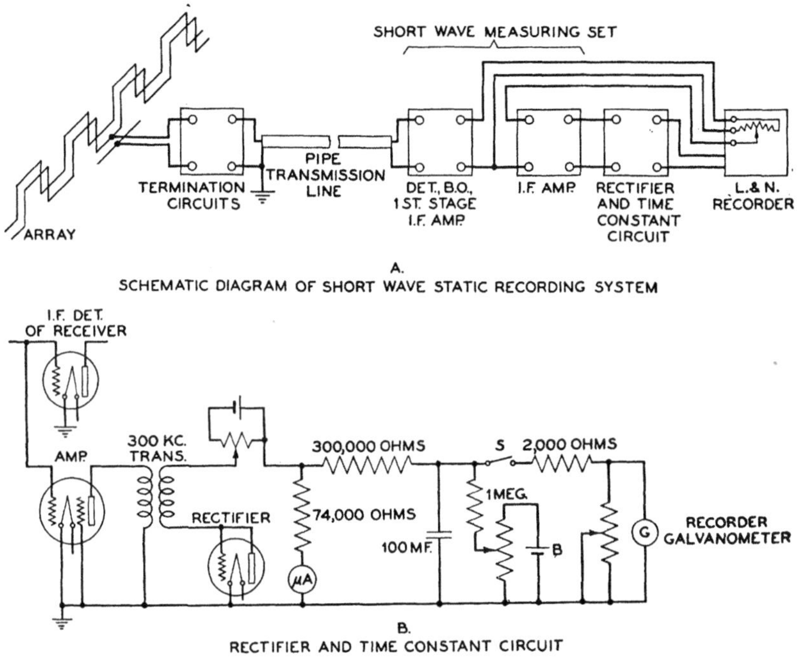

Fig. 1

The rotating antenna, a photograph of which is shown in Fig. 2, is a Bruce type broadside receiving array[3] two wavelengths long made of 3/4-inch brass pipe. The array was designed to operate on a wavelength of 14.5 meters. As shown in the photograph it is mounted on a wooden framework which in turn is mounted on a set of four wheels and a central pivot. The structure is connected by a chain drive to a small synchronous motor geared down so that the array makes a complete rotation once every twenty minutes.

[2] A detailed description of the measuring set and recorder is given in a paper by W. W. Mutch, Proceedings, this issue, pp. 1914–1919.

[3] A. A. Oswald, "Transoceanic telephone service, short wave equipment," *Jour. A.I.E.E.*, vol. 49, p. 267; April, (1930).* 3

Since static on short waves is extremely weak most of the time, the recording system had to be made very sensitive; so sensitive in fact that the first circuit noise of the receiver is recorded.[4] On account of interference which was found on 14.5 meters, it was necessary to operate the system on a wavelength of 14.6 meters. This, however, made

Fig. 2—Short-wave rotating antenna array.

little difference in the directivity of the array, the directional characteristic of which at this wavelength is shown in Fig. 3.

The array termination equipment is housed in a box mounted on the array and is connected to the measuring set in a small house about 275 feet away by means of a 3/8-inch copper concentric pipe transmis-

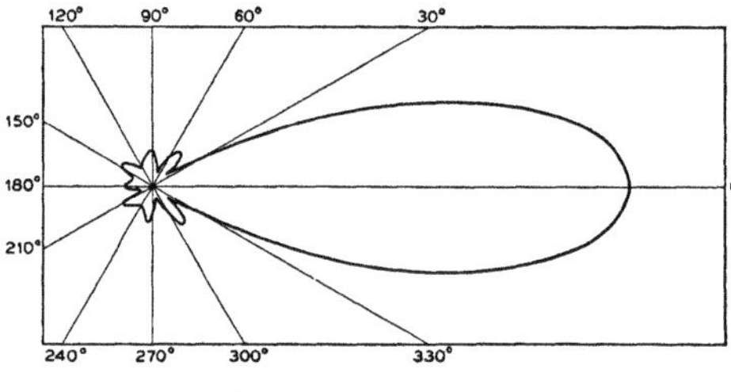

Fig. 3—Directional characteristic of array at 14.6 meters.

sion line buried about 6 inches in the ground. Fig. 4 is a schematic diagram of the array, the termination equipment and the copper pipe transmission line.

Fig. 5 is a photograph of the inside of the house, showing two receivers with their associated recorders. The apparatus on the right

[4] F. B. Llewellyn, "A study of noise in vacuum tubes and attached circuits," Proc. I.R.E., vol. 18, p. 243; February, (1930).

is the short-wave recording system. That on the left is a long-wave recording system, the records of which were used to compare with those

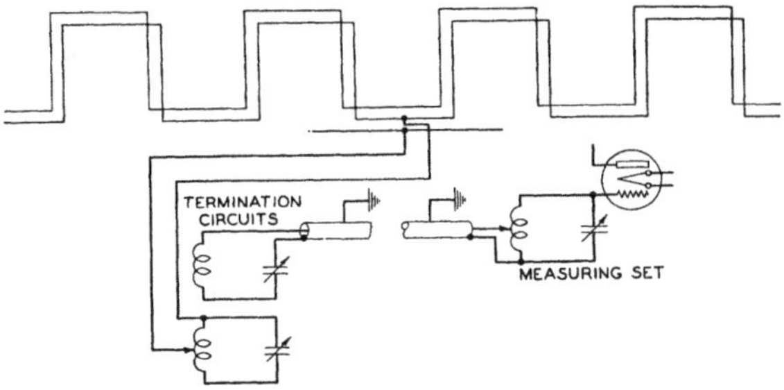

Fig. 4—Schematic diagram of array, termination, and pipe transmission line.

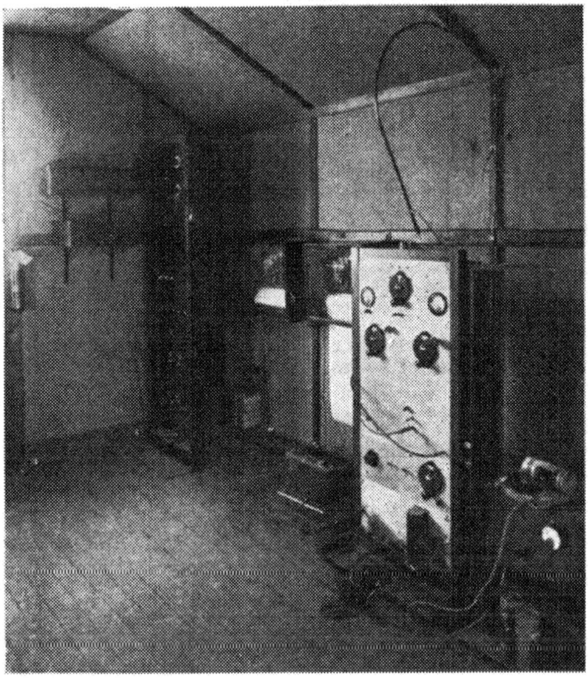

Fig. 5—Long- and short-wave static recording systems.

of the short-wave system. The long-wave antenna system consists of a rotating loop and an L type stationary antenna giving the familiar cardioid-shaped directional characteristic.

The receiver used is a short-wave field strength measuring set of the double detection type which was described some time ago.[5]

The output of this receiver is connected through a circuit with a long time constant to the Leeds and Northrup recorder the operation of which is discontinuous.[2] It automatically changes the gain of the receiver at the end of 10-second intervals in such a way that the output of the receiver is kept constant. The gain is changed by means of a noninductive potentiometer inserted in the intermediate frequency amplifier. This potentiometer replaces the slide wire found on the standard temperature recorders. The pen makes a continuous record of the position of the potentiometer arm and this record can be calibrated to give the field strength directly.

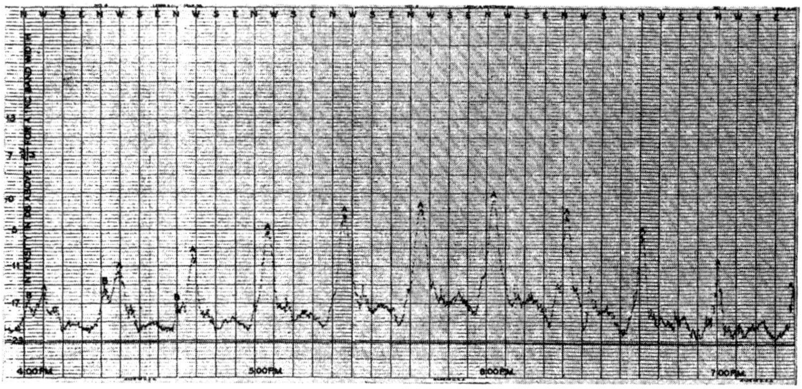

Fig. 6—Sample record of local thunderstorm static on short waves. August 27, 1931.

The operation of the recorder is as follows: For a period of 9 seconds the rectified output from the set charges the 100-microfarad condenser through the 300,000-ohm resistance, see Fig. 1B, the charge being proportional to the energy received from the static during the 9 seconds providing the rectifier is a square law device. During the same time the battery *B* charges the condenser in the opposite sense to the static. The battery *B* and associated resistance are adjusted so that if there is no change in the average static over the 9-second period the resulting charge on the condenser is zero. At the end of the interval the switch *S* is closed by a cam on the recorder shaft and the condenser is discharged through the recorder galvanometer. If the static level has not changed during the interval there will be no charge on the condenser

[5] H. T. Friis and E. Bruce, "A radio field strength measuring system for frequencies up to forty megacycles," Proc. I.R.E., vol. 14, p. 507; August, (1926).

and, hence the galvanometer will not be deflected and the gain of the receiver will remain unchanged. If the static level has increased or decreased the galvanometer will show a corresponding deflection and the recorder mechanism will decrease or increase the gain of the set accordingly.

In the system used the rectifier is not exactly a "square law" device being a two element rectifier in series with a resistance; however, as it was operated with a very small current (5×10^{-6} amperes) it approximated the square law sufficiently accurate for the present purpose.

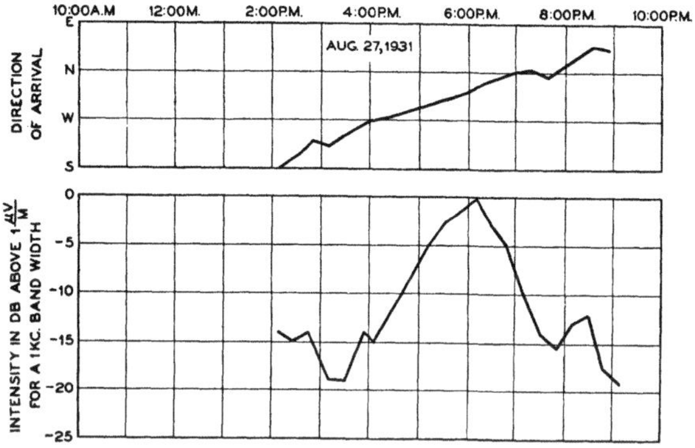

Fig. 7—Direction of arrival and intensity of local storm type static on 14.6 meters.

Results

From the data obtained it is found that three distinct groups of static are recorded. The first group is composed of the static received from local thunderstorms and storm centers. Static in this group is nearly always of the crash type. It is very intermittent, but the crashes often have very high peak voltages. The second group is composed of very steady weak static coming probably by Heaviside layer*refrac- 4
tions from thunderstorms some distance away. The third group is composed of a very steady hiss type static the origin of which is not yet known.

During the time that records have been taken, static of the first group arising from several local thunderstorms has been recorded and studied. The data from a few typical records of these storms have been replotted and are shown in Figs. 7, 9, 10, 11, and 12. In these figures the upper curve shows the direction of arrival of the main stream or streams of static plotted against time and the lower curve shows the intensity of these streams at the corresponding times.[6] In addition to the main streams shown there were usually other minor streams, but these are difficult to follow in detail due to interference from random static from local squalls which are generally present during these periods. Fig. 6 is a section of a typical record of this type of static. It is the record for August 27 of which Fig. 7 is the replot. The peaks marked *A* indicate the position of the main storm. Those marked *B* show the position of one of the minor storms.

Fig. 7 shows the data obtained from this record. It represents a severe electrical storm that passed Holmdel early in the evening. Dur-

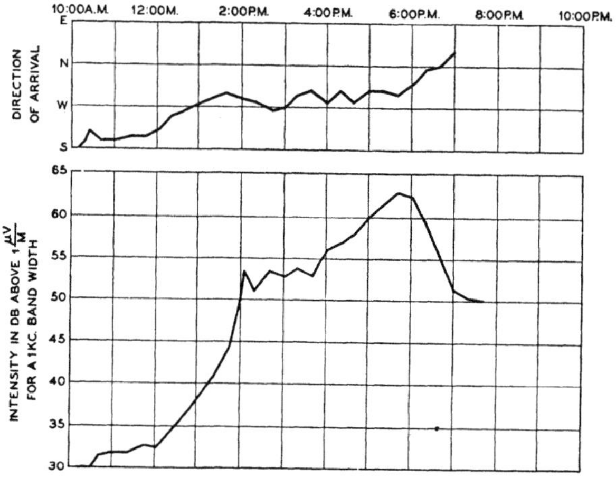

Fig. 8—Direction of arrival and intensity of local storm type static on 6936 meters. August 27, 1931.

ing the early afternoon hours the storm was preceded by several thunder squalls. The static from these squalls was recorded, but there were so many of them and the direction of each changed so fast (as could be observed visually) that it was not possible to follow them on the records. During the late afternoon and early evening hours the static from the south grew stronger than that from the local squalls indicating that a definite storm center was forming there or approaching within range of the receiver from that direction. From then on this storm center did not follow a straight path but, as shown by the records, circled around the receiver and disappeared in the northeast. The manner in which the intensity increased and decreased as the storm passed is clearly shown on the lower curve of the figure.

[6] The band width of the receiver used was 26 kc but before plotting, the data were reduced to the case of a receiver having a band width of 1 kc, i.e., the intensity values were reduced by a factor of $\sqrt{26}$.

For the purpose of comparison, Fig. 8 shows the replot of the long-wave record for the same day.[7] Note that the ratio of the intensity of the long-wave static in microvolts per meter for a 1-kilocycle band width to that of the short-wave static was 63 db when the storm was

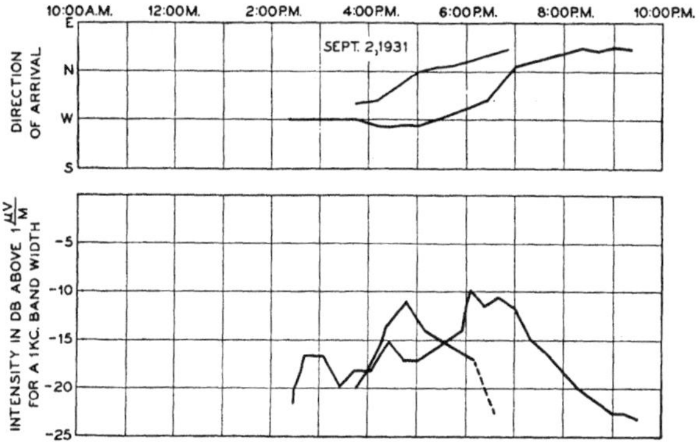

Fig. 9—Direction of arrival and intensity of local storm type static on 14.6 meters.

the severest. This ratio is probably a little too high because the rectifier device was not truly "square law." If we assume the inverse frequency law for the intensity of static this ratio should have been 53.5 db.

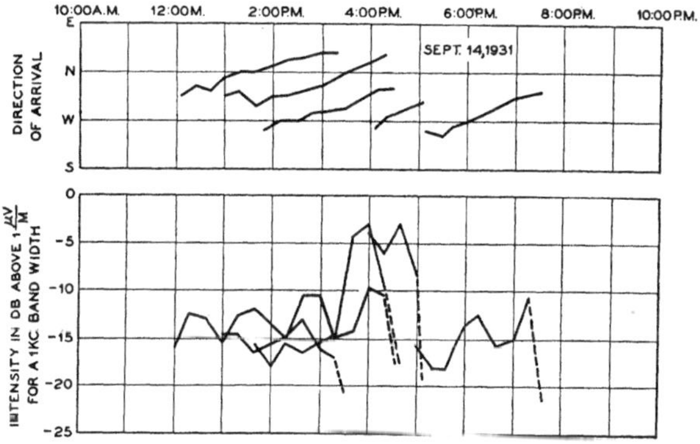

Fig. 10—Direction of arrival and intensity of local storm type static on 14.6 meters.

Fig. 9 shows the data obtained from the record of a well defined storm center that traveled in a straight or nearly straight line towards

[7] The long-wave data mentioned in this paper are all taken on a wavelength of 6936 meters.

the receiver. The static came from the west early in the afternoon, continued to come from that direction as the storm approached, and then as the storm passed Holmdel the direction shifted rapidly from the west through northwest and north to the northeast where it remained as the storm receded until the static no longer was strong enough to record. A minor storm preceded this main one by about an hour as is shown by the short curve preceding the main one. The main storm could clearly be seen passing Holmdel along the northern horizon, but at no time did it approach closer than 15 miles.

Fig. 10 shows the data obtained from the record of several small, but well defined storm centers that followed each other in rapid succession. On this day several small thunder squalls could be seen passing along the northern horizon.

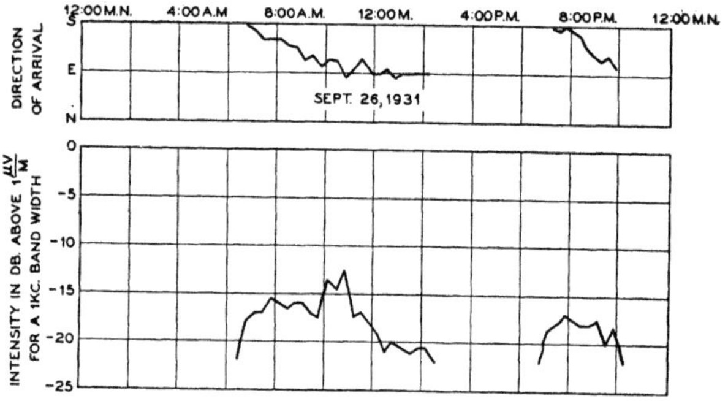

Fig. 11—Direction of arrival and intensity of local storm type static on 14.6 meters.

So far all of the records discussed have been of storms that approached from the southwest or west and passed northwest of the receiver. Fig. 11, on the other hand, shows the data from the record of two storms occurring the same day that approached from the south and passed the receiver on the southeast.

Finally Fig. 12 shows the data from a record of a storm that approached the receiver from the west and split, part of it passing to the north and part to the south of the receiver.

From these figures it is evident that on the average the thunderstorms were audible for four hours before and four hours after they reached Holmdel. Taking 35 miles an hour as the average velocity of a thunderstorm[8] this gives a distance of 140 miles that the storm centers were distant from Holmdel when the static could still be heard.

[8] *See* W. J. Humphreys, "Physics of the Air," p. 365. Also Ward, "The Climates of the United States," p. 322.

It is also worthy of note that by far the majority of storms came from the southwest and west and passed north of the receiver with only an occasional one passing south and southeast. The directions lying between southeast and northeast appear to be substantially free of this type of static at Holmdel and directional antennas built there to receive from those directions on short waves should be troubled with static only infrequently. Of course, this would not necessarily hold for other receiving locations. Locations in some sections, for example, would probably receive an equal amount of this type of static from all directions.

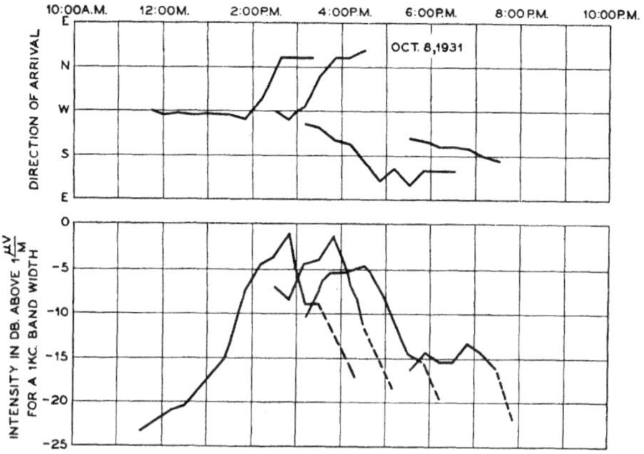

Fig. 12—Direction of arrival and intensity of local storm type static on 14.6 meters.

The static from the second group, which probably originates at long distances, is usually very weak on 14.6 meters. In fact, only occasionally is it strong enough to actuate the recorder. Because of this very few satisfactory records have been obtained of it. From the records that have been obtained, however, and from aural observations it has been determined that this static is of the crash and rumble type; its direction of arrival follows very closely that of the long wave static; and finally it is heard only when the long wave static coming from distant thunderstorms is very strong. It, therefore, probably comes from thunderstorms located some distance from the receiver. The most common directions of arrival of this static, as for the long wave static, are those directions lying between southeast and southwest.

On March 1, 1932, this kind of static was recorded by the short wave recorder from 2:30 P.M. to 3:50 P.M. The data obtained are shown and compared with those obtained on long waves in Table I. The direction of arrival for both the long and short wave static on this day was southeast. As shown in the table the difference between the intensity of the long and short wave static varied between 56 and 62 db. These values should also probably be somewhat lower because the rectifier is not a "square-law" device.

TABLE 1

Time	Intensity of static in db above 1 (microvolt per meter) For a 1—kc band width		Difference
	Long-Wave	Short-Wave	
2:30 P.M.	34.0	−22.0	56.0
2:50	37.5	−24.0	61.5
3:10	37.0	−22.5	59.5
3:30	35.5	−24.0	59.5
3:50	37.0	−25.0	62.0

Since this static is so weak that it cannot be recorded much of the time, the crash method[1] of measuring static as used by Potter could probably be used to great advantage to measure it.

The static of the third group is also very weak. It is, however, very steady, causing a hiss in the phones that can hardly be distinguished from the hiss caused by set noise. It is readily distinguished from ordinary static and probably does not originate in thunderstorm areas. The direction of arrival of this static changes gradually throughout the day going almost completely around the compass in twenty-four hours. It does not quite complete the cricuit, but in the middle of the night when it reaches the northwest, it begins to die out and at the same time static from the northeast begins to appear on the record. This new static then gradually shifts in direction throughout the day and dies out in the northwest also and the process is repeated day after day. Fig. 13 shows the direction of arrival of this static for three different days plotted against time of day. Curve 1 is for January 2, 1932, curve 2 is for January 26, 1932, and curve 3 is for February 24, 1932. Fig. 14 is a photograph of a section of one of the records.

This type of static was first definitely recognized only this last January. Previous to this time it had been considered merely as interference from some unmodulated carrier. Now, however, that it has been detected it is possible to go back to the old records and trace its position on them.

During the latter part of December and the first part of January the direction of arrival of this static coincided, for most of the daylight hours, with the direction of the sun from the receiver. (See curve 1, Fig. 13.) However, during January and February the direction has gradually shifted so that now (March 1) it precedes in time the direction of the sun by as much as an hour. It will be noticed that the curves

2 and 3 of Fig. 13 have shifted to the left.[9] Since December 21, the sun's rays have been getting more and more perpendicular at the receiving location causing sunrise to occur at the receiver earlier and earlier each day. It would appear that the change in the latitude* of

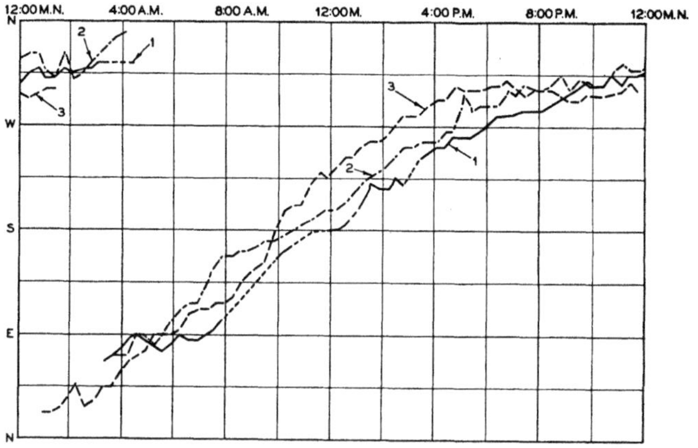

Fig. 13.—Direction of arrival of hiss type static on 14.6 meters.

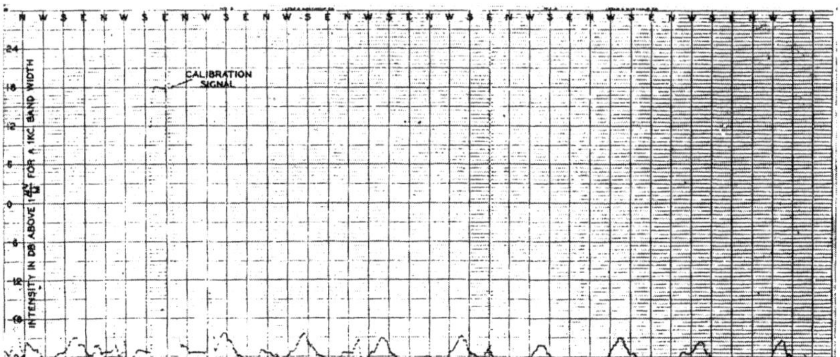

Fig. 14—Sample record of short-wave hiss type static.
February 24, 1932.

the sun is connected with the changing position of the curves. However, the data as yet only cover observations taken over a few months and more observations are necessary before any hard and fast deductions can be drawn.

The fact that the direction of arrival changes almost 360 degrees

[9] Since this paper was written the curve has shifted much further to the left. Now (May 25) it crosses south at 4:30 A.M.

during twenty-four hours and that the shift in the position of the curve
observed during the three months over which data has been taken cor-
5 responds to the change in latitude* of the sun affords definite indication
that the source of this static is somehow associated with the position
of the sun. It may be that the static comes directly from the sun or,
more likely, it may come from the subsolar point on the earth.

The intensity of this static is never very high. At no time during
the period that records have been taken has it exceeded 0.39 microvolts
6 per meter for 1-kilocycle band width.* As will be noticed from the
record (Fig. 14), however, its presence during otherwise quiet periods
is unmistakable.

The experiments which have been described in this paper were carried out at Holmdel, New Jersey. The writer wishes to acknowledge his indebtedness to Mr. Friis for his many helpful suggestions.

EDITOR'S NOTES AND AMENDMENTS

1. Corresponding to a frequency of 20.5 MHz.
2. B.O. = beating oscillator.
3. *Trans. Amer. Inst. Elec. Eng.*
4. *Heaviside layer* is a former term for the ionosphere.
5. *Latitude* here refers to the declination of the sun.
6. This value is much larger than those indicated in Figure 14; perhaps 0.39 μV is an error and 0.039 μV is intended.

ELECTRICAL DISTURBANCES APPARENTLY OF EXTRATERRESTRIAL ORIGIN*

By

Karl G. Jansky

(Bell Telephone Laboratories, Inc., New York City)

Summary—Electromagnetic waves of an unknown origin were detected during a series of experiments on atmospherics at high frequencies. Directional records have been taken of these waves for a period of over a year. The data obtained from these records show that the horizontal component of the direction of arrival changes approximately 360 degrees in about 24 hours in a manner that is accounted for by the daily rotation of the earth. Furthermore the time at which these waves are a maximum and the direction from which they come at that time changes gradually throughout the year in a way that is accounted for by the rotation of the earth about the sun. These facts lead to the conclusion that the direction of arrival of these waves is fixed in space; i.e., that the waves come from some source outside the solar system. Although the right ascension of this source can be determined from the data with considerable accuracy, the error not being greater than ±7.5 degrees, the limitations of the apparatus and the errors that might be caused by the ionized layers of the earth's atmosphere and by attenuation of the waves in passing over the surface of the earth are such that the declination of the source can be determined only approximately. Thus the value obtained might be in error by as much as ±30 degrees.

The data give for the coördinates of the region from which the waves seem to come a right ascension of 18 hours and a declination of −10 degrees.

Introduction

DURING the progress of a series of studies that were being made at Holmdel, N. J., on the direction of arrival of atmospherics at high frequencies,[1] records were obtained that showed the presence of very weak but steady electromagnetic waves of an unknown origin. The first indications of these waves were obtained on records taken during the summer and fall of 1931. However, a comprehensive study of them was not begun until January, 1932. The first complete records obtained showed the surprising fact that the hori-

* Decimal classification: R114. Original manuscript received by the Institute, June 22, 1933. Presented under a different title at the meeting of the American Section of the U.R.S.I., Washington, D. C., April 27, 1933. Revised for publication and presented before Eighth Annual Convention, Chicago, Illinois, June 27, 1933.

[1] Karl G. Jansky, "Directional studies of atmospherics at high frequencies," Proc. I.R.E., vol. 20, p. 1920; December, (1932). The waves referred to are those of group three in the above paper

Reprinted, with permission, from the 'Proceedings of the Institute of Radio Engineers' **21**, 1387–1398 (1933).

zontal component of the direction of arrival of these waves changed nearly 360 degrees in 24 hours, and at that time this horizontal component was approximately the same as the azimuth of the sun. These facts led to the assumption that the source of these waves was somehow associated with the sun.

Records of these waves have now been taken at frequent intervals for a period of more than a year. The data obtained from these records, contrary to the first indications, are not consistent with the suppositions made above relative to the source of the waves, but indicate that

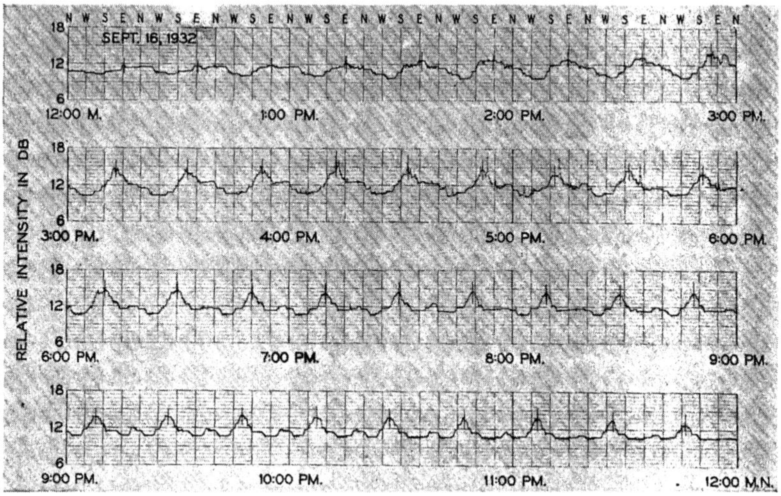

Fig. 1—Sample record of waves of extraterrestrial origion.

the direction of the phenomenon remains fixed in space, that is to say, its right ascension and declination remain constant.

Apparatus

The apparatus used and the type of records obtained were described in detail in a former paper.[1] Briefly, however, the apparatus consists of a rotating antenna array, a short-wave measuring set, and an automatic intensity recorder. The array is highly directive in the horizontal plane and is rotated about a vertical axis so that data obtained with the system, like that obtained with a loop rotating on a vertical axis, give the horizontal component of the direction of arrival of signals, but tell nothing directly about the angle the direction of arrival makes with the horizontal plane. The operation of the recorder is synchronized with that of the rotating array so that the records show directly the horizontal component of the direction of arrival of signals as well as their intensity. The apparatus was tuned to a wavelength of 14.6 meters during all the experiments.

Results

Fig. 1 shows a sample record of the waves of unknown origin obtained with this apparatus. The time at which the array was pointed in the direction from which the unknown waves come is clearly indicated on the record by the humps in the curve. The direction of the waves at those times can be determined from the scale along the top of the record.

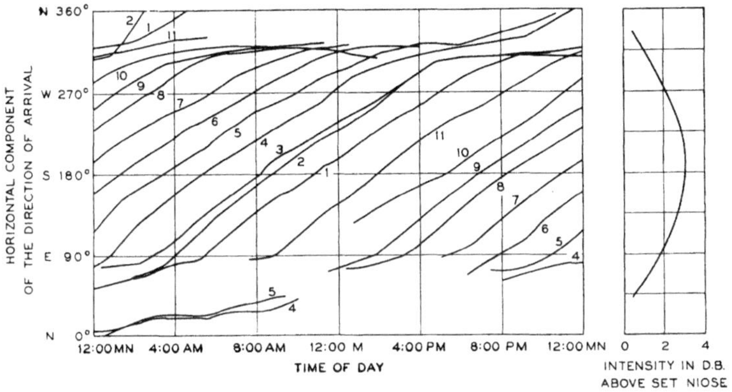

Fig. 2—Direction of arrival of waves of extraterrestrial origin.

1. Jan. 21, 1932
2. Feb. 24, 1932
3. March 4, 1932
4. April 9, 1932
5. May 8, 1932
6. June 11, 1932
7. July 15, 1932
8. Aug. 21, 1932
9. Sept. 17, 1932
10. Oct. 8, 1932
11. Dec. 4, 1932

If, now, the horizontal component of the direction of arrival is plotted against the time of day a curve similar to one of those of Fig. 2 is obtained. Thus, data from the record just mentioned constituted part of that from which curve 9 of this figure was obtained. The figure shows curves for eleven different days spaced approximately one month apart during the year 1932. There is no curve for the month of November. These curves were obtained by averaging the data taken over several consecutive days so as to eliminate the errors made in measuring the records. The day assigned to a given curve is the middle day of the group over which the data for that curve were obtained. The curve at the right in the figure shows the variation in intensity of the waves plotted against the direction of arrival.

This figure shows: first, that the horizontal component of the direction of arrival changes nearly 360 degrees in 24 hours, and, then that there is a uniformly progressive shift of the curves to the left from month to month which at the end of one sidereal year brings the curve back to its initial position. These facts show that the waves come, not from the sun, but from a direction which remains constant throughout the year.

Discussion

To show that this is the necessary conclusion it will be necessary to digress a little from the subject and discuss the celestial sphere and celestial coördinates.

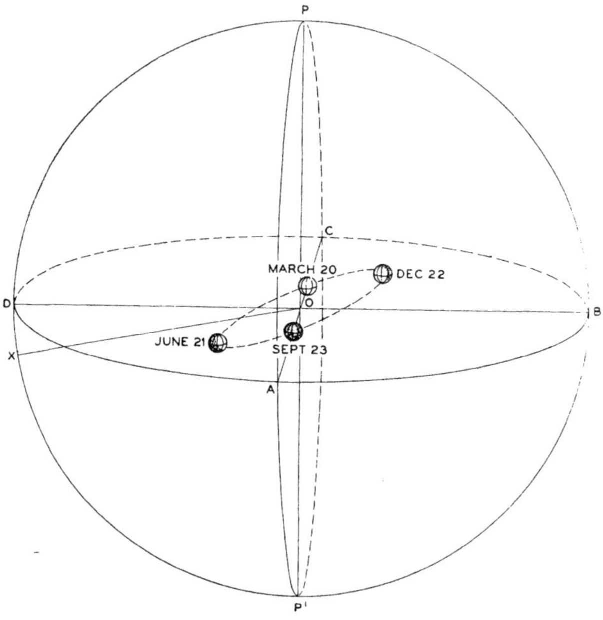

Fig. 3—Graphical representation of the celestial sphere.

The celestial sphere is that hypothetical sphere surrounding the earth upon which all the stars, whatever their distances, appear to be located. The direction of a star, then, is described by giving its apparent position on the celestial sphere, and its position on the sphere is given in terms of a pair of angles called the right ascension and declination of the star. Fig. 3 shows a graphical representation of the celestial sphere with the earth in its orbit around the sun at the center. The plane $ABCD$ represents the plane of the celestial equator and POP' the axis of the celestial sphere.[2] Right ascension is measured in hours

[2] For a more complete explanation of the system of coördinates used in this discussion see Russell-Dugan-Stewart, "Astronomy," vol. 1, chap. 1 or the opening chapters of any textbook on astronomy.

around the circle *ABCD* eastward from the line *OA* as reference. The line *OA* is determined by the direction of the sun from the earth at the time it crosses the equator on the first day of spring. Thus the line *OA* lies at 0 hours, *OB* at 6 hours, *OC* at 12 hours, and *OD* at 18 hours; 24 hours of right ascension being equal to 360 degrees. Declination is measured in degrees above or below the equatorial plane, plus, if it is above the plane and minus, if below. The positions of the earth with

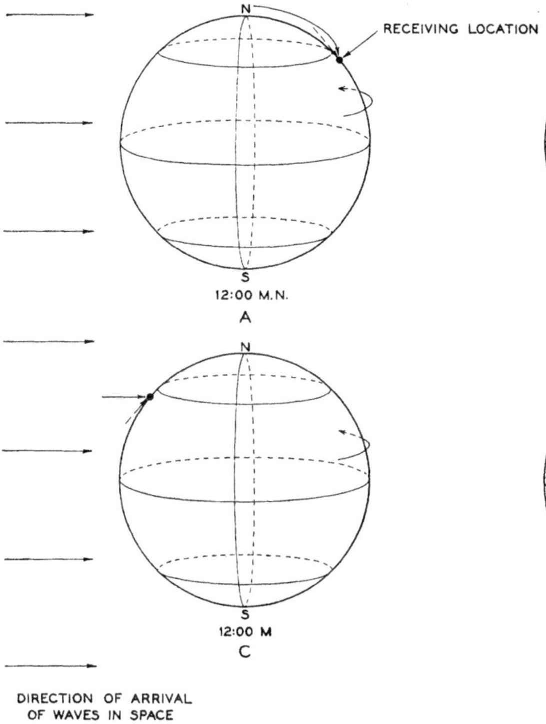

Fig. 4—The effect of the daily rotation of the earth on the direction of arrival as measured at the receiving location.

respect to the sun for the first day of each season are shown. Since the diameter of the earth's orbit is so small with respect to the distances to the stars, it is assumed that the earth is always at the center of the sphere at 0 and the rest of this discussion is based on that assumption. Accordingly, the plane of the celestial equator coincides with that of the earth's equator and the axis of the celestial sphere coincides with the earth's axis.

Now, if there were radio waves coming from a direction fixed in space and from a source so far removed from the sun that the direc-

tions of propagation throughout the whole solar system were substantially parallel, if there were no distortion in direction suffered by the waves during their passage through the ionized layers of the earth, if these waves had the ability to bend around the earth, and if there were no other unexplained phenomena taking place, then, for this idealized case, the horizontal component of the direction of arrival as measured at the receiving location would change 360 degrees during
1 one complete rotation of the earth.★ Let us assume for the sake of argument that the right ascension of the direction of arrival of these idealized waves is 18 hours and its declination 0 degrees, that direction

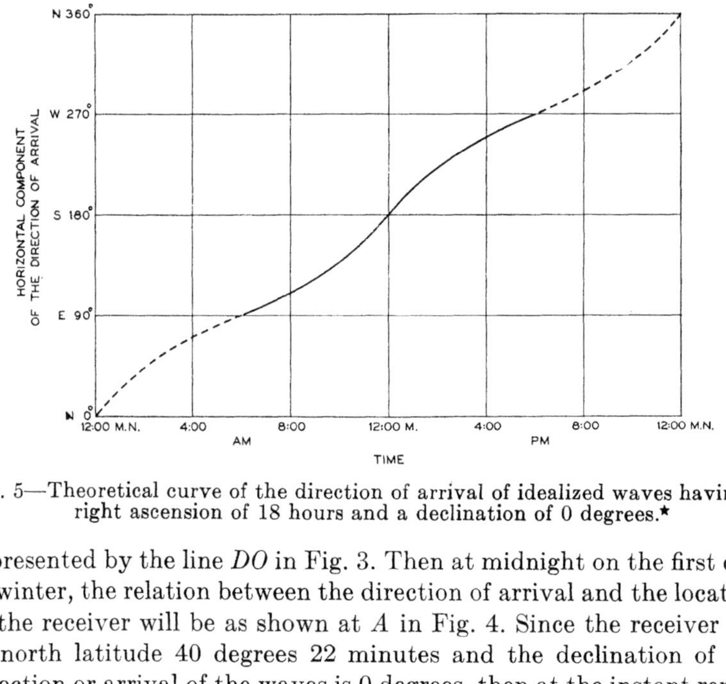

Fig. 5—Theoretical curve of the direction of arrival of idealized waves having a
2 right ascension of 18 hours and a declination of 0 degrees.★

represented by the line *DO* in Fig. 3. Then at midnight on the first day of winter, the relation between the direction of arrival and the location of the receiver will be as shown at *A* in Fig. 4. Since the receiver lies in north latitude 40 degrees 22 minutes and the declination of the direction or arrival of the waves is 0 degrees, then at the instant represented the horizontal component of the direction of arrival would be north as shown by the broken arrow. Six hours later the condition shown at *B* would exist and the horizontal component of the direction of arrival would coincide with the true direction and would be east. After another six hours, the direction of arrival would coincide with the meridian of the receiver, and its horizontal component would be south as shown by the broken arrow at *C*, after six hours more it would be west as at *D*, and finally after a complete rotation of the earth it would be back to north again. Or, if this horizontal component of the direction of arrival were plotted against time of day, a curve like that shown in Fig. 5 would be obtained. The curve is dotted for that portion of the time during which the earth would be between the source of the waves and the receiver. As will be seen from the figure, the horizontal component of the direction of arrival changes 360 degrees in about 24 hours, or in exactly 23 hours and 56.06 minutes since that is the time required for the earth to make one complete revolution with re-

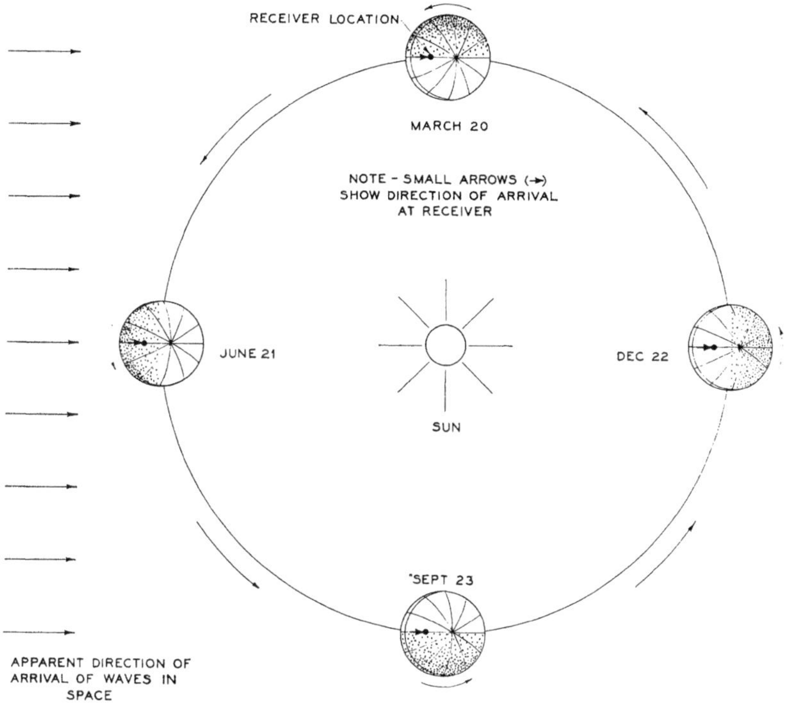

Fig. 6—The effect of the earth's orbital motion on the direction of arrival as measured at the receiving location.

spect to the stars. It is this difference between the length of the sidereal day and the mean solar day (3.54*minutes of solar time) that accounts 3
for the uniformly progressive shift of the curves of Fig. 2 to the left.

Fig. 6 will illustrate just how this shift takes place. This figure shows the earth in its orbit around the sun as seen from above. If, as has been assumed, the direction of arrival of the waves has a right ascension of 18 hours, they can be represented by some such group of arrows as shown at the left in the figure. When the earth is in the position shown for June 21 then, regardless of the declination of the direction of arrival of the waves, the time at which this direction of arrival will coincide with the meridian of the receiver will be at midnight. On September 23 this time of coincidence will occur six hours earlier at 6:00 P.M. On December 22 it will occur another six hours earlier or at 12:00 noon and on March 20 it will occur at 6:00 A.M. Consequently if a curve like the one of Fig. 5 is plotted for every month of the year a family of curves like that of Fig. 7 will be obtained where each curve occurs approximately two hours earlier than the one for the preceding month. Note the similarity between this family of curves and that of Fig. 2.

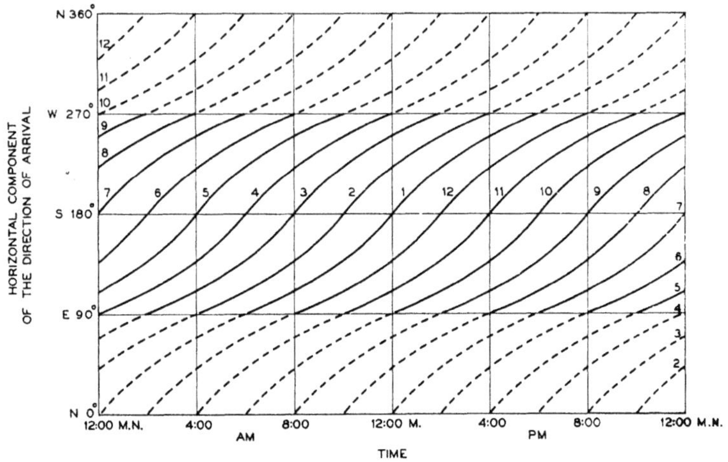

Fig. 7—Theoretical curves for twelve months of the direction of arrival of idealized waves having a right ascension of 18 hours and a declination of 0 degrees.

1. Dec. 22	5. April 21	9. Aug. 22
2. Jan. 21	6. May 21	10. Sept. 23
3. Feb. 21	7. June 21	11. Oct. 23
4. March 20	8. July 22	12. Nov. 22

It will be shown later that for idealized waves having a direction
4 of arrival the declination of which is between −40 degrees, 22 minutes* and +40 degrees, 22 minutes the horizontal component of the direction of arrival is south when, and only when, the direction of arrival coincides with the meridian of the receiver. Therefore, the times at which the curves of Fig. 7 cross the line whose ordinate is 180 degrees (south) are spaced approximately two hours apart, and if these times are plotted against the day of the year represented by each curve the points will all lie on a straight line[3] the slope of which will be 365.25/24 days per hour.

[3] Strictly speaking the line should not be exactly straight. Of the many reasons why this is so the most important is that the earth's motion in its orbit

It will be shown later that the declination of the direction of arrival of the waves detected by the measuring equipment is between the values of -40 degrees, 22 minutes* and $+40$ degrees, 22 minutes. Con- 4
sequently, if the right ascension remains constant then points obtained from Fig. 2 in a manner exactly similar to that just explained should fall on a straight line the slope of which should be 365.25/24 days per hour. It was in this manner that the points of Fig. 8 were obtained. The

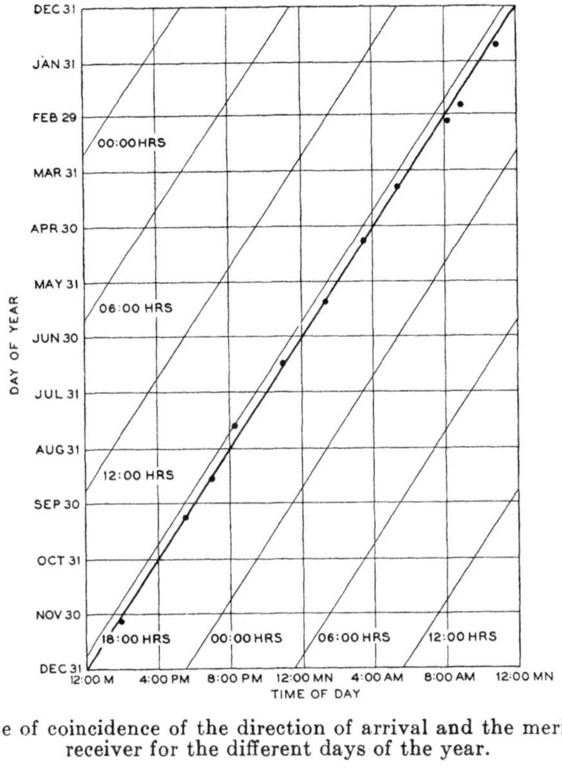

Fig. 8—Time of coincidence of the direction of arrival and the meridian of the receiver for the different days of the year.

correspondence of the points with the heavy line, the slope of which is 365.25/24 days per hour, cannot be accidental and proves that the right ascension of the direction of arrival of the waves is constant. The position of this heavy line on the graph is determined by the value of the right ascension. Thus the position of the curves corresponding to a right ascension of 0 hours, 6 hours, 12 hours, and 18 hours are shown by the light diagonal lines on the figure.

is not uniform. However, the effects are all so small that the greatest deviation would be scarcely perceptible on the curve so they will not be considered in this discussion.

From the relative positions of the heavy line and the light lines it
will be seen that the measured direction of arrival occurs at a right
ascension of approximately 18 hours, 30 minutes; however, because
the mechanism of the recorder takes a finite time to record the field
strength values, the directions measured on these records lag behind
5 the true directions by a value varying from 4 degrees to 9 degrees.* If
the measured values are corrected for this error the right ascension of
the direction of arrival becomes approximately 18 hours.[4]

Referring to Fig. 3, if the direction of arrival has a right ascension of 18 hours then it must lie in that half of the plane $PDP'B$ to the left of PP'. One such direction is represented by the line XO in the figure.

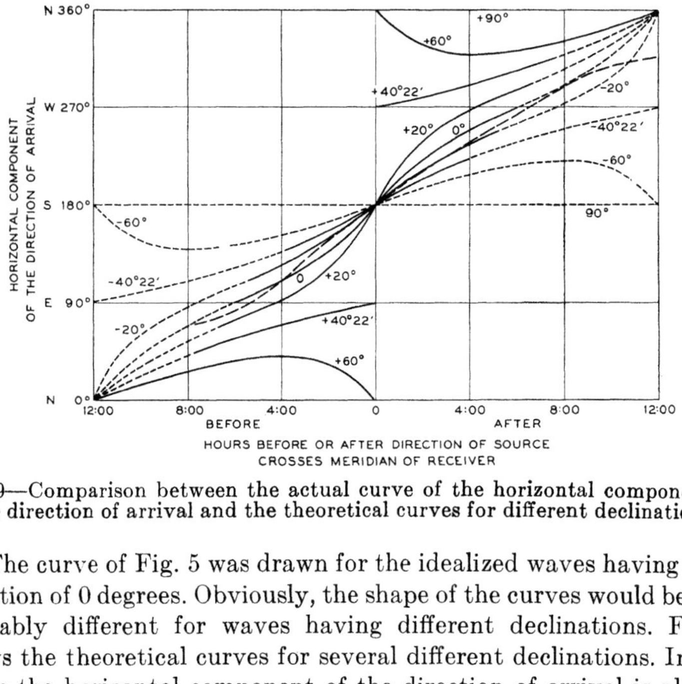

Fig. 9—Comparison between the actual curve of the horizontal component of
6 **the direction of arrival and the theoretical curves for different declinations.***

The curve of Fig. 5 was drawn for the idealized waves having a declination of 0 degrees. Obviously, the shape of the curves would be considerably different for waves having different declinations. Fig. 9 shows the theoretical curves for several different declinations. In this figure the horizontal component of the direction of arrival is plotted against time, but here the time is given not in terms of the hours of the day but in terms of the time interval before and after the direction of arrival coincides with the meridian of the receiver. The values of declination used for the different curves are given in the figure. As before, the curves are dotted for that portion of the time during which

[4] The limit of error has not been exactly determined but is certainly not greater than ±7.5 degrees, which is equivalent to ±30 minutes of right ascension.

the earth would be between the source of the waves and the receiver.

For the purpose of making a comparison between these curves and those of Fig. 2, an average of the curves of Fig. 2 is shown in Fig. 9 by the broken line. It will be seen that for the greater part of the time during which the direction of arrival is above the horizon it lies between the curves for a declination of 0 degrees and -20 degrees, giving a value of roughly -10 degrees for the declination of the direction of arrival of the waves. In Fig. 3 the line XO is drawn with a declination of -10 degrees and right ascension of 18 hours so that it represents the apparent direction of arrival of the waves.

Beyond the point where the direction of arrival drops below the horizon, the average curve is not at all similar to the theoretical curves of Fig. 9. However, for that portion of the curves the intensity is very weak (see the curve at the right in Fig. 2) and the directions cannot be measured very accurately. Furthermore, as the time interval before or after the direction of arrival coincides with the meridian of the receiver is increased, the waves must travel through an increasing thickness of the earth's atmosphere so that any bending of the waves caused by the ionized layers would increase also. At the time the direction of arrival coincides with the meridian of the receiver this bending is confined to the plane determined by the right ascension of the direction of arrival, and will therefore cause no error in the measurement of the right ascension if the data used for its determination are taken at this time, as has been done. It may, however, affect the measurement of the declination. At all other times, whatever bending the waves suffer will cause errors in both measurements and this bending might be the cause of the difference between the theoretical and actual curves noted.

It may very well be that the waves that reach the receiver instead of coming from a single point fixed in space originate in the earth's atmosphere, but are secondary radiations caused by some primary rays of unknown character, coming from a source or sources fixed in space, and striking the earth's atmosphere. If this is so the disturbance measured by the receiver is probably the summation of very many waves of various intensities coming from secondary sources in the earth's atmosphere that are scattered over a considerable area. In this case the declination of the source of the primary rays may be considerably different from that obtained from the curves; however, the measurement of its right ascension would not be affected appreciably if made in the manner described above.* 7

On the other hand it may be that the waves that reach the receiver are the primary waves themselves coming from a great many sources scattered throughout the heavens. In this case the direction measured would be the direction of the center of activity, and as before, the value of the right ascension would be accurate in spite of the bending of the rays in the ionized layers, and the declination would be in error by an amount equal to that for a single source at the center of activity.

From a consideration of the data and the method of interpretation it is believed that, in spite of the possible errors mentioned in the above cases, the declination of the source or center of activity, if there is more than one source, as measured would be accurate within an error not greater than ± 30 degrees.

The apparent direction of arrival of the waves has not as yet been definitely associated with any region fixed in space; however, there are two such regions that should be seriously considered. The point on the celestial sphere of right ascension 18 hours and declination -10 degrees, the direction from which the waves seem to come, is very near the point where the line drawn from the sun through the center of the huge galaxy of stars and nebulae of which the sun is a member would strike the celestial sphere. The coördinates of that point are approximately right ascension of 17 hours, 30 minutes, declination -30 degrees (in the Milky Way in the direction of Saggitarius[5]). It is also very near that point in space towards which the solar system is moving with respect to the other stars. The coördinates of this point are right ascension 18 hours and declination $+28$ degrees.[6] Whether or not the actual direction of arrival of the primary rays coincides with either of these directions cannot be determined definitely until some method of accurately measuring their declination is devised and the measurements made.

In conclusion, data have been presented which show the existence of electromagnetic waves in the earth's atmosphere which apparently come from a direction that is fixed in space. The data obtained give for the coördinates of this direction a right ascension of 18 hours and a declination of -10 degrees.

The experiments which are the subject of this paper were performed at Holmdel, N. J. (Latitude 40° 22′ N and Longitude 74° 10′ W) during the year 1932.

Acknowledgment

The writer wishes to acknowledge the help of Mr. A. M. Skellett, also of the Bell Telephone Laboratories, in making some af the astronomical interpretations of the data.

[5] H. Spencer Jones, "General Astronomy," pp. 358, 359; Forest Ray Moulton, "Astronomy," pp. 479, 504–509.

[6] Russel-Dugan-Stewart, "Astronomy," vol. 2, p. 661.

EDITOR'S NOTES AND AMENDMENTS

1. In other words the azimuth of the source changes by 360° even if the altitude becomes negative, i.e., the source sets, during part of the 24 hours.
2. Curve applies to 22 December.
3. Should be "3.94 minutes".
4. This declination should be −49°38′.
5. This lag corresponds to an instrumental time constant of ~13 sec combined with the varying angle of the antenna beam sweep with respect to lines of constant right ascension. It is not clear why the 9 sec quoted in Paper 1 seems inconsistent with the value implied here.
6. Actual curve: ----; theoretical curves: -- - -—-—- - - - .
7. The thought here is that the "secondary radiations" may not come from the same direction as the "primary rays", and that in this case it would probably be their altitude rather than their azimuth which would be primarily altered. This then would create a larger error in declination than in right ascension.

A NOTE ON THE SOURCE OF INTERSTELLAR INTERFERENCE*

By

Karl G. Jansky

(Bell Telephone Laboratories, Inc., New York City)

Summary—*Further consideration of the data obtained during observations on interstellar interference has shown that these radiations are received any time the antenna system is directed towards some part of the Milky Way system, the greatest response being obtained when the antenna points towards the center of the system. This fact leads to the conclusion that the source of these radiations is located in the stars themselves or in the interstellar matter distributed throughout the Milky Way.*

Because of the similarity in the sound produced in the receiver headset, it is suggested that these radiations might be due to the thermal agitation of charged particles.

IN FORMER papers,[1,2] it was explained how interstellar interference was first observed with an automatic field strength recording system making use of a highly directional rotating antenna. It was pointed out that the directions of arrival were fixed in space and that there seemed to be only a single direction of arrival having a
1 right ascension of eighteen hours and a declination of -20 degrees.*

Some of the data did not check very accurately the theory of a single direction of arrival and it was suggested that a possible explanation of the discrepancies might be found in refraction of the waves during their passage through the ionized layers of the atmosphere.

Since the publication of the above papers further consideration of the data has led to some very interesting conclusions and speculations. The data obtained from the system are in the form of a continuous record of the output for all hours of the day and, since the antenna rotates continuously, for all directions as well. If we examine a typical day's record of the disturbance, for example that for September 16, 1932, given in Figs. 1 and 2, the following facts are evident. Besides varying gradually in height throughout the day the peaks obtained for each revolution of the antenna also change decidedly in shape. In Fig. 2 from 12:00 M to 3:00 P.M. the peaks are very broad, in fact one peak covers the time taken up by one complete revolution of the

* Decimal classification: R113.5. Original manuscript received by the Institute, April 10, 1935. Presented before Tenth Annual Convention, Detroit, Mich., July 3, 1935.

[1] Karl G. Jansky, "Directional studies of atmospherics at high frequencies," Proc. I.R.E., vol. 20, p. 1920; December, (1932).

[2] Karl G. Jansky, "Electrical disturbances apparently of extraterrestrial origin," Proc. I.R.E., vol. 21, p. 1387; October, (1933).

Reprinted, with permission, from the 'Proceedings of the Institute of Radio Engineers' **23**, 1158–1163 (1935).

antenna. From 3:00 P.M. on, the peaks gradually get narrower and narrower until at 10:00 P.M. they are only one quarter the previous

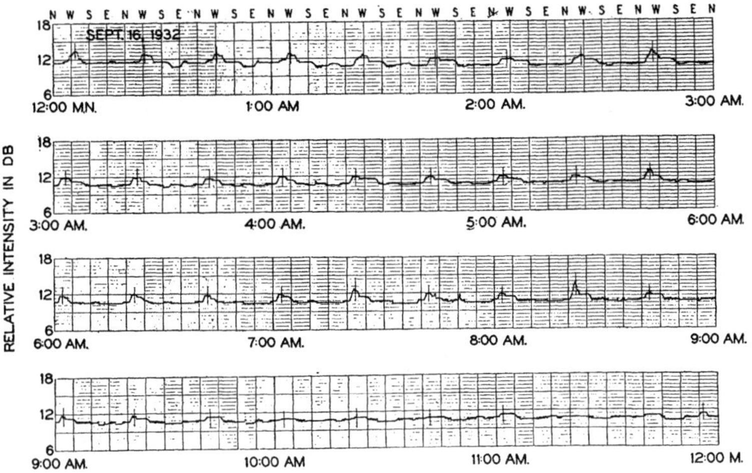

Fig. 1—Sample record of waves of interstellar origin. Record taken 12:00 midnight to 12:00 noon on September 16, 1932.

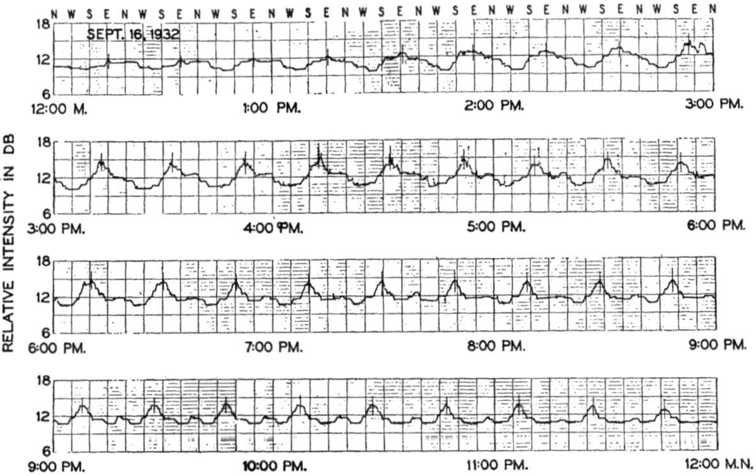

Fig. 2—Sample record of waves of interstellar origin. Record taken 12:00 noon to 12:00 midnight on September 16, 1932.

width. During the same time a much smaller and weaker peak begins to appear on the record. At 9:20 P.M. it is very clear.

Upon determining the direction towards which the antenna system points in space at these times, it is discovered that when the peaks are broad, the antenna is so located in space that it sweeps along the Milky Way and the maximum response is obtained when it points in the direction of the center of the Milky Way. When the large sharp peaks and the alternate small peaks are obtained the antenna is so located in space that it sweeps across the Milky Way, the large peak being obtained when that section of the Milky Way nearest the center is crossed and the small peak when that section farthest from the center is crossed.

If we consider the belief now held by astronomers that the Milky Way is a large galaxy of stars having the same general shape as a huge discus or grindstone with the solar system, and therefore the earth, located at some distance from the center and almost in the galactic plane, then the phenomena described above would seem to indicate that the disturbances recorded are due to radiations emanating from the stars themselves. The various heights and widths of the peaks obtained on the record would then be explained in the following manner.

If the axis of rotation of the antenna were perpendicular to the plane of the Milky Way the antenna would rotate so that it always pointed at some part of the Milky Way and therefore would always receive some energy. This energy should reach a maximum value when the antenna points in the direction of the center of the Milky Way System, for the greatest number of stars would then be included within the angle of reception of the antenna. As the antenna rotates the number of stars included within this angle would very gradually decrease until the antenna points in just the opposite direction when the number of stars within the angle would be a minimum. As the antenna rotates further the number of stars within the angle would again increase until the maximum was again reached, etc. Thus the energy received at such a time would show a gradual decrease and increase with one maximum and one minimum for a single rotation of the antenna.

Actually the axis of rotation of the antenna is never exactly perpendicular to the plane of the Milky Way, but approaches that condition closely when the meridian of the receiving location has a right ascension of twelve hours and forty minutes. Translated in terms of the time on the records this occurs about five hours and twenty minutes before the azimuth of the direction of arrival of the disturbances is south. For the record shown in the figures this time would be about 1:50 P.M. at which time the angular distance between the axis of rotation and the perpendicular to the plane of the Milky Way is twelve degrees and twenty minutes. However, due to the facts that the Milky Way has a very appreciable width, and the vertical directional characteristic of the antenna is very broad, the discussion given above is still applicable and explains the type of record that should be and is obtained at this time. Furthermore, at this time, the center of the Milky Way System as seen from the receiving location has an azimuth very slightly south of east which checks exactly the direction of the maximum disturbance as obtained from the curves.

Since the direction of the axis of rotation of the antenna changes as the earth rotates, the above condition exists for only a short time. Thus, after seven hours and forty-seven and one-third minutes (when the right ascension of the meridian of the receiving location is twenty-two hours and twenty-seven and one-third minutes)* the axis of rota- 2
tion will lie in the plane of the Milky Way instead of being perpendicular to it. In this position the antenna will sweep across the Milky Way twice for every revolution and we would expect two peaks on the record where previously we had only one, a large peak when the antenna sweeps across that part of the Milky Way nearest the center and a smaller one when it crosses that part farthest from the center. These peaks should be relatively sharp because the number of stars within the angle of reception of the antenna changes rapidly. The first case occurs when the antenna points in a southwesterly direction and the second when it points north-northeast.

Turning now to Fig. 2 we see that at $9{:}37\frac{1}{3}$ P.M. (seven hours and forty-seven and one-third minutes after 1:50 P.M.) we have two definite peaks on the record for every revolution of the antenna, the larger of which is obtained when the antenna is pointing southwest and the smaller when it is pointing north-northeast, checking the predictions exactly.

After another eight hours and twenty-five and one-third minutes, or at $6{:}02\frac{1}{3}$ A.M. on the record given, the axis of rotation of the antenna again lies in the Milky Way, but this time the two points where the antenna sweeps across the Milky Way are both some distance from the center so that neither peak should be very large. At this time the two points have directions of northwest by north and southeast by south, the former being the nearest to the center of the Milky Way. Turning again to the figures we find that at $6{:}02\text{-}\frac{1}{3}$ A.M. the peaks are much weaker than at $9{:}37\text{-}\frac{1}{3}$ P.M. and that they have the directions predicted.

A more detailed analysis of the data has shown that every time the antenna points towards some part of the Milky Way the record shows an increase in the energy received, and also every time the record shows an increase of energy received the antenna is found to be pointing towards some part of the Milky Way.

As said before, the most obvious explanation of these phenomena is one that assumes that the stars themselves are sending out these radiations and that the direction of arrival at the receiving location, instead of being confined to a single direction as was formerly intimated, include all directions, a greater indication being obtained for those directions confined to the Milky Way because of the greater star density there.

Another plausible explanation is one based on an hypothesis previously suggested,[2] that the waves which reach the antenna are secondary radiations caused by some form of bombardment of the atmosphere by high speed particles which are shot off by the stars.

Upon examining the characteristics of these radiations for further clues as to their source, one is immediately struck by the similarity between the sounds they produce in the receiver headset and that produced by the thermal agitation of electric charge.[3] In fact the similarity is so exact that it leads one to speculate as to whether or not the radiations might be caused by some sort of thermal agitation of charged particles. Such particles are found not only in the stars, but also in the very considerable amount of interstellar matter that is distributed throughout the Milky Way, which matter, according to Eddington[4] has an effective temperature of 15,000 degrees centigrade. If the radiations come from such particles one would expect the response obtained to depend upon the directional characteristic and gain of the antenna and the way it is pointed relative to the Milky Way, an expectation which agrees with the observed facts.

Attempting to explain the radiations in question on the basis of any of the above hypotheses immediately raises a serious question as to the effect of the sun. Since the sun is a star, although in comparison with some, a rather insignificant star, and since it is much closer to the earth than the stars of the Milky Way, so close in fact that the energy received from it in the form of light and heat is many times the combined energy received from all the stars and all the interstellar matter of the Milky Way, it would naturally be expected that radiations similar to those in question would be found coming from the sun with an intensity much greater than the intensity of those coming from the other sections of the Milky Way. So far no such solar radia-

[3] F. B. Llewellyn, "A study of noise in vacuum tubes and attached circuits," PROC. I.R.E., vol. 18, p. 243; February, (1930).

[4] Arthur Stanley Eddington, "Stars and Atoms," pp. 66–69, Yale University Press, 1927.

tions have been detected, and, although a possible explanation for their absence might be based on a supposition that the temperature of the sun is such that the ratio of the energy radiated by it on the wavelengths studied to that radiated in the form of light and heat is much less than for some other classes of heavenly bodies found in the Milky Way, the question will have to remain unanswered until more data have been taken.

EDITOR'S NOTES AND AMENDMENTS

1. Reference 2 (Paper 2) actually gives $\delta = -10°$; the revised value is explained in Jansky (1933c).
2. By comparison with the previous case, it can be seen that this value should be twenty hours and twenty-seven and one-third minutes. Calculations indicate, however, that these two cases imply that the zone time which Jansky here quotes is Eastern *Daylight* Time. Yet it is quite certain from a detailed comparison of Figures 1 and 2 in this paper and Figure 2 of Paper 2 that all Figures are, in fact, labelled with Eastern *Standard* Time.

SECTION B

FIRST MAPS OF THE GALAXY

The following papers by Grote Reber, all entitled "Cosmic Static", span a decade of scientific investigation by a most remarkable man. Reber single-handedly carried on the science of radio astronomy, despite the fact that he could only do research during his free hours at home. During the day he was a radio engineer in Chicago while at night a radio astronomer in his backyard in the suburb of Wheaton, Illinois. For the details of his story, the reader is referred to a delightful and informative blow-by-blow description of his trials and joys set down by Reber himself in 1958. Suffice it to say here that he recognized the fundamental importance of Jansky's work, but also realized that further progress would only be possible with much improved equipment. Thus, he took it upon himself to design and build a 31 ft (9 m) paraboloid reflector illuminated by a novel feed system.[1] The magnitude of his engineering skill and ingenuity can only be appreciated by realizing that this part-time, backyard project produced what was undoubtedly the largest antenna of its kind in the world.

Reber's first experiments were in the spring of 1938 at 3300 MHz, a frequency as high as commercially available tubes would allow him to go. His motivation towards high frequencies was soundly reasoned. Not only would he obtain higher angular resolution than Jansky, but he also expected a much stronger signal if the radiation were similar to that of a black body ($I_\nu \propto \nu^2$). Imagine his disappointment, then, when he could detect no radiation at 3300 MHz in early 1938, nor any with a second receiver at 910 MHz later that same year. These negative results certainly established that the extraterrestrial radiation was not thermal in origin, but they were nevertheless quite discouraging. Not to be defeated, he persisted and went to a yet lower operating frequency – 160 MHz – and, in 1939, was finally able to confirm Jansky's discovery.

Paper 4 (1940) is Reber's first publication dealing with the "cosmic static". He presents relatively crude data showing the run of radio brightness at 160 MHz with galactic longitude[2]. The quoted intensities in this paper, as well as in succeeding ones, unfortunately are in error, due to incorrectly estimated or measured beamwidths, antenna efficiencies, and/or receiver bandwidths. Table I summarizes the situation. The quoted half-power beamwidths, being

TABLE I
Reber's antenna and receivers

	Papers 4, 5	Paper 6	Paper 7
Frequency (MHz)	160	160	480
Wavelength (m)	1.85	1.85	0.62
Half-power beamwidth (deg)†	3 (~12.5)	6 x 8 (~12.5)	3 x 3 (~4.0)
Antenna efficiency†	0.42 (0.8*)	0.85 (0.8*)	– (0.8*)
Bandwidth (MHz)†	1.0 (0.19)	0.19	8.0
Total derived correction factor for quoted intensities	0.12	0.25	0.41
T_b (K) corresponding to quoted $10^{-22} \frac{\mathrm{W}}{\mathrm{cm}^2\ (\mathrm{circ.\ deg})\ \mathrm{MHz}}$	64	129	24

† The first value in each instance is that given in the original paper; those values in parentheses are the best estimates available from other considerations (see text).
* This is an assumed probable value for the main-beam efficiency.

much less than λ/D, must of course be underestimates; a value of 12°.5 for the 160 MHz data is not only consistent with the antenna size, but can also be derived by measuring (a) the apparent size of the point source Cas A, recognizable today on the contour map in Paper 6, or (b) the width of the Sun deducible on the tracings given in Figure 5 of Paper 6. We have used a main-beam efficiency of 0.8 as the most probable value, while the bandwidth error is noted by Reber himself in a 1942 paper. We should not, however, let these errors detract from the genius in Reber's work. In his first paper he takes Eddington's 15-year-old model of the interstellar medium and becomes the first to suggest that the mechanism of radiation is free-free transitions by electrons in a hot plasma. It turns out, however, that he over-corrects for the effects of stimulated emission (by a factor of ~50) and ignores the effects of opacity; thus is he misled to the conclusion that his observations can be reasonably explained using free-free radiation. This suggestion, as well as his use of a disk model for the galaxy to aid in understanding the run of intensity with longitude, is nevertheless illuminating.

Three months after the submission of his first paper, Reber submitted his results to an astronomical journal. Otto Struve, the editor of *The Astrophysical Journal*, and the other astronomers at the Yerkes Observatory of course had never seen such data and were perplexed as to how to gauge its reliability. Struve finally sent a delegation from the University of Chicago, including Gerard Kuiper and Phillip Keenan, to Wheaton. It was this visit which convinced the optical astronomers that Reber knew what he was doing *experimentally.* Apparently they felt that the *theoretical* interpretation, however, should best be left to "real" astronomers, for Reber's 1940 *Astrophysical Journal* article (Paper 5) is very similar to Paper 4, except for the omission of the entire section on free-free radiation and galactic structure. And immediately following this paper, Struve inserted Paper 8 by Henyey and Keenan, which contained a more rigorous and correct development of the free-free radiation theory.

Reber's 1942 paper in the *Proceedings of the Institute of Radio Engineers* is omitted from this collection due to its great length. In that paper he presented many details and photographs of his antenna and receiver, as well as typical strip chart recordings. He also went through a rather thorough circuit analysis in order to estimate the sensitivity of his receiver. His intuitive grasp for the nature of his measurements is well illustrated by the last sentence in the Summary: "It is suggested that cosmic static is the equivalent of thermal agitation in which all space is the conductor and the input terminals of the detecting equipment are projected by means of an antenna system to some far-distant part of space." What a marvellous conception! He further noted that he had estimated solar coronal radiation to be measurable, but had not been able to try for a detection, presumably as a result of daytime interference from automobile ignitions.

By this time, Reber could see that what was really needed was an all-sky survey. To this end he automated his data recording, yet again improved his receiver stability and sensitivity (to a system temperature of ~11 000 K) and, in Paper 6, presents the first radio contour maps of the sky. His basic map was not improved upon until six years later in the Southern Hemisphere (Bolton and Westfold, 1950; Allen and Gum, 1950) and a full 12 years later in the North (Dröge and Priester, 1956). The Milky Way is beautifully limned and the influences of the following discrete sources (only later recognized as such) and regions can clearly be seen: Cas A, Cyg A and Cyg X, Sgr A and the galactic center region, and Vela X.[3] Table II will assist the reader in locating the constellations mentioned in Papers 6 and 7. Reber's picture is now one of spiral structure consisting of a central disk, which is the Milky Way proper and from which protuberances project.

Modesty tucked away at the end of the paper is the first *published* radio detection of the Sun, although Hey (Paper 16) and Southworth (Paper 17) had both earlier detected the Sun, but had been prevented by wartime secrecy from openly publishing. Reber notes that the measured solar intensity is "rather surprising", but does not indicate that his measurements imply an

TABLE II
Approximate positions of constellations mentioned in Reber's papers

Constellation	Abbreviation	α	δ
Sagittarius	Sgr	17^h30^m	$-30°$
Lyra	Lyr	19 00	+35
Aquila	Aql	19 30	+10
Cygnus	Cyg	20 30	+40
Cassiopeia	Cas	0 00	+55
Perseus	Per	2 30	+55
Orion	Ori	5 30	+05
Canis Major	CMa	7 00	–20
Puppis	Pup	8 00	–30

astounding apparent brightness temperature of 9×10^5 K for a disk the size of the optical Sun.

Paper 7 reports the culmination of Reber's work with his homemade dish. His map of the galaxy at 480 MHz shows fine detail on a large scale and was not improved upon in a major way for almost a decade (Piddington and Trent, 1956). The higher observing frequency causes the thermal complex labelled Cyg X to be comparable in flux density to the non-thermal source Cyg A. Reber correctly deduces not only the differing spectra of these two sources, but also their vastly differing angular extents[3]. The "minor maximum" in Aquila is the North Polar Spur becoming detectable, but the projection from the galactic center region at $\delta = -20°$ is not confirmed in later surveys. Finally, with higher resolution a section of the Milky Way in Orion and Taurus can now be seen to be breaking up into three main components, now recognized as the Crab Nebula, IC 443, and the Rosette Nebula.

NOTES

[1] This reflector has been preserved and is, in fact, still used for monitoring interference at the National Radio Astronomy Observatory in Green Bank, West Virginia.

[2] All galactic coordinates in this volume, except Figure 3, are in the old (l^I, b^I) system. In this system, the plane of the galaxy is defined as 0° in latitude (b^I), while longitude (l^I) increases easterly, being 327° in the direction of the galactic center and 57° in the direction of the Sun's orbit about that center.

[3] For the positions of these, see Table III in the introduction to Part IV.

Cosmic Static*

GROTE REBER†, ASSOCIATE, I.R.E.

INTRODUCTION

IN 1932 Jansky[1] published the first of a series of papers[2,3,4] indicating that a certain type of static appears to come from space and in particular from the plane of the Milky Way. Very few other[5] data are available on the disturbance. Various[6] hypotheses have been advanced to account for the phenomenon but all have failed under quantitative calculation.

160-MEGACYCLE TESTS AT WHEATON, ILLINOIS

The writer became interested in this work about three years ago. It was decided to make measurements at various frequencies with equipment of high resolving power. The apparatus shown in Fig. 1 is really a transit telescope adapted to work at radio frequencies. The mirror is 31 feet in diameter and has a focal

* Decimal classification: R114. Original manuscript received by the Institute, September 8, 1939.

† Wheaton, Ill.

[1] K. G. Jansky, "Directional studies of atmospherics at high frequencies," PROC. I.R.E., vol. 20, pp. 1920–1932; December, (1932).

[2] K. G. Jansky, "Electrical disturbances of extraterrestrial origin, PROC. I.R.E., vol. 21, pp. 1387–1398; October, (1933).

[3] K. G. Jansky, "A note on the source of interstellar interference," PROC. I.R.E., vol. 23, pp. 1158–1163; October, (1935).

[4] K. G. Jansky, "Minimum noise levels obtained on short-wave radio receiving stations," PROC. I.R.E., vol. 25, pp. 1517–1530; December, (1937).

[5] H. T. Friis and C. B. Feldman, "A multiple unit steerable antenna for short-wave reception," PROC. I.R.E., vol. 25, pp. 841–917; July, (1937); *Bell Sys. Tech. Jour.*, vol. 16, pp. 337–419; July, (1937).

[6] Greenstein and Whipple, "The origin of interstellar radio disturbances," *Proc. Nat. Acad. Sci.*, vol. 23, pp. 177–181; March, (1937).

Reprinted, with permission, from the 'Proceedings of the Institute of Radio Engineers' **28**, 68–70 (1940).

Fig. 1—Antenna system used for the investigation of cosmic static.

length of 20 feet. Details of operation have been given elsewhere.[7]

The output of the amplifier is indicated by the meter at the right of Fig. 2. When no radiation is present the reading is constant at some predetermined level. Any radiation captured will cause the meter to move in a downward direction. Readings are taken visually at minute intervals. A typical set of points is shown in Fig. 3. The dotted line across the dip in the curve is the condition that would have obtained had

[7] G. Reber, "Electric resonance chambers," *Communications*, vol. 18, pp. 5–8; December, (1938).

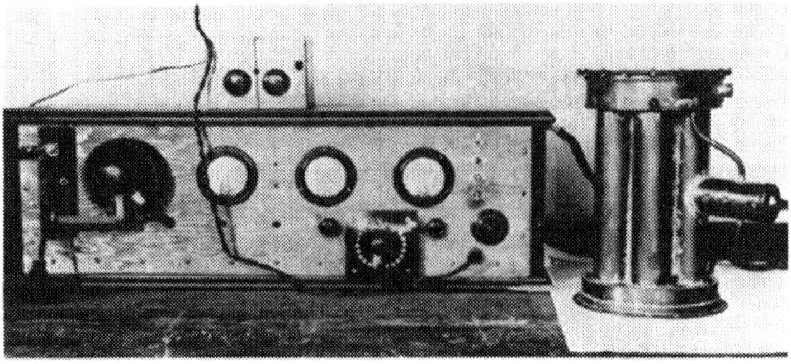

Fig. 2—Control and indicating equipment. The amplifier is shown at right.

no radiation been intercepted. The input to the receiver is

$$\text{input} = \left[\left(\frac{\text{reference}+\text{deflection}}{\text{reference}}\right)^2 - 1\right] 8\times10^{-18}$$
$$\text{watt/kc band} \quad (1)$$

where "reference" is the number of microamperes corresponding to 8×10^{-18} watt per kilocycle band[4,8] and "deflection" is the number of microamperes represented by the dip in the curve. The maximum recorded input is 15 per cent of the reference-level power and the
1 band width of the receiver is about 10^3 kilocycles.* The available input power, therefore, is 1.2×10^{-15} watt. The drum efficiency is 50 per cent, reflector efficiency 85 per cent, and area of mirror 7×10^5 square centimeters. This gives a maximum radiation intensity of 4×10^{-21} watt per square centimeter or equivalent to the energy arriving from a star of 22.1 bolometric magnitude which indicates some of the technical difficulties encountered.

The acceptance cone of the mirror is about 3 degrees

[8] F. B. Llewellyn, "Limits to Amplification," *Bell Sys. Tech. Jour.*, vol. 14, pp. 85–95; January, (1935).

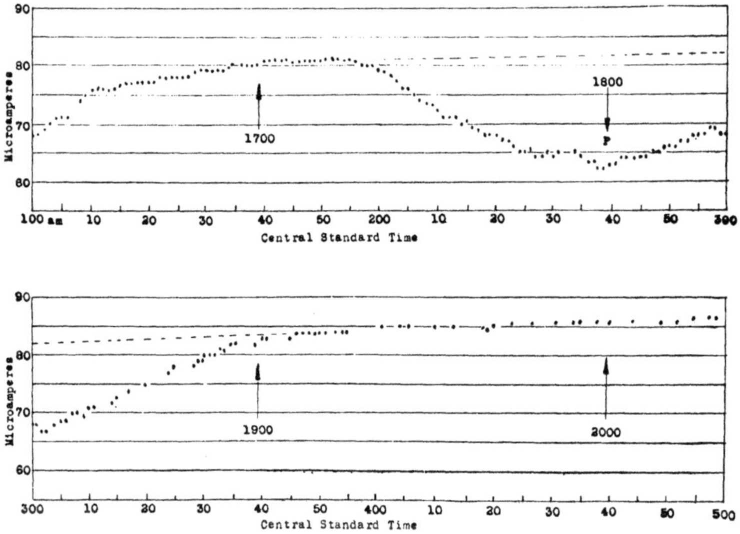

Fig. 3—Microamperes versus C.S.T. on May 11, 1939, for declination −20 degrees. Hours right ascension indicated by arrows. *P* marks the plane of the Milky Way. Galactic longitude for this declination is 337.2 degrees. The at first rapid, and later gradual, increase of reading is due to the warming up of the equipment.

in diameter* so the maximum absolute intensity of 2
radiation becomes 4.5×10^{-25} watt per square centimeter per circular degree per kilocycle band. When intensity is plotted versus galactic longitude the points of Fig. 4 result. The rest of the directions around the Milky Way will be measured during the coming year as they arrive in good position.

The amplifier of Fig. 2 consists of five stages of acorn tubes tuned by concentric lines to the signal frequency. The last stage works into a diode detector which actuates a direct-current amplifier with a microammeter in the plate circuit. An increase in input will cause an increase in bias and a resultant decrease in the microammeter reading. In actual operation the amplifier is on the end of the drum in Fig. 1 and the

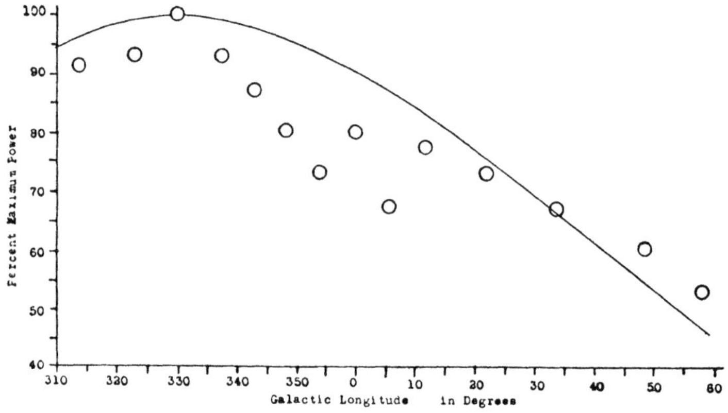

Fig. 4—Intensity versus Galactic longitude l. Circles indicate measured points. Solid line computed from equation (7).

direct bias voltage is fed down the cable to control the equipment.

Perhaps it is well to point out that cosmic static is in no way connected with cosmic rays. The latter are high-speed particles and have no directional effect except a 1 per cent increase on the leading face of the earth as it moves through space.

Radiation of Charged Particles

Considerable material is being thrown off by the stars. This expands and tends to fill up the region between the stars with a very tenuous gas. Since the gas is at a much lower density than the best artificial vacuum the particles are relatively far apart. Under the action of starlight these atoms of gas are ionized or free electrons are ejected with a velocity dependent upon the color of the light involved. Therefore space is filled with positive particles and high-speed electrons (of a few volts energy).

These particles may encounter in a variety of ways. The one of interest here is a free-free transition; that is an electron approaches a positive charge from one direction and is acted upon to leave in another direction. In 1923 Kramers[9] deduced from classical electric theory that when an electron encounters an ion in such a free-free transition energy will be radiated at the expense of the velocity of the electron. Eddington[10] applied Kramers' theory to the material inside a star. At that time he argued a certain correction due to Einstein should not be applied to Kramers' formula. This correction takes into account the stimulation of such radiation due to the transition occurring in equilibrium with an electromagnetic field. Later Eddington[11] applied Kramers' theory to the material of interstellar space with the specific argument that the average velocity loss of an electron due to free-free transitions could not exceed 12 per cent. In 1930 Gaunt[12] rederived Kramers' formula on the basis of quantum mechanics. He decided that Kramers' work is correct for visual range, too high for the X-ray range and too low for the low frequencies. Since no application was known he did not go into the low-frequency case in detail but definitely stated Kramers' formula does not include the effect .of stimulated radiation. Partly on this authority and partly to fit the theory to measurements this correction is included in what follows.

[9] H. A. Kramers, "Theory of the continuous X-ray spectrum," *Phil Mag.*, vol. 46, pp. 836; November, (1923).

[10] A. S. Eddington, "Absorption of radiation inside a star," Monthly Notices of Royal Astronomical Society, vol. 84, p. 104, (1924).

[11] A. S. Eddington, "Diffuse matter in interstellar space," *Proc. Royal Soc.*, vol. 111, pp. 424, (1926).

[12] Gaunt, "Continuous absorption," *Trans. Royal Soc.*, vol. 229, pp. 163, (1930).

Quantitative Calculations

Krămers gives the amount of energy liberated as

$$\rho = \frac{32\ \pi^2 Z^2 e^6}{3\sqrt{3}\ C^3 m^2 U}\, ns\Delta\nu \text{ ergs/cm}^3\text{/sec} \qquad (2)$$

where

n and s are numbers of electrons and ions per cubic centimeter

e, m, and U are the charge, mass, and velocity of the electrons

3 Z is the atomic number*

C is the velocity of light

$\Delta\nu$ is the band width

The correction[10] for stimulated radiation may be reduced to incremental form such that true radiation becomes

$$\rho' = \frac{kT}{h\nu}\, \rho \text{ ergs/cm}^3\text{/sec} \qquad (3)$$

where

k is Boltzmann's constant

h is the Planck's constant

T is the effective temperature of the electromagnetic field (3.2 degrees absolute for space)

ν is the frequency in question

Eddington[11] evaluated ρ as $\rho = 8.7 \times 10^{-42}\ \Delta\nu$ erg per

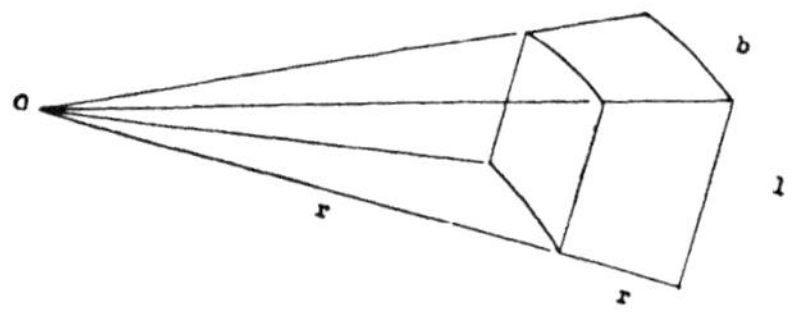

Fig. 5—Incremental volume used in calculation of radiation intensity.

cubic centimeter per second*; therefore ρ' becomes 4

$$\rho' = 5\ 84 \times 10^{-31} \frac{\Delta\nu}{\nu} \text{ ergs/cm}^3\text{/sec.} \qquad (4)$$

Consider now the situation shown in Fig. 5. The incremental volume will be (when the angles are small)

$$\Delta V = r\Delta b\, r\Delta l\Delta r = r^2\Delta b\Delta l\Delta r. \qquad (5)$$

The incremental energy arriving at 0 from this incremental volume is

$$\Delta I = \frac{\rho'}{4\pi r^2}\Delta V = \frac{\rho'}{4\pi}\Delta b\Delta l\Delta r. \qquad (6)$$

Integrating from zero to r centimeters gives an absolute intensity at 160 megacycles of $7\times10^{-41}r$ erg per second per square centimeter per circular degree per kilocycle band.

Astronomical Interpretation

Consider the geometry of Fig. 6 where

$$r = d\cos l + (d^2(\cos^2 l - 1) + c^2)^{1/2} \qquad (7)$$

c = radius of the Milky Way

d = distance from the sun to the center

l = galactic longitude

r = distance from the earth to the edge of the galaxy.

The way r varies with l depends upon the ratio of d to c. The relation giving best fit to points is when $d = 2c/3$ and is shown by a solid line in Fig. 4. Dividing the observed maximum absolute intensity by the

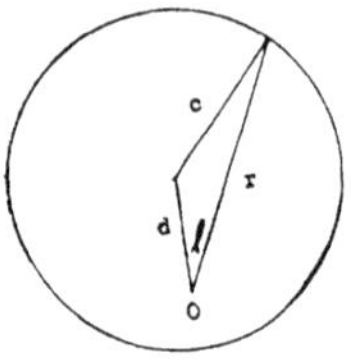

Fig. 6—Plan view of the Milky Way. Earth and sun at point O.

absolute intensity due to radiation from charged particles gives a maximum r of 6.5×10^{22} centimeters or 2.1×10^{4} parsecs. The radius of the galaxy then becomes 12,600 parsecs.

Other Data

Various bright stars such as Vega, Antares, etc.,
have been tried but no measurable response was ob-
tained. Repeated attempts to measure radiation from
Mars were also of no avail. While some fine structure
of radiation pattern seems to be present the existing
5 equipment is not adequate for its accurate resolution*.
No conclusive measurements have been made on any
of the extragalactic nebulae; however, these are to be
worked on next. Tests on the sun are always under
6 rather unfavorable conditions* but if radiation occurs
from this body it is of intensity below 4×10^{-24} watt-
per-square-centimeter-per-kilocycle band at 160 mega-
cycles.

In the beginning it was hoped (from the theory of black-body radiation) that the intensity would increase as the square of frequency. Therefore the first tests were made at 3300 megacycles with rather crude equipment. Nothing was found at the sensitivity limit of 10^{-20} watt per square centimeter per circular degree per kilocycle band. Better apparatus for a frequency of 900 megacycles also produced no positive results at a limit of 10^{-22} watt per square centimeter per circular

degree per kilocycle band.* These negative results point 7
to a function for ρ' similar to the above.

The only other data amenable to calculation is that of Jansky's original paper. Taking into consideration the ground reflection[13] and assuming the Milky Way to have a thickness through the center of one tenth the diameter his maximum measured intensity will produce a radius of 13,100 parsecs and an eccentricity of 3/4 radius for the earth which is a good check on the inverse-frequency function of ρ'. It is also worthy of note that Plaskett in a recent paper[14] deduces a most probable radius of 15,000 parsecs for the Milky Way so these figures are in fair agreement.

[13] C. B. Feldman, "The optical behavior of the ground for short radio waves," PROC. I.R.E., vol. 21, pp. 764–801; June, (1933).

[14] J. S. Plaskett, "Modern conception of the stellar system" *Pop. Astronomy*, p. 293, May, (1939).

EDITOR'S NOTES AND AMENDMENTS

1. Actual bandwidth was 190 kHz; see Table I in the introduction to Section B.
2. Actual beamwidth was $\sim 12\overset{\circ}{.}5$; see Table I in the introduction to Section B.
3. Z is more precisely defined as the effective charge of the ions.
4. Eddington used the values $n = s = 0.045\ \mathrm{cm}^{-3}$, $Z = 3$ and $T_e = 10{,}000$ K (implying $U = 620\ \mathrm{km\ s}^{-1}$).
5. The meaning here is that the chart record hints of possible signals from Mars, but only barely above the receiver noise level and not repeatably.
6. The "unfavorable conditions" during daylight hours were mainly caused by interference from auto ignitions. Curiously, Reber's stated upper limit for any solar radiation is about one-quarter of the eventually detected level of intensity, as derived from Paper 6.
7. These limits may well be in error if Reber also used incorrect values for his beamwidth at these frequencies when converting received power to specific intensity.

COSMIC STATIC

Several papers[1] have been published which indicate that an electromagnetic disturbance in the frequency range 10–20 megacycles arrives approximately from the direction of the Milky Way. It has been shown[2] that black-body radiation from interstellar dust particles is not the source of this energy.

The antenna system shown in Figure 1 was constructed for the investigation of this phenomenon.[3] The receiver can be set at the desired declination by rotating along the meridian on the circular tracks at each side. Readings at a fixed declination are taken over an interval of several hours, the rotation of the earth providing the change in right ascension.

The drum at the focal point is an artificial black body described elsewhere.[4] The entire receiving system has an effective cone of ac-
1 ceptance approximately 3° in diameter.*

The output is indicated by a microammeter so connected that any intercepted energy will cause the readings to decrease. A few typical records are shown in Figure 2, in which the individual points are omitted because they lie too close together. The magnitude of the dip in the curve gives a measure of the intensity of the received energy. Over a long period, as the apparatus warms up, the zero level will gradually rise. The dotted line indicates the run which would have been obtained had no radiation been captured.

The results of preliminary measures of the variation of the static

[1] K. G. Jansky, *Proc. I.R.E.*, **20**, 1920, December, 1932; **21**, 1387, October, 1933; **23**, 1158, October, 1935; **25**, 1517, December, 1937. H. T. Friis and C. B. Feldman, *Bell Tech. J.*, **16**, 337, July, 1937.

[2] Whipple and Greenstein, *Proc. Nat. Acad. Sci.*, **23**, 177, 1937.

[3] For details of the instrumental design and the method of reduction of the data see G. Reber, *Proc. I.R.E.*, **28**, 68, 1940.

[4] G. Reber, *Communications*, **18**, 5, December, 1938.

Reprinted, with permission, from 'Astrophysical Journal' 91, 621–624 (1940).

disturbance as a function of galactic longitude are shown in Figure 3, in which each point represents the central intensity of the dip on one

FIG. 1.—Antenna system viewed from the north. The resonance chamber can be seen supported at the focal point of the parabolic reflector.

night's record. The magnitude of the systematic error may be ±50 per cent, but the general order of the disturbance is correct. The plane from which this energy arrives is tilted about 5° south of the plane of the galaxy in the vicinity of longitude 150°.

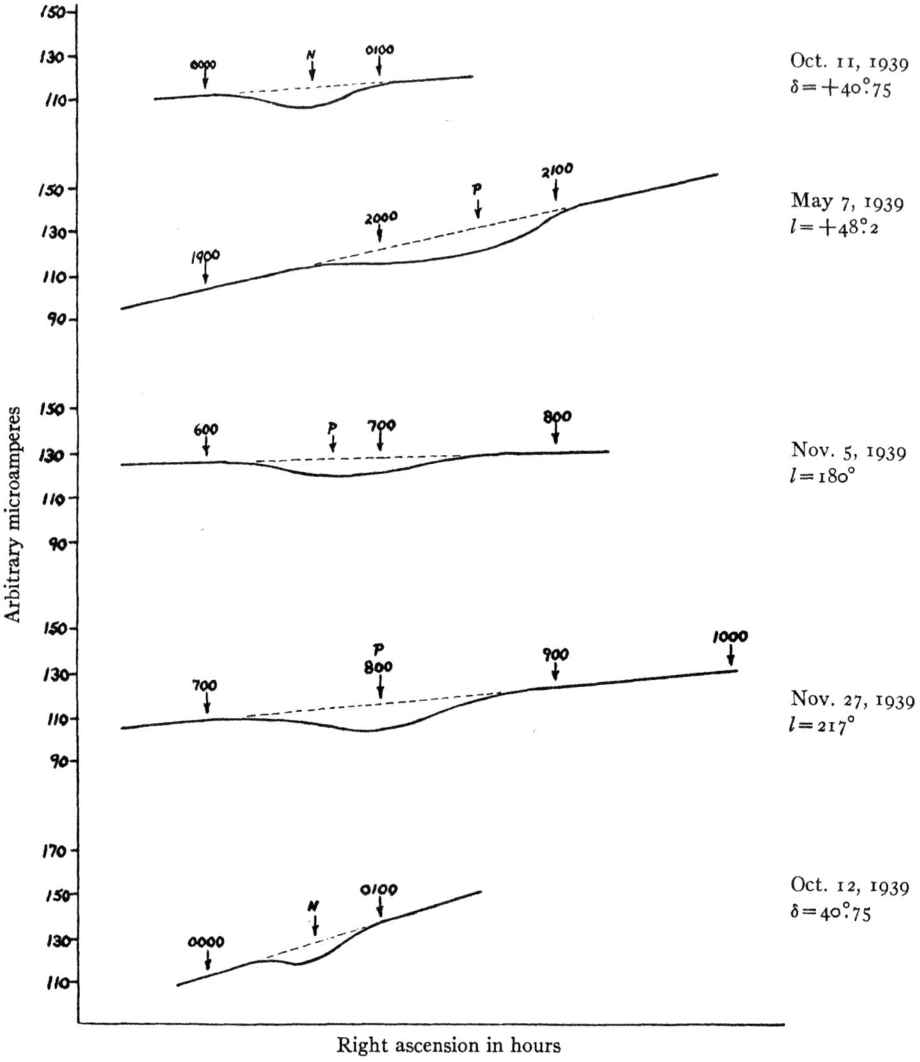

FIG. 2.—Typical records of observation. *P*—plane of Milky Way; *N*—center of Andromeda Nebula.

Since the theory of black-body radiation predicts an intensity proportional to the square of the frequency in this range, the first tests were made at 3300 megacycles. Nothing was found at the sensitivity limit of 10^{-20} watts per square centimeter per circular degree per kilocycle band width. Improved equipment for the frequency of 900 megacycles gave no results at the limit of 10^{-22} watts per square centimeter per circular degree per kilocycle band width. The data of Figures 2 and 3 were obtained at 162 megacycles.* 2

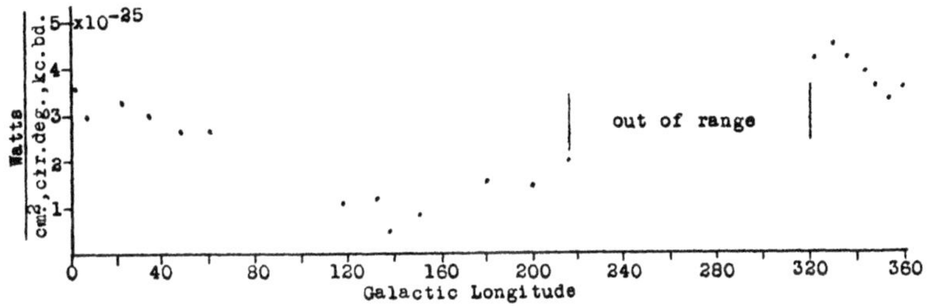

FIG. 3.—Intensity of maximum incident radiation as function of galactic longitude.

A few bright stars, such as Vega, Sirius, Antares, Deneb, and the Sun, gave negative results. Mars and the Orion nebula also gave no readable indication. If radiation is present from any of these objects, the intensity is below 10^{-25} watts per square centimeter per circular degree per kilocycle band width at 162 megacycles.* The 3,4
only other positive results are from the great nebula in Andromeda, with a mean of four readings, giving a maximum intensity of 8×10^{-26} watts per square centimeter per circular degree per kilocycle band width.* 5

The foregoing observations confirm previous evidence that radiation in the radio spectrum is apparently coming from the direction of the Milky Way. The intensity is a function of galactic longitude.

GROTE REBER

212 WEST SEMINARY AVENUE
WHEATON, ILLINOIS
December 1939

EDITOR'S NOTES AND AMENDMENTS

1. Actual beamwidth was $\sim 12^\circ\!.5$; see Table I in the introduction to Section B.
2. The operating frequency is given as 162 MHz in this paper and as 160 MHz elsewhere.
3. See Table I in the introduction to Section B for a summary of the probable errors in all these quoted specific intensities.
4. The upper limit given here for the sun is about one-quarter of that given in Paper 4 and ~5% of the eventually detected intensity as derived from Paper 6.
5. As van de Hulst first pointed out in 1944 (Paper 34), these results for the Andromeda Nebula must be spurious since the traces in Figure 2 indicate "signals" only one-half as wide as the true half-power beamwidth of the antenna.

COSMIC STATIC

GROTE REBER
Wheaton, Illinois
Received May 8, 1944

ABSTRACT

Cosmic static is a disturbance in nature which manifests itself as electromagnetic energy in the radio spectrum arriving from the sky. The results of a survey at a frequency of 160 megacycles per second show the center of this disturbance to be in the constellation of Sagittarius. Minor maxima appear in Cygnus, Cassiopeiae, Canis Major, and Puppis. The lowest minimum is in Perseus. Radiation of measurable intensity is found coming from the sun.

Experiments on the measurement of electromagnetic energy at radio wave lengths arriving from the sky have been conducted at Wheaton, Illinois, for a number of years. Preliminary results have already been published.[1] During the year 1943 new and improved apparatus was put into operation and considerably better data were obtained.

The electromagnetic energy is captured by the mirror shown in Figure 1 and is directed to the mouth of the drum at the focal point of the mirror. Within the drum are a pair of cone antennae. These convert the electromagnetic energy into alternating current. This current is fed from the tips of the cones up a parallel-wire transmission line to the receiver-mounting on the end of the drum. On this mounting is attached a five-stage amplifier of about 90-decibels gain over the frequency range 156–164 megacycles per second.[2] It uses type 954 acorn tubes and a circuit arrangement involving transmission-line elements. The output of the amplifier is rectified by a type 9006 diode, and the resulting direct-current voltage is fed down a concentric cable into the house, where the associated power supply and recorder equipment shown in Figure 2 are located.

It is well known that considerable amounts of random-charge voltage are generated in the grid and plate circuits of the first tube of a multistage amplifier. This background random-charge voltage will be rectified by the diode and appears as a constant direct-current voltage, V_{on}, at the end of the concentric cable. Its magnitude has been the point of much investigation and serves as a reference level for the calibration of the intensity of cosmic static phenomena. The value of V_{on} is noted on a meter before and after a set of data is taken. During the actual taking of data, V_{on} is canceled out by a small battery so that the cosmic static may be more easily perceived.

The mirror of Figure 1 is mounted on an east-west axis so that it may be pointed to any angle of declination between the limits of $-32\overset{\circ}{.}5$ and $+90^\circ$ along the north-south meridian. One of the circular tracks is calibrated in degrees. The mirror is set to point at the desired declination; and then, as the earth rotates, the mirror sweeps out a band in the sky along this particular declination. Whenever the mirror passes over a cosmic static disturbance, energy is collected, and the voltage at the end of the concentric cable increases by an amount ΔV_{on} The recorder of Figure 2 is set to a chart speed of 6 inches per hour. If no cosmic static is intercepted, the recorder will draw a straight line on the chart. If cosmic static is encountered, the pen will move up by an amount ΔV_{on}. Figure 3 shows sample charts taken at various declinations.

If no cosmic static had been intercepted, the recorder would have drawn a line of some constant slope. The magnitude of this slope is unimportant, provided the line remains

[1] G. Reber, *Proc. I.R.E.*, **28**, 68, 1940; *Ap. J.*, **91**, 621, 1940; *Proc. I.R.E.*, **30**, 367, 1942.

[2] G. Reber, *Electronic Industries*, **3**, 89–92, 1944.

Reprinted, with permission, from 'Astrophysical Journal' **100**, 279–287 (1944).

FIG. 1.—Sheet-metal mirror, 31.4 feet in diameter, 20 feet in focal length, used for collecting cosmic static

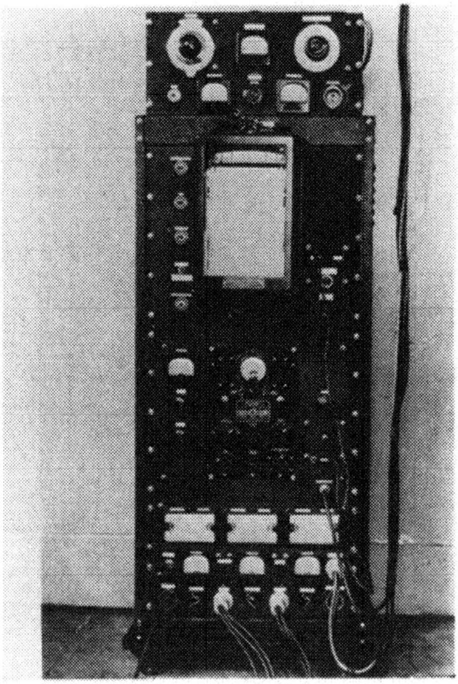

FIG. 2.—Power supply and automatic recorder for cosmic-static investigation

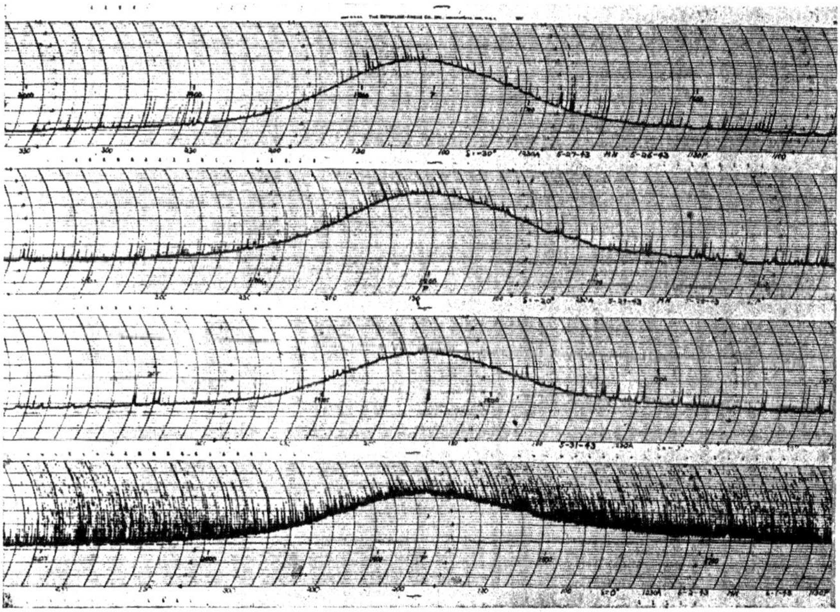

FIG. 3a

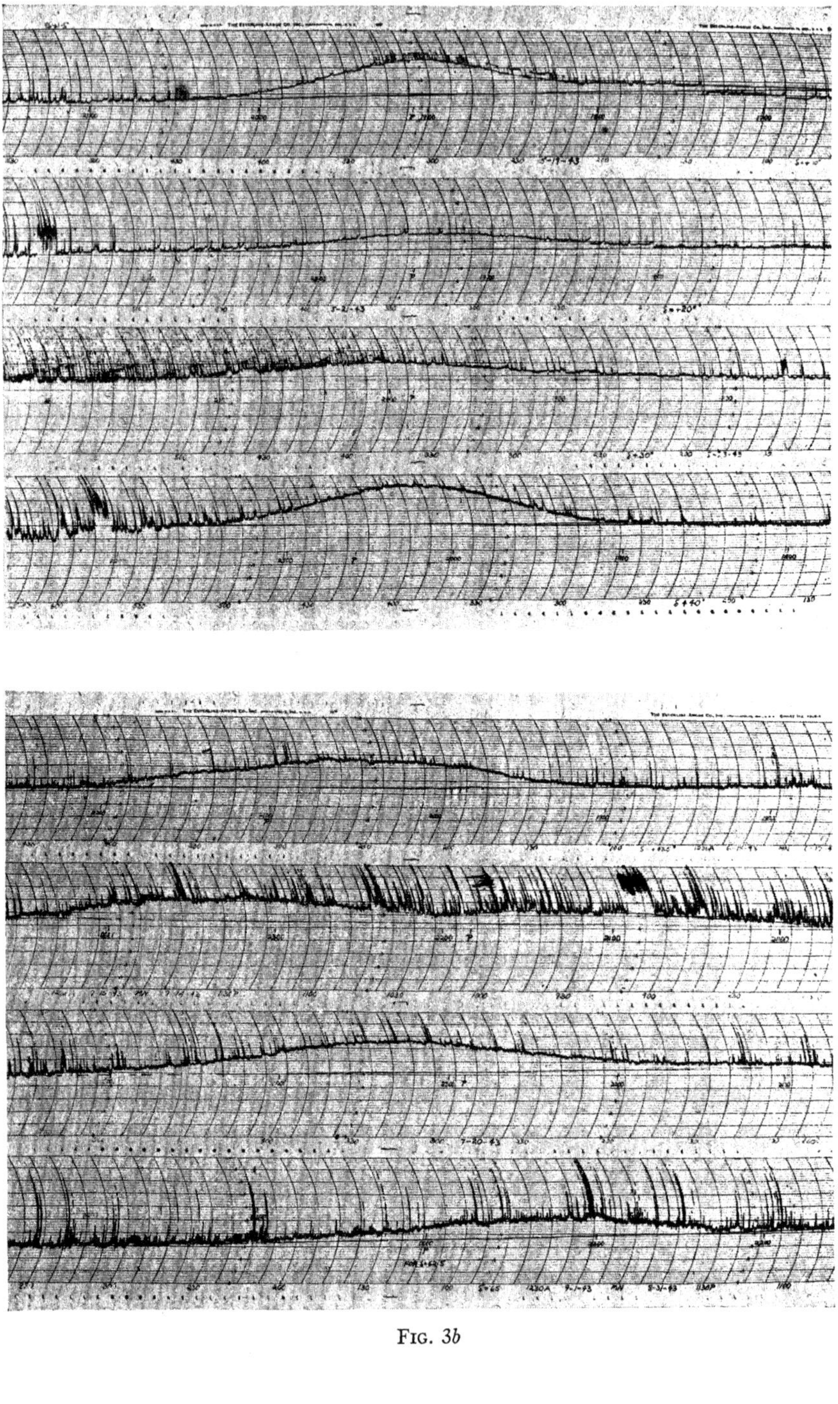

FIG. 3b

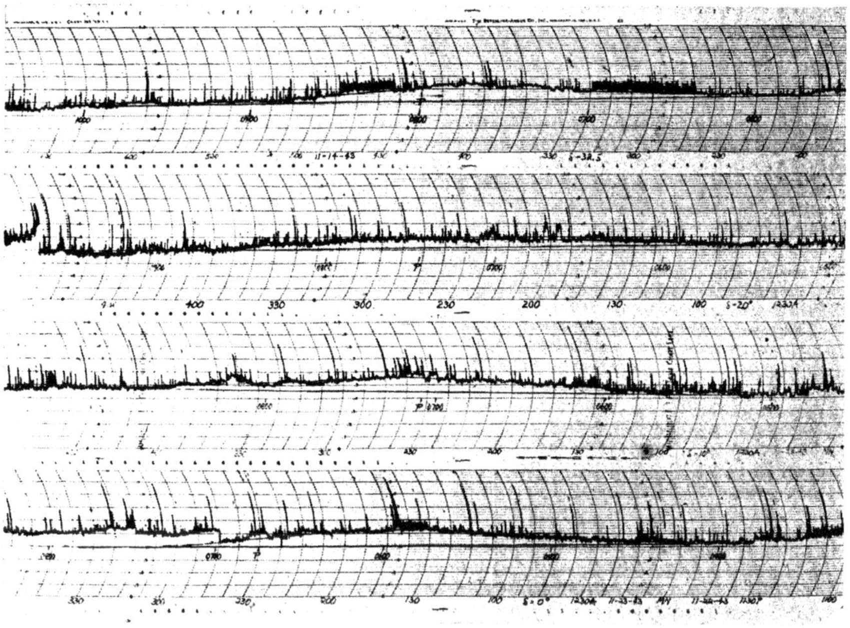

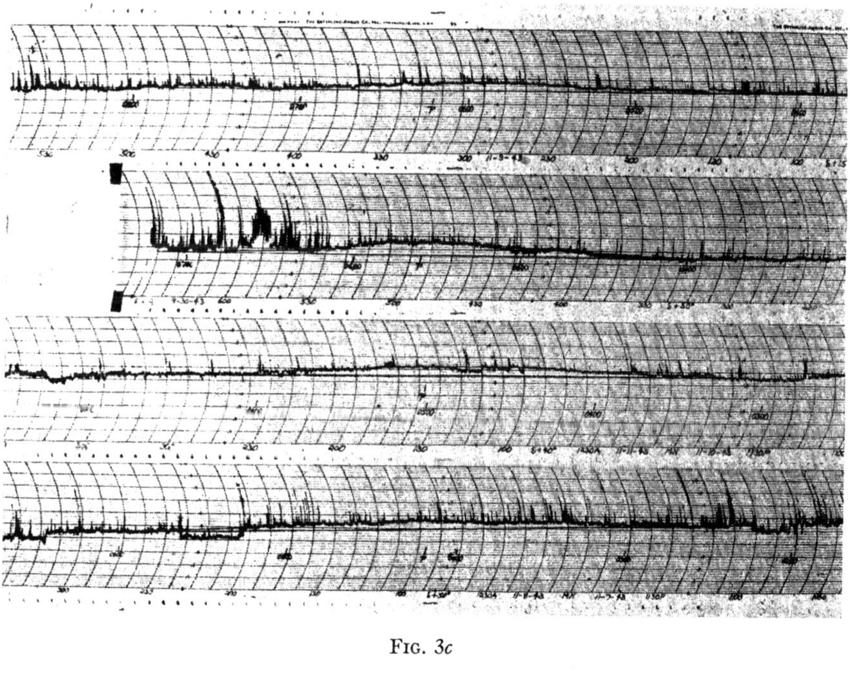

Fig. 3*c*

on scale for the required length of time. On the charts of Figure 3 this slope has been drawn in. The rise of the curved line above the straight one indicates cosmic static, and its magnitude is ΔV_{on}. The small peaked rises are due to the ignition noise from passing automobiles. The sharp spikes are caused by opening and closing of switches. Central Standard time is noted in the margin. Right ascension in hours is marked by short lines on the chart. The plane of the galaxy is indicated by the mark at P.

Referring to a previous paper,[3] the intensity of cosmic static is

$$I = \frac{2U}{E_d E_m A \phi_e \phi_m}, \qquad (1)$$

where

1 $U = 15.1 \times 10^{-14} \Delta$ watts/M.C. band*,

$E_d = 1.00 =$ Drum efficiency ,

$E_m = 0.85 =$ Mirror efficiency ,

$A = 7 \times 10^5$ sq. cm. $=$ Mirror area ,

$\phi_e = 8° =$ Resolving-power in plane of electric vector ,

$\phi_m = 6° =$ Resolving-power in plane of magnetic vector ,

and

$$\Delta = \frac{\Delta V_{on}}{V_{on}}.$$

Assembling these values gives

2 $I = 10.6 \times 10^{-21} \Delta$ watts/sq. cm., cir. deg., M.C. band*. (2)

Several charts are taken at each declination under consideration. All charts of like declination are then worked up, the right ascension for a given intensity is determined, and the results are averaged to give mean right ascension for this intensity. About two hundred charts were obtained in 1943. The final results plotted on a flattened globe are shown in Figure 4, *a* and *b*, for the two hemispheres of the sky. The constant-intensity lines are in terms of 10^{-22} watts/sq. cm., cir. deg., M.C. band. The small numbers at the centers of major maxima are the top values in this direction. The points at declination $-48°$ in Figure 4, *a* are beyond the normal range of the collector machine. They were obtained as the result of an accident when the machine was run off the end of the track in a heavy snowstorm. It lodged with the mirror nearly vertical and the drum somewhat out of focus and resting in the service tower. These points show the general trend of the phenomena but are not of a high order of accuracy and, consequently, are connected with dashed lines.

Too little is known about the cause of this phenomenon to read a great deal from Figure 4. However, it is suggested that this disturbance is in some way connected with the amount of material in space. Since the wave length is long (1.87 meters), the absorption caused by dust is small. Therefore, the intensity is roughly indicative of the amount of material between us and the edge of the Milky Way. On this basis the various maxima point to the directions of projections from the Milky Way. These projections may be similar to the arms often photographed in other spiral nebulae. In the case of the Milky Way this general picture would call for the center toward Sagittarius, and arms in the
3 directions of Cygnus, Cassiopeiae, and Canis Major.* A minimum occurs in Perseus, indicating that we are nearest the edge of the galaxy in that direction. The maximum in lower Puppis is possibly a general rise toward the center. This region from Puppis to

[3] *Proc. I.R.E.* **30**, 376, eq. (20), 1942.

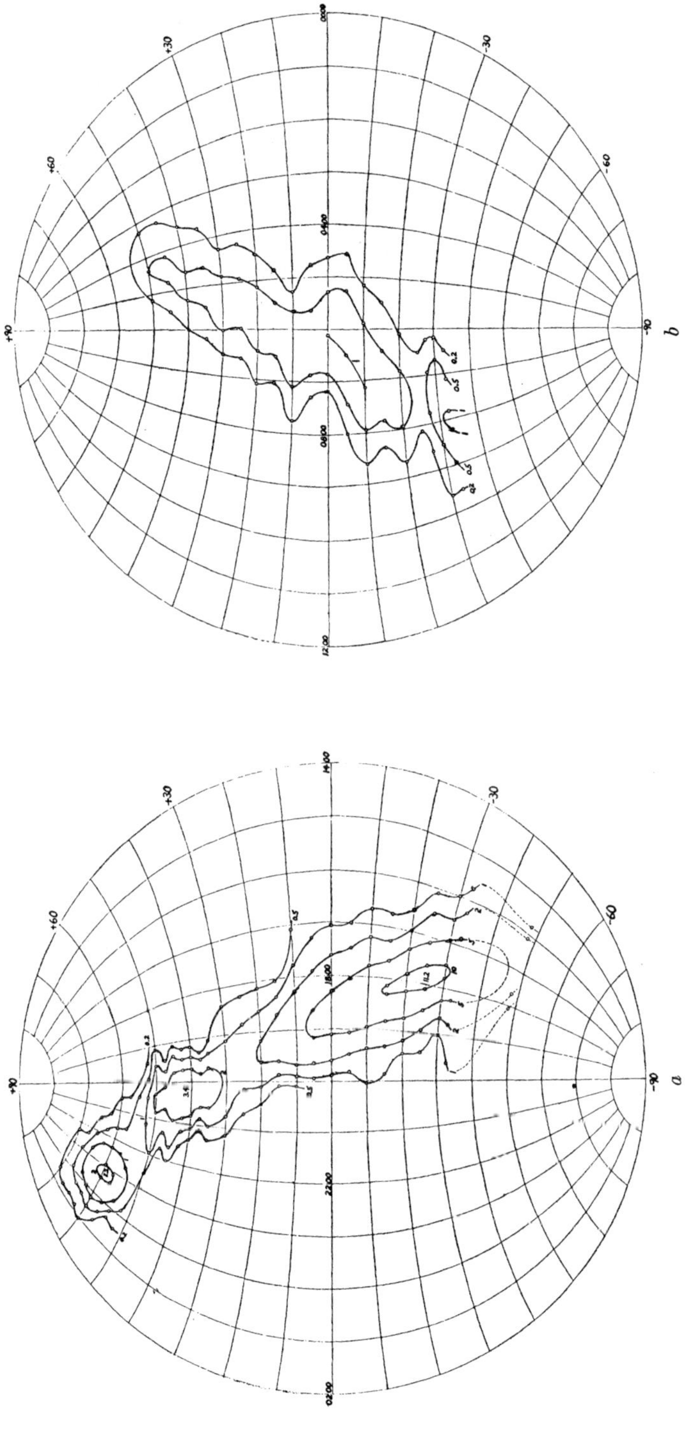

FIG. 4.—Constant intensity lines in terms of 10^{-22} watt/sq. cm./cir. deg./M.C. band

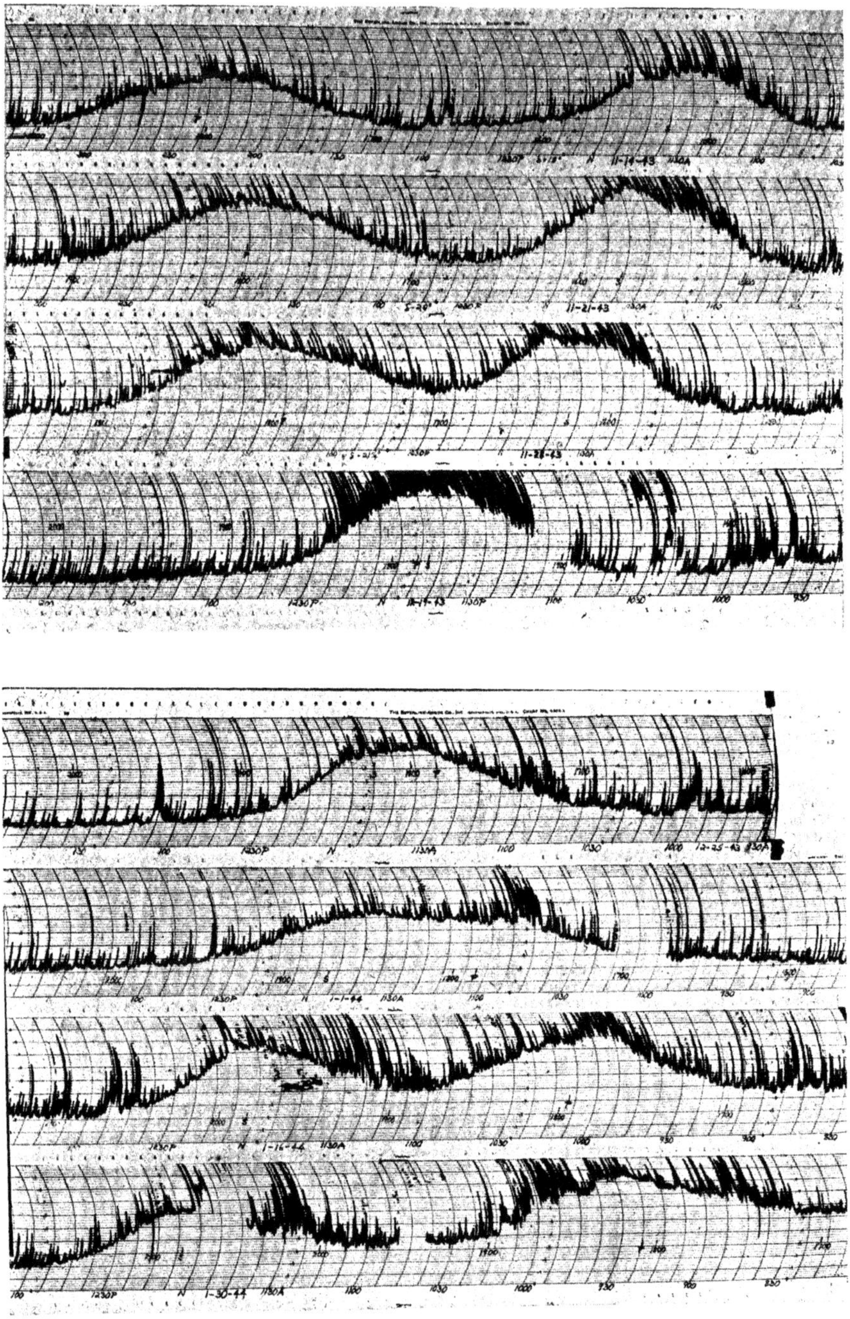

FIG. 5

Scorpius is out of reach at the latitude of Wheaton. However, it is deserving of study, as it may contain other minor maxima.

It has been suggested that this long-wave radiation could be set up in the corona of the sun. Until recently no positive evidence was available. However, the new apparatus was sufficiently stable to allow some data to be taken during the day in spite of severe automobile noise. Figure 5 shows a series of charts taken over several months as the sun passed through the center of the Milky Way. V_{on} is different on the various charts; hence they are not strictly comparable. In any case the sun had the rather surprising center intensity of 10×10^{-22} watts/sq. cm., cir. deg., M.C. band. In spite of the apparent great strength from the sun, this source must be greatly discounted when explaining the origin of cosmic static. If it were the source and the Milky Way were made of average stars like the sun, a very large area in Sagittarius would have a visible intensity equal to that of the sun. Since this is not the case, some other cause must be found to make up the difference
of 20 or 30 mag.* 4

A few other objects, such as the moon, the Pleiades, the Orion Nebula, and the spiral in Andromeda have been tested without conclusive results. The last object seems to be at the threshold of sensitivity of the present apparatus, namely, 0.2×10^{-22} watt/sq. cm.,
cir. deg., M.C. band.* Further improved sensitivity will probably pick up a variety of 5
other objects, and an increase in resolving-power will probably show more detail in the galactic structure.

EDITOR'S NOTES AND AMENDMENTS

1. This equation implies, in modern terms, a system temperature of 10,900 K.
2. See Table I in the introduction to Section B for the probable correction factor which should be applied to all quoted specific intensities.
3. See Table II in the introduction to Section B for the locations of these constellations.
4. Stated differently, the thought here is that the ratio of radio to optical intensity from the Milky Way is $\sim 10^{10}$ times that same ratio for the sun.
5. Although the wording here is more cautious and the claimed intensity is lower than in Paper 5, this possible signal from the Andromeda Nebula still amounts to $\sim$ 6 times the value accepted today.

Cosmic Static*

GROTE REBER†, SENIOR MEMBER, IRE

Summary—**A brief description of the apparatus is given. The results of a survey of the galaxy made at a frequency of 480 Mc are compared to a previous survey made at 160 Mc. The principal new findings are a projection from Sagittarius in the direction of the north galactic pole; a supplementary small rise in Aquila; and a splitting of the maxima in Cygnus and Orion each into two parts.**

INTRODUCTION

THE RESULTS of a survey of the sky at a frequency of 160 Mc made during 1943 have been published.[1] The equipment was then remodeled, and a survey of the sky at a frequency of 480 Mc was conducted during the years 1946 and 1947. The same mirror and general techniques which were used in the first survey were also used in the second survey. Fig. 1 shows the mirror with new focal-point equipment. The large drum used in the first survey has been replaced by a hemisphere, the details and theory of which have been described elsewhere.[2] New electronic equipment was developed and mounted in the cylindrical container above the hemisphere. Fig. 2 shows the rf amplifier. The antenna transmission-line connection is at the bottom left and couples into the first stage, which consists of a pair of 2C42 tubes connected in push-pull. The output of this stage is coupled by a balanced-to-unbalanced transformer to the input of the second stage, which uses a single 446B tube. A total of six cascade stages using 446B tubes terminate in a 9005 diode. All seven stages operate in a grounded-grid fashion, producing 115 db gain with an 8-Mc bandwidth. The noise figure of the amplifier is about 6.5 db above theoretical KTB. The same

* Decimal classification: R113.414. Original manuscript received by the Institute, March 15, 1948.

† 212 W. Seminary Ave., Wheaton, Ill.

[1] Grote Reber, "Cosmic static," *Astrophys. Jour.*, vol. 100, pp. 279–287; November, 1944.

[2] Grote Reber, "Antenna focal devices for parabolic mirrors," PROC. I.R.E., vol. 35, pp. 731–734; July, 1947.

Fig. 1—Sheet-metal mirror 31.4 feet in diameter, 20-foot focal length, on meridian transit mounting, with 480-Mc focal apparatus.* 1

recorder system was used as before, although a new power supply was constructed for the above amplifier.

The mirror of Fig. 1 is mounted on an east-west axis so that it may be pointed to any angle of declination between the limits of $-32°$ and $+90°$, along the north-south meridian. One of the circular tracks is calibrated in degrees. The mirror is set to point at the desired declination; then, as the earth rotates, the mirror sweeps out a band in the sky along this particular declination. Whenever the mirror passes over a cosmic-static disturbance, energy is collected, and the voltage at the end of the amplifier increases. If no cosmic static is intercepted, the recorder will draw a straight line on the chart. If cosmic static is encountered, the pen will move up by an amount depending upon the intensity of the cosmic static. The small spikes and "fur" on the traces shown in Figs. 4, 5, and 6 are due to local interference of impulsive nature. Most of this interference is due to the ignition systems of passing automobiles.

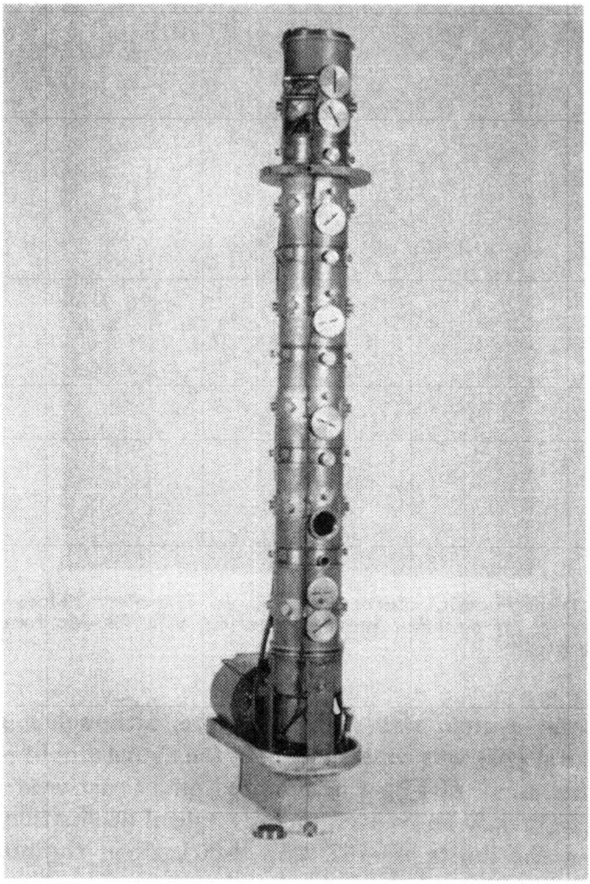

1 Fig. 2—A seven-stage rf amplifier at 480 Mc, using lighthouse tubes.*

Results

Approximately 200 traces were secured. Quite a number of these were made in regions near the galactic poles to determine whether or not any radiation could be found coming from these places. All positive data were organized as before, and the results plotted on two flattened globes. Fig. 3 shows the end results in the form of constant-intensity contours. These levels are in terms of 10^{-22} watt per square centimeter per circular degree per

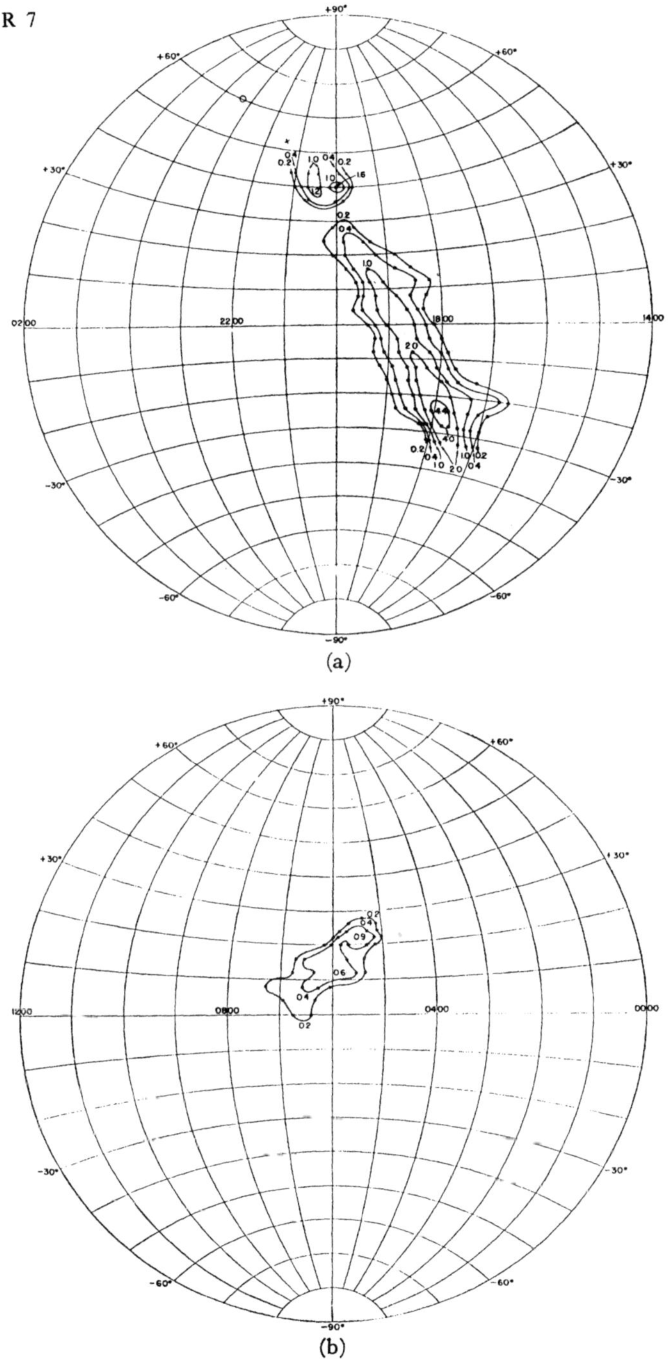

Fig. 3—Constant-intensity contours at 480 Mc in terms of 10^{-22} watts per cm² per circular degree per Mc bandwidth in summer (a), and winter (b).

Mc bandwidth, and are comparable with results of the 160-Mc survey.

Inspection of Fig. 3 and comparison with the results of the first survey show that the Milky Way appears much narrower at 480 Mc than at 160 Mc, and considerably more detail is present at the higher frequency. This may be due, in part, to the narrower width of the
2 antenna acceptance cone, which is only about $3° \times 3°$*at 480 Mc. The higher resolving power should show more detail, if such be present. However, there is no reason to suppose that the outline of the Milky Way will be the same at widely different frequencies, or that the same high resolving power at 160 Mc would produce contours similar to Fig. 3.

Summer Milky Way

3 The projection off to the right of Sagittarius*at declination $-20°$ is real, as evidenced by the trend of the data at $-16°$ and $-24°$. There is no similar projection even faintly apparent at 160 Mc. Peculiarly enough, the maximum in Sagittarius is very close to declination $-25°$, right ascension 17h 50m at both frequencies. This is quite close to the position of the maximum intensity of radiation at 1 micron, as measured by Stebbins and Whitford.[8] Fig. 4 shows two sample traces. The cosmic static appears as the smooth bump in the curve. The quick irregularities or "fur" on the curve represents interference from the ignition systems of passing automobiles.

The saddle in Lyra is much more pronounced at 480 Mc, and an additional minor maximum can be seen in Aquila, as demonstrated by the swelling of the contour lines.

The maximum in Cygnus is split into two parts. Fig. 5 shows a sample trace taken in this region. To make sure of the effect, six traces were taken, and all confirmed its existence. If the average intensity of the two peaks in

[8] J. C. Stebbins and A. E. Whitford, "Infrared radiation from region of the galactic center," *Astrophys. Jour.*, vol. 106, pp. 235–242; November, 1947.

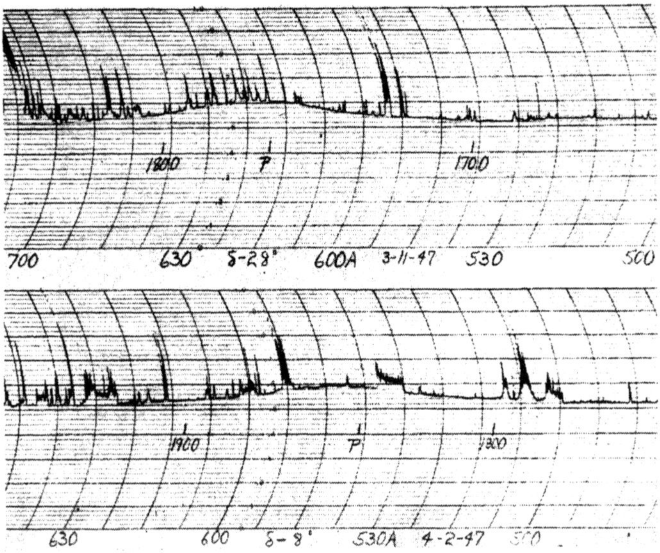

Fig. 4—Sample traces showing rise in drift curve due to cosmic static.

Cygnus is divided by the intensity in Sagittarius, a ratio will be obtained which is closely the same as the corresponding ratio of intensities at 160 Mc.

The 1.6 level maximum at approximately 2000RA and +40° declination agrees quite well in position with a place of high intensity at 60 Mc and 200 Mc located by Bolton.[4] Using Lloyds-mirror technique over the sea, he secures very high effective resolving power and gives the position at 19h 58m 47s plus or minus 10s right ascension, and plus 41° 47m plus or minus 7m. It is interesting to note that apparently this phenomenon extends down to 9.51 Mc.[5,6]

For some reason, the left-hand peak of 1.2 level does not seem to radiate as strongly on the lower frequencies.

[4] J. G. Bolton and G. J. Stanley, "Variable radiations from Cygnus," *Nature*, vol. 161, pp. 312–313; February 28, 1948.

[5] H. T. Friis and C. B. Feldman, "A multiple unit steerable antenna for short-wave reception," PROC. I.R.E., vol. 25, pp. 897, 911, 912, and Table VIII; July, 1937.

[6] Grote Reber and J. L. Greenstein, "Radio investigations of astronomical interest," *Observatory* (London), vol. 67, p. 16, February, 1947.

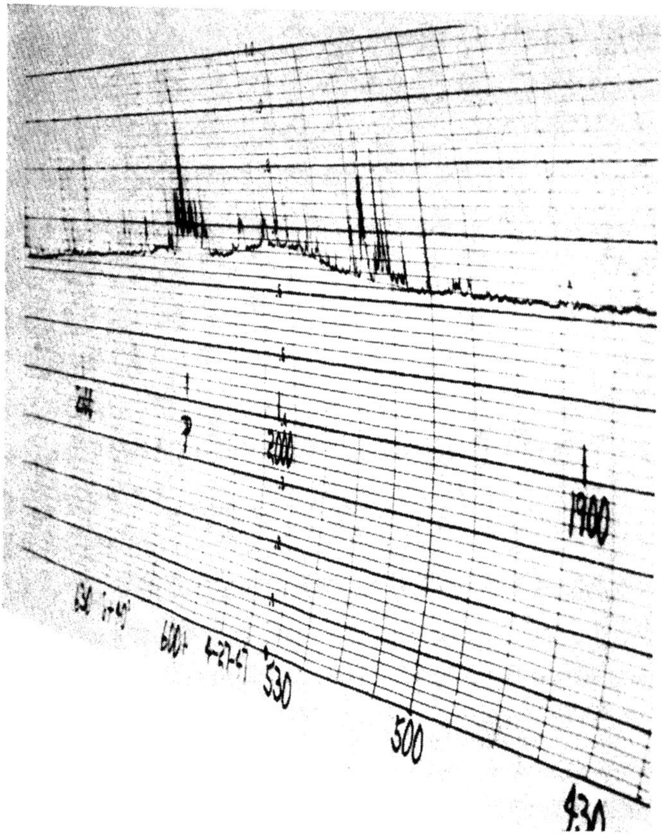

Fig. 5—Sample trace showing cosmic static at declination +40° in Cygnus.

If it had, the Lloyds-mirror technique would have been confused by two sources close together. Thus, it may be assumed that the intensity versus frequency distribution of two sources relatively close together is rather different. Another interpretation is that the 1.6 level peak is really a very-high-intensity area of very small solid angle (perhaps only a minute of arc in diameter), and the 1.2 level peak is a broad source (perhaps a degree in diameter) of low intensity. The integrated product of surface brightness times solid angle might be nearly the same for both sources. Thus the present equipment could not differentiate between these two types of sources. Conversely, the Lloyds-mirror technique would be insensitive to the one of low surface brightness and large solid angle, while the one of reverse characteristics would stand out clearly.

Unfortunately, the sky above declination +44° could not be covered in detail due to lack of time. Since winter weather was closing in and the equipment was scheduled to be moved to Virginia, only a few exploratory traces were secured. At declination +52° the Milky Way was very faint. An *X* marks the spot on Fig. 3(a). At +60° declination considerable energy was found near right ascension 23h 15m. This point is marked by a circle on Fig. 3(a), and is in the region of Cassiopeia where a similar maximum was found at 160 Mc.

WINTER MILKY WAY

The form of the intensity contours shown in Fig. 3(b) is similar, but much smaller than at 160 Mc. It appears that the region of Orion is also split into two parts. Fig. 6 shows a sample trace taken at declination +20°. If the contours had been taken at slightly higher levels, the contour lines would have closed at about the 0.5 level, and the two separate sources would have appeared more clearly.

Several traces showed some energy to be present very faintly up to declination +36° and down to declination −32°.

Too little is known of the cause of the phenomena to read a great deal from Figs. 3(a) and (b). However if the energy is generated by clouds of interstellar gas, and these clouds are of discrete nature as Oort[7] and others have suggested, then perhaps the directions in which these clouds may be located are now becoming apparent.

[7] J. H. Oort, "Interstellar Matter," *Monthly Notices Royal Society*, vol. 106, pp. 159–179; no. 3, 1946.

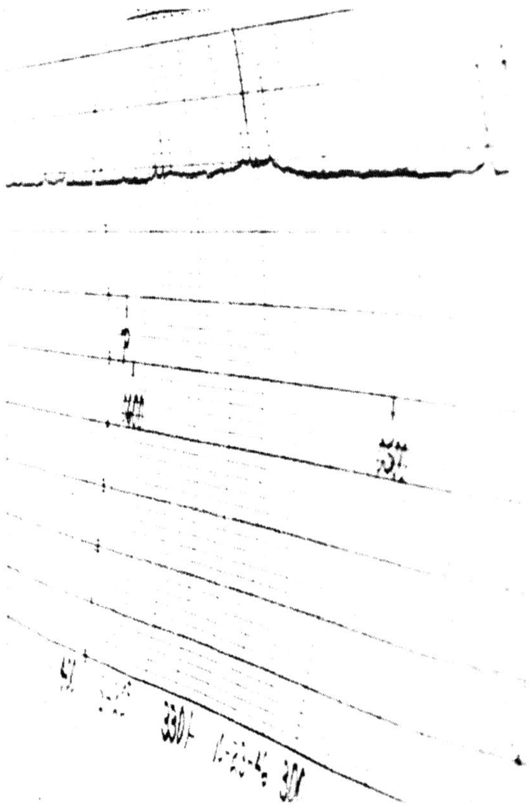

Fig. 6—Sample trace showing cosmic static at declination +20° in Orion.

Other Objects

A few traces showed some faint disturbance to be present near right ascension 12h between declination 0° and −20°. The situation on the Andromeda nebula is the same as at 160 Mc; namely, inconclusive. If the sensitivity of the electronic equipment could be increased by an order, or the diameter of the mirror tripled, then it is quite likely that this object would be readily apparent and easily measurable at both fre-
4 quencies.*

Considerable data on solar radiation was secured and reported upon.[8,9]

[8] Grote Reber, "Solar radiation at 480 Mc," *Nature*, vol. 158, p. 945; December 28, 1946.
[9] Grote Reber, "Solar intensity at 480 Mc," PROC. I.R.E., vol. 36, pp. 88; January, 1948.

EDITOR'S NOTES AND AMENDMENTS

1. The photographs shown in Figures 1 and 2 here are substantially, but not exactly, the same as those in the original article. The man in Figure 1 is unidentified.
2. This beamwidth is probably somewhat of an underestimate and thus all quoted specific intensities are somewhat high (see Table I in the introduction to Section B).
3. See Table II in the introduction to Section B for the location of all mentioned constellations.
4. The Andromeda Nebula (as well as Virgo A) was in fact just about at Reber's limit of detection.

SECTION C

FREE-FREE EMISSION THEORY

With the strong encouragement of Reber and his data, L. G. Henyey (1910–1970) and P. C. Keenan set out to check Reber's calculations of the expected free-free emission from interstellar gas at radio wavelengths. Today, the calculation seems extremely straightforward, but one must always tread carefully when extrapolating physical laws. Henyey and Keenan apply the theory of radiative transfer and atomic recombination to wavelengths fully seven orders of magnitude longer than previous astronomical experience. And in fact they do it correctly. If one uses their parameters in modern theory, one obtains a spectrum with intensities only ~20% lower than those plotted in their Figure 1. The fact that the computed spectrum agrees with Reber's 160 MHz measurement is only a coincidence, since we now know that only a small portion of the observed radiation at low frequencies is thermal by nature. They are nevertheless able to draw the important and correct conclusion that Jansky's observed brightness temperatures of order 100 000 K at 20.6 MHz cannot in any way be reconciled with the notion of thermal emission from a gas at a temperature of ~10 000 K. This problem remained outstanding until 1950 when Kiepenheuer suggested the synchrotron mechanism (Paper 9).

INTERSTELLAR RADIATION FROM FREE ELECTRONS AND HYDROGEN ATOMS

ABSTRACT

The intensity of the radiation produced by electrons and hydrogen ions is considered.
It is shown that for an electron temperature of 10,000° and a density of one electron
per square centimeter* we may expect from interstellar space an intensity of radiation at 1
a wave length of 1.65 meters of 5.8×10^{-18} ergs per square centimeter per second per
square degree per kilocycle band width. This is in excellent agreement with the observational value given by Reber. The computations have been extended over the range 10
meters to 4400 A.

In this note we wish to consider the possible contribution of atomic or ionic processes to the radiation in the radio region of the spectrum. The observations of Jansky[1] and of Reber[2] indicate that there exists radiation, in this region, of the order of 10^{-17} ergs per second per square centimeter per kilocycle coming from a square degree in the Milky Way. Whipple and Greenstein[3] attempted to explain this radiation in terms of the thermal emission by interstellar dust particles, with no success, since the maximum temperature which they could attribute to these particles was about 30°, leaving a discrepancy in intensity of the order of 10,000. Reber[4] proposed an explanation of the radiation in terms of the so-called free-free transitions of electron and proton systems in interstellar space. While his computations were very rough and did not allow for all possibilities, they did indicate that the order of magnitude approaches that of the actual observed intensity. This success may be ascribed principally to his use of an electron temperature of 10,000°, a value which is probably of the right order.

In addition to the free-free transitions we shall take account of the transitions between continuous and bound levels near the continuous limit in hydrogen. The transfer of radiation is governed also by electron and particle scattering. On the other hand, interactions between two free electrons will give rise only to quadrupole radiation, which can be neglected in the spectral range under consideration.

[1] *Proc. I.R.E.*, **20**, 1920, 1932; **25**, 1387, 1937.

[2] *Ap. J.*, **91**, 621, 1940.

[3] *Proc. Nat. Acad. Sci.*, **23**, 177, 1937.

[4] *Proc. I.R.E.*, **28**, 68, 1940.

Reprinted, with permission, from 'Astrophysical Journal' 91, 625–630 (1940).

To compute the contributions of each of these processes we must have some information concerning the velocity distribution in the electron gas. No adequate theory exists at the present time which predicts the nature of the distribution function, so that as a substitute we must make use of a Maxwellian one. While this is certainly not the exact function, it should predict the correct order of magnitude for the intensities of the light emitted and absorbed. There is only one possible source of serious error. One can allow for the stimulated emission, in the case of thermal equilibrium, by multiplying the true absorption coefficient by the correction factor

$$1 - e^{-h\nu/kt} .$$

In the far infrared the exponential is very nearly unity, and the correction factor is usually very small. It is evident that one has here a delicate balance of the two mechanisms that may be seriously disturbed if the condition of thermal equilibrium, or the condition of a Maxwellian velocity distribution in our case, is only approximately true.

The emission and absorption of radiation due to the free-free transitions progress according to the conditions of local thermodynamic equilibrium. If we denote the volume absorption coefficient by κ_ν, the emission is $\kappa_\nu B_\nu$. The bound-free transitions produce only emission, because of the high dilution of the exciting radiation, which may be written $\kappa_\nu' B_\nu - \mu_\nu' I_\nu$, where κ_ν' is the absorption coefficient that would obtain in thermal equilibrium and μ_ν' is the coefficient of stimulated emission. We have

$$\mu_\nu' = \kappa_\nu' \frac{1}{e^{h\nu/kt} - 1}$$

by virtue of the thermal distribution of the emitting particles. In addition we introduce σ, the coefficient of scattering by electrons. Since the density of radiation in space is low, we use σ only in computing the extinction and neglect its contribution to the *Ergiebig-*
2 *keit.**

Under the conditions which we have outlined, the equation of transfer for a ray in the plane of the Milky Way becomes

$$-\frac{dI_\nu}{ds} = (\kappa_\nu + \sigma_\nu - \mu_\nu')I_\nu - (\kappa_\nu + \kappa_\nu')B_\nu . \qquad (1)$$

Putting

$$- d\tau_\nu = d(\kappa_\nu + \sigma_\nu - \mu'_\nu) , \quad (2)$$

it follows that

$$\frac{dI_\nu}{d\tau_\nu} = I_\nu - \frac{\kappa_\nu + \kappa'_\nu}{\kappa_\nu + \sigma_\nu - \mu'_\nu} B_\nu . \quad (3)$$

If we suppose that the radiation finds its origin along the line of sight between the earth and a limiting optical depth τ_ν, we can deduce from (3) that

$$I_\nu = \int_0^{\tau_\nu} \frac{\kappa_\nu + \kappa'_\nu}{\kappa_\nu + \sigma_\nu - \mu'_\nu} B_\nu e^{-\tau_\nu} d\tau_\nu ,$$

or, replacing the *Ergiebigkeit* by a mean value,

$$I_\nu = \frac{\kappa_\nu + \kappa'_\nu}{\kappa_\nu + \sigma_\nu - \mu'_\nu} B_\nu(1 - e^{-\tau_\nu}) . \quad (4)$$

The wave-mechanical equations for the absorption coefficients given by Menzel and Pekeris[5] were used in carrying out the numerical computations. For each spectral region the appropriate expressions for the Gaunt correction factors were inserted in the integrals. The bound-free contributions were taken care of by means of the smoothed formula, and for this reason the resulting curve does not display the discontinuities at the heads of the various series. For electrons at 10,000° the intensity at the head of the Balmer series jumps by a factor of 17, while the Paschen discontinuity is 2.6; but it is only in the immediate vicinity of the series heads that the smoothed formula is seriously in error.

The value of 10,000° adopted for the electron temperature is certainly of the right order of magnitude, since the principal source of the ionizing radiation is found in the B-type stars. The optical depth, τ_ν, was computed on the assumption that the thickness of the emitting layer is about 16,000 parsecs, or 5×10^{22} cm.* 3

The number of electrons per cubic centimeter may be estimated from three different lines of approach. Struve and Elvey[6] have ob-

[5] *M.N.*, **96**, 77, 1935.

[6] *Ap. J.*, **88**, 364, 1938.

served the intensity of $H\alpha$ as produced in interstellar space in the directions of Cygnus and Cepheus. Their value for the absolute intensity of $H\alpha$ may be used to determine N_iN_e, the product of the numbers of electrons and of hydrogen ions per cubic centimeter. Using the anaysis of the recombination of hydrogen given by Baker and Menzel[7] (their case B), we have computed that $N_i = N_e = 0.6$.

Struve[8] and Dunham[9] have estimated the electron density from the observed relative intensities of interstellar *Ca* I and *Ca* II absorption lines. They find 30/cc and 10/cc, respectively. Dunham points out that the density may well be a tenth of his determination because of the uncertainties of the problem.

The intensity of the recombination spectrum, both continuous and discrete, is proportional to N_iN_e multiplied by the thickness of the emitting layer. This intensity in the photographic region, centered around $\lambda = 4400$ A and including the continuum and the lines $H\gamma$ and $H\delta$, must be less than the observed intensity of diffuse light in the Milky Way, since it is likely that the larger portion of this light arises from starlight scattered by interstellar dust. The observed intensity is about 60 tenth-magnitude stars per square degree.[10] Using
4 a thickness of the emitting layer of 1600* parsecs and an electron temperature of 10,000°, we have computed that this intensity implies that

$$N_i = N_e < 1\ .$$

This determination of an upper limit must be given considerable weight since it is based on observational material whose uncertainty lies in the possibility that the true intensity may be smaller than 60 but is certainly not larger by more than the photometric uncertainty of about 20 per cent. We have, therefore, adopted as an upper limit the rounded value

$$N_i = N_e = 1\ ,$$

which is compatible with all the determinations.

[7] *Ap. J.*, **88**, 52, 1938.

[8] *Proc. Nat. Acad. Sci.*, **25**, 39, 1939.

[9] *Proc. Amer. Phil. Soc.*, **81**, 277, 1939.

[10] Henyey and Greenstein, unpublished.

The results of the computation are indicated in Figure 1. The broken line gives the trend of the black-body intensity in the long-wave region. The full curve shows the computed intensities for the various parts of the spectrum. In the photographic region the course

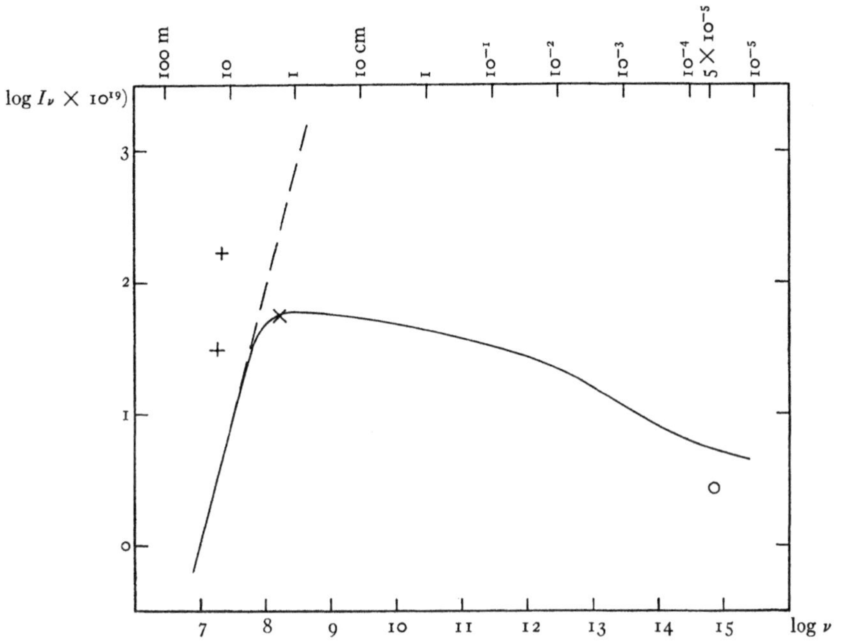

FIG. 1.—The full curve represents the total radiation from free electrons in the galaxy. Ordinates are $19 + \log_{10} I$, where the unit of I is ergs per cm² per sec per square degree per kc band width. Broken curve shows black-body radiation for 10,000°. Observational data are entered as follows:

o—Photographic intensity of diffuse light from Milky Way (Henyey and Greenstein)
+—Radio measures by Jansky (reduced by Reber)★ 5
×—Radio measures by Reber

of the curve is fixed essentially by the bound-free transitions, but to the red of 20μ they become negligible in comparison to the free-free processes. It can be seen that the two curves begin to deviate one from the other at a wave length of about 5 meters. This results from the fact that the value of τ_ν changes at this wave length from an amount much greater than unity to less than unity, or, physically, at this wave length the radiating layer ceases to be opaque and becomes effectively transparent.

The crosses in the long-wave region represent the intensities measured on radio receivers by Jansky and by Reber. The predictions of the theory are in satisfactory agreement with Reber's observation, but in the case of Jansky's data the discrepancy is serious. A change in the value of τ_ν could, at most, lower the computed curve very slightly, since the layer is practically opaque at this wave length, but could never raise it above the black-body intensity. In order to produce the intensity observed by Jansky, it would be necessary to have a source at a temperature of the order of 150,000°.

The observation indicated in the photographic region is the one of relatively high precision discussed above.[10] It will be seen that this observation falls considerably below the computed curve. In reality, if we allow for the probable existence of scattered starlight, the curve should pass well below the point. This apparent discrepancy originates from the large thickness of emitting matter, 16,000 parsecs, which we have used. The high value of the absorption coefficient in the photographic region, amounting to 1 mag. per kiloparsec, means that we cannot receive light from an equivalent thickness much greater than 1500 parsecs. This requires that the intensity at λ 4400 must be cut by a factor of more than 10, in agreement with our previous conclusions.

We conclude that interaction between free electrons and ions in interstellar space probably produces radiation of observable intensity in the radio and possibly the visual region of the spectrum.

L. G. Henyey
Philip C. Keenan

Yerkes Observatory
February 1940

EDITOR'S NOTES AND AMENDMENTS

1. Should be "*cubic* centimeter".
2. *Ergiebigkeit* means "source function".
3. The origin of the value of 16 kpc is not clear; it probably corresponds to a typical path length for a line of sight through the central Milky Way.
4. Should be "16,000 parsecs".
5. The higher Jansky datum refers to 1932 measurements at a wavelength of 14.6 m (Paper 1); the lower is derived from measurements at 16.7 m (Jansky 1937). The Reber datum refers to his measured peak intensity (Paper 4).

SECTION D

SYNCHROTRON RADIATION AND COSMIC RAYS

Inspired by Alfvén and Herlofson's suggestion, only a few months earlier (Paper 27), of the synchrotron mechanism as the cause of the emission in radio sources, K. O. Kiepenheuer (1910–1975) then became the first to directly link the cosmic rays and the general galactic radio emission. It nevertheless was several more years before this proposal became generally accepted[1] and thus resolved the long-standing battle between the proponents for emission from an interstellar gas and those for an integrated effect from unresolved radio stars (e.g., Unsöld (1949), Ryle (1950), Westerhout and Oort (1951)). The clinching evidence for most astronomers was provided by the later demonstration of optical and radio polarization in discrete sources and in the background (e.g., Dombrovsky (1954), Vashakidze (1954), Mayer *et al.* (1957)). The synchrotron theory, at least as applied to astronomy,[2] was not developed further in the West, but almost exclusively in the Soviet Union by V. L. Ginzburg, I. S. Shklovsky and their collaborators. They alone seemed to realize the importance of an intertwining of cosmic ray physics and radio astrophysics. It is interesting to note that it was not until 1960 that primary electrons, which were a necessary consequence of this entire theory, were definitely established to be a component of cosmic rays (Earl, 1961; Meyer and Vogt, 1961).

In Paper 9, Kiepenheuer demonstrates that reasonable values for the cosmic-ray gas in the galaxy yield a radio brightness not unlike that observed. In Paper 10, Ginzburg develops the theory for the galactic background in more detail and also correctly shows that, while Alfvén and Herlofson's suggestion for discrete sources is a fruitful one, the sources cannot in truth be anything like the similarity to our solar system which they imagine. Many other Soviet physicists quickly followed up this work. One of the more important of these studies was done by Getmansev (1952) in which the well-known relation was derived between the radio spectral index α and the electron energy spectral index γ, namely $\alpha = (1 - \gamma)/2$. Another was the suggestion by Shklovsky (1952a) of the existence of a spherical galactic corona of cosmic rays, which would account for the considerable radio brightness found at high galactic latitudes.

NOTES

[1] For instance, it is not mentioned at all in Pawsey and Bracewell's comprehensive textbook *Radio Astronomy*, which was published in 1955.

[2] The theory concerning radiative losses of an electron was, of course, already about 50 years old at this point; the actual (optical) radiation from relativistic electrons was first observed in the General Electric synchrotron by Elder *et al.* (1947).

Cosmic Rays as the Source of General Galactic Radio Emission

K. O. KIEPENHEUER
Yerkes Observatory, University of Chicago, Williams Bay, Wisconsin
June 20, 1950

THE galactic radio emission is not a thermal free-free radiation of interstellar gas, as was first believed. The electronic temperature would have to be of the order of 100,000° in contradiction to all spectroscopic evidence which gives values around 10,000°. Stars could be considered as sources only under very artificial assumptions. The observed intensities, which must come from the outermost layers of stellar atmospheres, could not be blackbody radiation[1] and might be understood only in terms of coherent plasma oscillations of extended regions. The formation and maintenance of these oscillations is hardly possible in stellar atmospheres.[2]

1 It will now be shown that the general cosmic radiation* of our star system is a high frequency source of sufficient power. In interstellar space, at least inside the interstellar clouds which occupy about 5 percent of space, the mean density of kinetic energy ought to be of the same order as the magnetic-field energy; therefore, fields of around 10^{-6} gauss are to be expected. An energetic electron with energy $W \gg m_0c^2$, which is circulating in this field, is radiating electromagnetic energy into a very narrow cone whose angular aperture is m_0c^2/W in the direction of motion. Therefore, an observer at rest receives very short pulses corresponding to a frequency which is very much higher than the classical Larmor frequency, ν_0. The mean spectral intensity distribution of this radiation will then be[3]

$$P(\nu) \approx (e^2/\pi R)(\nu/\nu_0)^{\frac{1}{3}}$$

for $\nu_0 \ll \nu < \nu_c$, where R is the radius of the electron's circular orbit
2 and $\nu_c = \frac{3}{2}\nu_0(W/m_0c^2)^3$.* If n_e is the number of electrons per cm^3 with energy W, the emissivity of high frequency radiation will be

$$\epsilon_\nu \Delta\nu = n_e P(\nu)\Delta\nu = (e^3H/\pi W)n_e(\nu/\nu_0)^{\frac{1}{3}}\Delta\nu \text{ ergs/cm}^3\text{/sec.}$$

This increases steadily with frequency until $\nu = \nu_c$ and then decreases rapidly. The observed distribution[1,4] within the frequency range of 10 to 3000 Mc seems rather to be $\propto \nu^{-0.3}$. We therefore expect to be already in a region with $\nu \geq \nu_c$. Also the involved interstellar magnetic field strengths might vary through space and so influence sensibly the spectral intensity distribution. If the thickness of the emitting layer is D, the intensity of the radiation becomes $I = \epsilon_\nu \Delta\nu D$. Let us suppose that there is one

Reprinted, with permission, from 'Physical Review' 79, 738–739 (1950).

electron per 100 cosmic-ray particles,[5] that is, $n_e \approx 3 \times 10^{-11}$ cm^{-3}. If we tentatively put $W = 10^8$ ev and $D = 1000$ light years (1/100 of the diameter of the galaxy), assuming $\nu = 100$ Mc, and $\Delta\nu = 10$ Mc, we get a radiation intensity of $I \approx 10^{-11}$ erg/cm^2/sec.* This agrees 3
in order of magnitude with the observations of Hey, Parsons, and Philips.[4]

The total radiation loss of an electron moving in a homogeneous magnetic field, H, is

$$-dW/dt = (4\pi/3)\nu_0(e^2/R)(W/m_0c^2)^4 \sim W^3H^2.$$

Since this radiation loss increases so rapidly with energy, electrons with energies greater than about 10^9 ev are not expected. It seems also possible that electrons are eliminated from cosmic rays in the vicinity of stars by collisions with thermal photons.[6] The composition of cosmic rays we observe at the outer boundary of the earth's atmosphere (being close to the sun) is therefore not an average sample of interstellar space.

The general radio emission shows some relation to the visible structure of the galaxy,[4] but does not seem to be directly correlated with stars or other galactic objects. It follows, therefore, that the sources of radio emission are more closely related to the general shape of the galaxy than to its visible components. This conclusion favors Fermi's hypothesis that the distribution of cosmic rays and of galactic matter is more or less the same, cosmic rays being created in interstellar space and not by the stars.

A relation between radio emission and cosmic rays has previously been suggested by Alfvén[7] for the special case of so-called radio stars (discrete centers of strong radio emission). This interesting suggestion, which has stimulated the above analysis, appears to the writer to be in need of re-examination.

I wish to thank Mr. Grote Reber for helpful discussions.

[1] A. Unsöld, Zeits. f. Astrophysik **26**, 176 (1949).
[2] M. Ryle, Proc. Phys. Soc. London **A62**, 483 (1949).* 4
[3] J. Schwinger, Phys. Rev. **75**, 1912 (1949).
[4] Hey, Parsons, and Philips, Proc. Roy. Soc. **A192**, 425 (1947).

[5] There must be a certain number of electrons in cosmic rays as a direct consequence of their ionization losses in interstellar space and of electron pair production by encounters of protons and thermal starlight photons. Also there is some indication of the existence of electrons in the primary radiation given by direct measurements (B. Rossi, Rev. Mod. Phys. **21**, **104** (1949)) and by the east-west asymmetry (L. Jánossy, *Cosmic Rays* (Oxford University Press, London, 1948)).

[6] E. Feenberg and H. Primakoff, Phys. Rev. **73**, 449 (1948).
[7] H. Alfvén and N. Herlofson, Phys. Rev. **78**, 616 (1950).

EDITOR'S NOTES AND AMENDMENTS

1. *Cosmic radiation* here refers to cosmic rays.
2. Exponent "3" should be "2".
3. This rough estimate for I comes from a calculation using the theory for $\nu<\nu_c$ and then adjusting downwards somewhat since it has been already argued that probably $\nu \geqslant \nu_c$.
4. Should be 194<u>8</u>.

COSMIC RAYS AS THE SOURCE OF GALACTIC RADIO EMISSION

(КОСМИЧЕСКИЕ ЛУЧИ КАК ИСТОЧНИК ГАЛАКТИЧЕСКОГО РАДИОИЭЛУЧЕНИЯ)

V. L. GINZBURG*
(Introduced by Academician M. A. Leontovich, Member of the Academy, 24 November 1950)

It has lately become evident that the galactic radio emission cannot basically be explained by thermal emission from interstellar electrons, as has been previously suggested (e.g., see the review paper by Unsöld [1]). Thus, it is particularly worthwhile considering the idea that the galactic radio emission is due to bremsstrahlung of relativistic electrons (i.e., of the electron component of the cosmic radiation) in interstellar and circumstellar magnetic fields [2, 3].

If we assume that the observed general radio emission of the galaxy or the emission from individual discrete sources is due to the said mechanism, and if we know the magnetic field intensity, it is possible to show what the intensity of the corresponding cosmic ray particles should be. The question of the acceptability or inacceptability of cosmic electron data obtained in this way, in fact, amounts to much the same as the problem of the origin of cosmic rays. Since the problem thus formulated, and seen by us as fundamental, is not considered by Alfvén and Herlofson [2] and is solved by Kiepenheuer [3] with inadequate correctness, it seems opportune for us to present here the results of such an analysis.

1. The emission spectrum of particles with energy $E \gg \mathrm{mc}^2$, moving in a homogeneous magnetic field, is given a most detailed consideration by Vladimirski [4]. The energy emitted per second per unit frequency is:

$$P(\nu) = 2\pi P(\omega) = 16\frac{e^3 H}{mc^2}\, p\left(\frac{\omega}{\omega_1}\right); \tag{1}$$

* Translated from the Russian in *Dokl. Akad. Nauk SSSR*, 76, 377 (1951) by E. R. Hope (translation T 132 R of the Defence Scientific Information Service (Canada), 1954, with modifications).

if

$$\frac{\omega}{\omega_1} \ll 1, \quad p\left(\frac{\omega}{\omega_1}\right) = 0.256\left(\frac{\omega}{2\omega_1}\right)^{1/3};$$

and if

$$\frac{\omega}{\omega_1} \gg 1, \quad p\left(\frac{\omega}{\omega_1}\right) = \frac{1}{16}\left(\pi\frac{\omega}{\omega_1}\right)^{1/2} e^{-2\omega/3\omega_1},$$

where $\omega_1 = (eH/mc)(E/mc^2)^2, \omega/2\pi = \nu =$ frequency of the emission, and H = the component of the magnetic field perpendicular to the particle velocity.

The function p, in the interval $0.18 < (\omega/\omega_1) < 16$, is tabulated by Vladimirski [4]; its maximum is at $\omega/\omega_1 = 0.5$, where $p(0.5) = 0.10$; further, $p(0.18) = 0.09$, $p(2) = 0.055$, $p(4.05) = 0.018$, $p(6.5) = 0.004$, $p(10.4) = 4 \times 10^{-4}$, $p(16) = 10^{-5}$. All these values are in round numbers, since we do not need an accuracy greater than a few percent. Outside the above interval we may use, with the same degree of accuracy, the limit-formulae indicated in Equation (1).[1]

At the maximum

$$P(\nu_{\max}) = 1.6\frac{e^3H}{mc^2} = 2.15 \times 10^{-22} H \frac{\text{erg}}{\text{sec Hz}},$$

and (2)

$$\nu_{\max} = 0.5\frac{\omega_1}{2\pi} = 1.4 \times 10^6 H\left(\frac{E}{mc^2}\right)^2 \text{Hz},$$

where in the passage to numerical values it is assumed that the radiating particle is an electron. As we shall see later, there is no reality in the assumption that the protons radiate, and we shall not even discuss it in more detail.

The specific intensity of the radio emission in the given interval is $I_\nu = 1/4\pi \int P(\nu)N \exp(-\kappa_\nu x)\,dx$, where x is the distance from the point of observation (i.e., the Earth), κ_ν is the absorption coefficient in the interstellar gas, and N is the density of the emitting particles (more precisely, $P(\nu)N = \int P(\nu, E)N_1(E)\,dE$, where $N_1\,dE$, is the particle density in the energy interval dE). In addition we assume that the radio emission from a volume element is in the mean isotropic, owing to the isotropy of the cosmic radiation and the random orientation of the field H along the line of sight. (These assumptions introduce an error in I_ν which, in general, may be characterized by a coefficient of order unity.) For wavelengths $\lambda \lesssim 15$ m, the Galaxy is still found to be optically quite thin, i.e., $\int \kappa_\nu\,dx \ll 1$ (Unsöld [1]). Hence,

$$I_\nu = \frac{P(\nu)}{4\pi} NR, \tag{3}$$

where $P(\nu)$ and N are averaged over the path and over the range of energy of the cosmic particles, while R is the size of the system in the given direction. The value of $I(\nu)$ will obviously be a maximum where the given frequency ν is the frequency $\nu_{\max}$ in Equation (2).

Thus we find

$$I_{\nu,\max} = 1.7 \times 10^{-23} HNR \frac{\text{erg}}{\text{sec cm}^2 \text{ Hz ster}},$$

and (4)

$$\nu_{\max} = 1.4 \times 10^6 H \left(\frac{E}{mc^2}\right)^2 \text{ Hz}.$$

It has been experimentally established that at $\lambda = c/\nu \lesssim 15$ m, the quantity I_ν decreases as the frequency increases. Assuming, for instance, that the emission is maximum at a wavelength of 20 m, then from Equation (4) we find that $H(E/mc^2)^2 \simeq 10$, i.e., for $H \sim 10^{-6}$ oersted we have $E \sim 10^9$ eV. As is known from experiment, the intensity of the radio emission at $\lambda \sim 10 - 15$ m corresponds to black-body radiation of temperature $T_{\text{eff}} \sim 1 \times 10^5$ ° to 2×10^5 °. For a black body we know

$$I_\nu = \frac{2kT_{\text{eff}}}{\lambda^2} = \frac{2.76 \times 10^{-16}\, T_{\text{eff}}}{\lambda^2 \text{ (in cm)}} \quad \frac{\text{erg}}{\text{sec cm}^2 \text{ Hz ster}}. \tag{5}$$

Comparing Equations (4) and (5), we see that with $T_{\text{eff}} = 10^5$ °, $\lambda = 10^3$ cm and $R = 10^{22}$ cm (1/10 of the diameter of the Galaxy), we have $HN \simeq 10^{-16}$. Hence for $H \sim 10^{-6}$ oersted we find $N \sim 10^{-10}$ cm^{-3}. As is evident from its nature, this calculation of ours gives the maximum admissible value of HN. At the boundary of the terrestrial atmosphere and in the vertical direction, the intensity of the cosmic radiation (basically protons) is

$$I_c \simeq 0.12 \frac{\text{particles}}{\text{cm}^2 \text{ sec ster}};$$

thus, for isotropic conditions, $N \simeq (4\pi/c) I_c \sim 5 \times 10^{-11}$ cm^{-3}. The number of electrons impinging upon the Earth is at least 100 times smaller (Rossi [5]), but in interstellar space a value of $N \sim 10^{-10}$ cm^{-3} for electrons of energy $E \sim 10^9$ eV is not inadmissible.* Thus the observed intensity of the 1
general galactic radio emission may be explained by the radiation of cosmic electrons, provided that independent considerations do not ultimately force

us to doubt our figures: $H \sim 10^{-6}$ oersted, a path of length $R \sim 10^{22}$ cm, and $N \sim 10^{-10}$ cm^{-3}.

Examination of a number of other possible sources of radio emission (bremsstrahlung and radio emission associated with electron scattering) shows that they cannot account for the observed picture. At the present time, then, we do not see any sources for the general galactic radio emission other than the mechanism discussed here. Any other mechanism actually amounts to a denial of the very existence of emission sources in interstellar space and reduces the general galactic radio emission to that from discrete sources associated with stars.

The origin of the radio emission from discrete sources is likewise still not clear; Alfvén and Herlofson [2] refer to the mechanism considered here, namely radiation from relativistic electrons in the magnetic field surrounding a star.

2. Turning to the question of the discrete sources, we remember that the total intensity corresponding to these discrete sources, at $\lambda \sim 3 - 5$ m, attains a value (Unsöld [1])

$$J_\nu = I_\nu \, \Delta\Omega \sim 10^{-19} \frac{\text{erg}}{\text{cm}^2 \text{ sec Hz}}, \tag{7}$$

where $\Delta\Omega$ is the solid angle subtended by a source. All we know from experiment is that the angular sizes of the sources are less than $6 - 8'$. Under the most favorable circumstances, i.e., using Equation (4), we have

$$J_\nu \simeq \frac{P(\nu)}{4\pi} N \, \Delta\Omega \, D = 1.7 \times 10^{-23} HN \, \Delta\Omega \, D \,, \tag{8}$$

where D is the diameter of the source.

Making use of Equation (7) and supposing that $\Delta\Omega = (6')^2 = 3 \times 10^{-6}$ ster and that the distance to the source is $R = 5 \times 10^{22}$ cm (whence $D = R \cdot 6, \simeq 10^{20}$ cm), we have

$$(HN)_{\text{min}} \sim 10^{-11}, \tag{9}$$

from which N_{min} is $\sim 10^{-7}$ cm^{-3} even for $H \sim 10^{-4}$ oersted. If, however, we assume, as do Alfvén and Herlofson [2], that $D \sim 0.1$ light year $\simeq 10^{17}$ cm, then with $H \sim 10^{-4}$ oersted we have $N_{\text{min}} \sim 10^{-4}$ cm^{-3}. In other words, the required electron density is 10^6 times greater than the density of all cosmic [ray] particles in the vicinity of the Earth. It is thus clear that the discrete sources of radio emission contemplated by Alfvén and Herlofson cannot possibly be similar to our "solar cosmic [radiation] system" characterized by $D \sim 10^{17}$ cm, $H \sim 10^{-5}$ oersted, N(electrons) $< 5 \times 10^{-13}$ cm^{-3}, and N(protons) $\sim 5 \times 10^{-11}$ cm^{-3} (Alfvén [6]).

From the point of view of testing the correctness of this mechanism of cosmic ray generation, it is particularly important to explain the spectrum of the general galactic emission, primarily at $\lambda > 10$ m, and also the radio emission spectra and sizes of the discrete sources (determination of the sizes is particularly difficult, but certain possibilities in this direction do exist (Getmansev and Ginzburg [7])).

3. In connection with the data which we have been considering, there naturally arises the question of radio emission from cosmic rays in the terrestrial magnetic field of $H \sim 0.5$ oersted. It is easy to see that the emission from a single particle is in this case too feeble to make its observation possible (the total radiated energy is not greater in order of magnitude than $P(\nu_{\max})\rho/c \sim 10^{-24}$ erg/Hz, where ρ is the radius of the Earth). Also too feeble is the radio emission produced in the retardation [bremsstrahlung] and generation of cosmic particles. In the latter case (see, e.g., Ginzburg and Frank [8]) we have

$$P(\nu) = \frac{2e^2\beta^2}{c}\left[\frac{\beta^2+1}{2\beta^3}\log_e\frac{1+\beta}{1-\beta} - \frac{1}{\beta^2}\right] \simeq \frac{4e^2}{c}\left(\log_e\frac{2E}{mc^2} - \frac{1}{2}\right), \qquad (10)$$

where $\beta = v/c$ and the use of the second expression is valid when $E \gg mc^2$. Thus, even in a large Auger shower, in which, let us say, 10^8 particles are formed, there is radiated an energy of only $\sim 10^{-20}$ erg Hz^{-1} distributed over an area of the order of the shower area.[2]

However, both of the above effects may be important in explaining "bursts" of sky radio emission (Covington [10]) and radio emission con- 2
nected with the aurorae (Forsyth *et al.* [11]).* Thus, for instance, Equation (10) indicates that a stream of any kind of particles with unit charge and with $\beta \sim 1$ will (for a single abrupt deceleration of each particle of the stream) radiate at a wavelength $\lambda = 10$ cm an energy corresponding to a rise of $\sim 30°$ in the sky radio emission temperature (see Equation (5) and Covington [10]) if there are 10^{13} particles per second falling upon each cm^2 of the surface of the atmosphere, corresponding to a current density of $\sim 10^{-6}$ A cm^{-2}. In the case of sufficiently high-speed electrons emitting in the terrestrial magnetic field, the same increase of the effective temperature will be produced if the particle stream amounts to $\sim 10^8$ electrons cm^{-2} sec^{-1}. Considering all other geophysical data, the question of the admissibility of such stream rates

is one requiring a special analysis which has not yet been carried out. It is also necessary to bear in mind that other sources of sky and auroral radio emission, beside those mentioned, are conceivable. The problem is also complicated by the extreme scantiness of the relevant experimental data.

For this reason, we shall limit ourselves to just one further remark.

Forsyth *et al.* [11] state that the emission during an auroral display arrives in the form of individual "bursts" of length $\Delta t \sim 1 \times 10^{-6}$ to 5×10^{-6} sec and, it seems, at intervals $T \sim 10^{-3}$ sec one from the other. In this connection it is curious to note that an electron (or a sufficiently small cloud of relativistic electrons) moving in a magnetic field will radiate in just this fashion. In fact, the electrons radiate only in directions within an angle $\theta_0 \sim mc^2/E$ of their instantaneous velocity vectors. Hence, a motionless observer located in the plane of the orbit will "see" separate bursts of duration $\Delta t \sim \theta_0/\omega_0 = mc/eH$, at intervals of $T = 2\pi/\omega_0$, where $\omega_0 = (eH/mc)(mc^2/E)$ = rotation frequency in the magnetic field. In a magnetic field of intensity $H \sim 0.1$ oersted we should have, for emission at a frequency $\nu_{max} \sim 3 \times 10^9$ Hz (see Equation (2); Forsyth *et al.* [11] observed at $\lambda = c/\nu = 10$ cm), $E/mc^2 \sim 10^2$. If so, $\Delta t \sim 10^{-6}$ sec and $T \sim 3 \times 10^{-4}$ sec.

Of course, the fact that we obtain in this way figures of the right order of magnitude is still very far from proving that the observed auroral radio emission is due to emission from electrons in the terrestrial magnetic field. But it seems to us that the proposed mechanism merits consideration.

P. N. Lebedev Institute of Physics,
Academy of Sciences of the U.S.S.R.
Received: 31 October 1950

NOTES

[1] The error of the calculation by Kiepenheuer [3] is primarily due to a mistake in evaluating $P(\nu)$; a formula valid only for $\omega \ll \omega_1$ is used for $\omega > \omega_1$.

[2] We note in this connection that it is erroneous to talk of the possibility of observing the reflection of radio signals from Auger showers as do Blackett and Lovell [9]. More precisely, they have overestimated the reflection coefficient by a factor of 10^5.

LITERATURE CITED

1. Unsöld, A., 1949, *Zs. f. Astrophys.* **26**, 176.
2. Alfvén, H. and Herlofson, N., 1950, *Phys. Rev.* 78, 616.
3. Kiepenheuer, K. O., 1950, *Phys. Rev.* **79**, 738.
4. Vladimirski, V. V., 1948, *Zh. Eksp. Teor. Fiz.* 18, 392.
5. Rossi, B., 1949, *Usp. Fiz. Nauk.* **38**, 222.
6. Alfvén, H., 1950, *Phys. Rev.* 77, 375.
7. Getmansev, G. G. and Ginzburg, V. L., 1950, *Zh. Eksp. Teor. Fiz.* **20**, 347.
8. Ginzburg, V. L. and Frank, I. M., 1946, *Zh. Eksp. Teor. Fiz.* **16**, 15.
9. Blackett, P. M. S. and Lovell, A. C. B., 1941, *Proc. Roy. Soc.* **A177**, 183.
10. Covington, A. E., 1950, *J. Geophys. Res.* **55**, 33.
11. Forsyth, P. A., Petrie, W., and Currie, B. W., 1949, *Nature* **164**, 453.

EDITOR'S NOTES AND AMENDMENTS

1. Earl (1961) measured a value of $N \sim 1.3 \times 10^{-12}$ cm^{-3} above the Earth's atmosphere; however, Ginzburg's value for the product *HNR* is still close to that accepted today with $R \sim 5 \times 10^{23}$ cm and $H \sim 10^{-6}$ oersted.
2. Later observations of radio emission associated with aurorae give conflicting results as to the reality and nature of the phenomenon – the solar cycle may also have an influence on this emission.

PART II

TECHNIQUES

The overall development of radio astronomy can fairly be viewed as a continuous quest for yet greater sensitivity and better angular resolution. The improving sensitivity over the years came through a combination of lower-noise receivers and larger collecting areas. Greater resolution at a given wavelength can, of course, only result from building larger antennas or longer baseline interferometers, and indeed up until the mid-1950s, it was this problem of resolution which was of greater concern. This is primarily because the attention of early radio astronomy was directed towards the longer decimeter and meter wavelengths where the typical signal-to-noise ratio was often quite workable, but the available resolution seemed appalling, especially when a radio researcher tried to talk to an optical astronomer! Only in later years did receiver technology and telescope design become such that the range of wavelengths less than 10 cm were amenable to detailed study of non-solar objects.

The fact that Part II only includes two papers may, at first sight, seem a gross injustice to a science born of technology such as radio astronomy. In a sense, almost every observational paper included in this volume relies for its results on some new instrument or technique. Promiment examples include the paraboloidal reflector (the Reber papers), the Michelson interferometer (Paper 18), the sea-cliff or Lloyd's interferometer (Papers 19, 24 and 26), the Dicke radiometer (Paper 22), the total-intensity interferometer (Paper 33), the phase-switched interferometer (Paper 32), and the technique of frequency-switching (Paper 36). This section is limited, however, to those early papers in which observations are not discussed in detail, but which represent work which nevertheless became vital to the further development of radio astronomy.

Nevertheless, mention should be made of several important techniques which were beginning in the early 1950s, but which are not represented in this Part:

(1) *The Mills Cross*, discussed at the end of the introduction to Paper 12.

(2) *The concept of aperture synthesis mapping.* This technique, whereby fringe amplitudes (and later phases) are measured with an interferometer of variable spacing, was extensively developed at Cambridge, at first mainly on the Sun (Stanier, 1950; Machin, 1951; O'Brien, 1953), and later for non-solar applications (Blythe, 1957; Ryle, 1957; Ryle and Hewish, 1960). At the same time, Christiansen and Warburton (1955) (also see Christiansen, 1953) were simultaneously using two crossed grating interferometers (the "Chris-cross") in conjunction with the rotation of the Earth to produce an unambiguous two-dimensional map of the Sun. The full technique of aperture synthesis only eventually became practicable as a result of two technological advances: (1) the advent of the large digital computer, allowing the millions of computations necessary in the Fournier transform to be accomplished within a reasonable time, and (2) improvement in phase stability, allowing a good comparison of relative phase over a period of several hours. The beauty of the method lies in its economy[1] and in its flexibility. One need sample only the number of antenna spacings required for the experiment at hand; moreover, one can later manipulate the data in countless ways in the computer, further optimizing the resulting map of sky brightness.

(3) *The total-intensity interferometer*, as developed at Jodrell Bank by Hanbury Brown and Twiss (1954). In this technique, the revolutionary idea was advanced that basic information on source size and structure could be gained from correlation of signals *after detection.* The notion was then not only proved in the field (see Papers 30 and 33), but also eventually in the laboratory at optical wavelengths after many theoretical objections that such a scheme was in violation of the very laws of physics! In fact, it was in the study of stars at optical wavelengths that Hanbury Brown's technique was later to prove most productive. See Hanbury Brown's own 1974 book for an account of the historical development of this technique.

(4) *The design and use of large paraboloidal reflectors* (see Hanbury Brown and Lovell, 1955). These were, of course, extensively used from the start. In the early 1950s the 218 ft Jodrell Bank fixed spherical dish was the largest at meter wavelengths, while the 50 ft reflector of the U.S. Naval Research Laboratory was most sensitive at centimeter wavelengths.

NOTE

[1] For the university researcher, time is usually cheaper, or at least more available, than steel.

SECTION E

THE DICKE RADIOMETER

One of the products of the wartime development of radar in the United States was the following work of Robert H. Dicke. In Paper 11, he describes both basic thermodynamic principles and practical hardware concerned with microwave measurements. In addition, accurate and detailed derivations are given for the influences on overall system sensitivity of various receiver parameters such as bandwidths and integration times.

It turns out that the noise power from an astronomical source at these relatively high frequencies is usually considerably less than that arising in the receiver itself. In this situation, one is extremely susceptible to the slightest time variations of receiver gain. Dicke's ingenious solution is simply to install a switch (in its initial realization consisting of a rotating, slotted wheel) which yields an effective modulation of the desired signal. When the "switch-off" signal, which ideally is of a similar magnitude to the "switch-on" signal, is subtracted from the latter, the resulting difference is much less affected by long-period gain variations. It is pointed out that this is so because, in practice, the most serous instabilities of microwave radiometers possess characteristic frequencies much lower than the switch frequencies. This scheme thus produces greatly-improved accuracy and effective sensitivity, as evidenced in the pioneering solar and lunar measurements reported in Paper 22. Dicke does not, however, explicitly discuss the two primary drawbacks to the technique: (1) insertion loss from the switch itself, and (2) the fact that the sensitivity of the radiometer is reduced by a factor of ~ 2 from what it would otherwise be (if perfectly gain stable). This latter factor arises from the combination of time not spent on-signal and the subsequent differencing of two noisy signals. Nevertheless, the advantages of the technique are such that it is still used in a wide variety of applications, involving comparison not only with thermal absorbers, but also with different sky positions, adjacent frequencies, differently polarized horns, etc.

The Measurement of Thermal Radiation at Microwave Frequencies

R. H. DICKE*
*Radiation Laboratory, Massachusetts Institute of Technology, Cambridge, Massachusetts***
(Received April 15, 1946)

The connection between Johnson noise and blackbody radiation is discussed, using a simple thermodynamic model. A microwave radiometer is described together with its theory of operation. The experimentally measured root mean square fluctuation of the output meter of a microwave radiometer (0.4°C) compares favorably with a theoretical value of 0.46°C. With an r-f band width of 16 mc/sec., the 0.4°C corresponds to a minimum detectable power of 10^{-16} watt. The method of calibrating using a variable temperature resistive load is described.

INTRODUCTION

SINCE radio waves may be considered infra-red radiation of long wave-length, a hot body would be expected to radiate microwave energy thermally. In order to be a good radiator of microwaves, a body must be a good absorber and the best thermal radiator is the "blackbody."

Although their discoveries were historically unconnected, there is a very close connection between "Johnson noise" of resistors and thermal radiation. The thermal fluctuations of electrons in a resistor set up voltages across the resistor. These "noise voltages" are of such a magnitude that a noise power per unit frequency of kT can be drawn from the resistor. k is Boltzmann's constant; T is the absolute temperature of the resistor.

The connection between thermal radiation and Johnson noise can best be shown by considering the system of Fig. 1.

An antenna is connected to a transmission line which is in turn terminated by a resistor. The radiation impedance of the antenna is assumed to be equal to the characteristic impedance of the coaxial line, i.e., the antenna is "matched" to the line. Also the resistor is assumed to "match" the line. When a transmission line is terminated by a "matched" load, the running waves in the transmission line incident on this load are completely absorbed without reflection. The antenna is completely surrounded by black

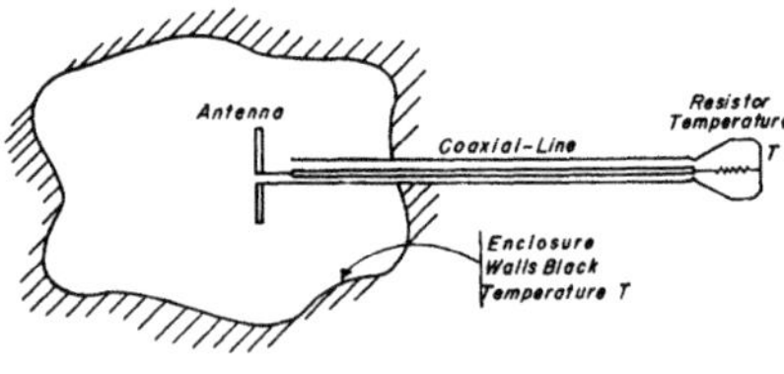

FIG. 1. Antenna system in black enclosure.

* Now at Palmer Physical Laboratory, Princeton University, Princeton, New Jersey.

** This paper is based on work done for the Office of Scientific Research and Development under contract OEMsr-262 with the Massachusetts Institute of Technology.

Reprinted, with permission, from 'The Review of Scientific Instruments' 17, 268–275 (1946).
(1946).

walls of an enclosure. The black enclosure and the resistor are assumed to be at the same temperature T.

Thermal radiation emitted by the walls of the blackbody is picked up by the antenna and transmitted down the line where it is absorbed by the resistor. Johnson noise in the resistor causes a noise power to be transmitted down the line which, passing out the antenna, is absorbed by the black walls. If these two powers were unequal the resistor would either lose or gain energy, resulting in a violation of the second law of thermodynamics. Thus, the available Johnson noise power from a resistor must be equal to the power picked up by an antenna pointed at a blackbody at the same temperature. It is evident that an antenna pointed at surroundings in thermal equilibrium at a temperature T behaves like a resistor at this temperature terminating the coaxial line. It is convenient to describe the radiation intercepted by an antenna in terms o an "antenna temperature." An antenna is said to have a certain temperature when it is receiving an amount of radiation equal to that from a resistor at the temperature in question.

The above arguments were made for a coaxial transmission line, but it is easily seen that they are valid for an antenna fed by any type of single mode transmission line.

The frequency dependence of blackbody radiation differs from that of electrical noise, and this difference may at first glance seem paradoxical. It will be remembered that Planck's blackbody radiation formula reduces to that of Rayleigh-Jeans at the long wave-length end, and in this region the radiation per unit area, per unit solid angle, and per unit frequency of a blackbody is proportional to frequency squared. However, the available noise power from a resistor per unit frequency is frequency independent. The paradox is resolved as follows: in general, the lobes of an antenna pattern become more directive at a higher frequency. It accepts noise from a smaller solid angle; in fact, it can be shown that the average absorption cross section of an antenna is $\lambda^2/4\pi$. The factor λ^2 from the antenna pattern and $1/\lambda^2$ from the blackbody radiation formula just cancel each other, and the total absorbed power is frequency independent.

MEASUREMENT OF THERMAL RADIATION

The detection and measurement of thermal radiation at radiofrequencies requires techniques differing somewhat from those commonly employed for the monitoring of c.w. signals. A c.w. 1
signal, for instance a c.w. carrier, is characterized by a spectrum of small band width, and a receiver for monitoring such a signal is designed with a narrow band. The narrow band rejects noise from frequencies for which there are no

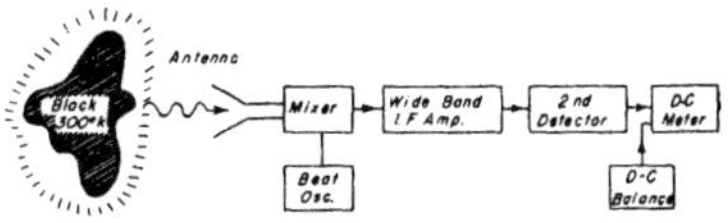

FIG. 2. Wide band receiver used as radiometer.

signal components, and a good signal to noise ratio is obtained.

Thermal radiation, on the other hand, is usually characterized by a spectrum which is very wide. The amplitudes, and phases of different frequency components, are completely independent and are characterized by a complete randomness in the phases. A receiver for the measurement of r-f★noise should employ a wide 1
intermediate frequency (i-f) band.

Figure 2 is a block diagram of a wide band receiver used as a radiometer. A microwave receiver somewhat similar to Fig. 2 has been used by Southworth[1] in an interesting series of measurements on microwave radiation from the sun.

It will be assumed for purposes of definiteness that the radiometer of Fig. 2 has an i-f band width of 10^7 cycles/sec., that the noise figure[2] of the receiver is 20,★and that the response time of 2
the d.c. output meter is 1 second. The operation of the radiometer is roughly as follows:

[1] G. C. Southworth, J. Frank. Inst. **239**, 285 (1945).

[2] The noise figure of a receiver is defined by the following operation: A resistor whose absolute temperature is 300°K is connected to the input terminals of the receiver. This resistor is heated until the available noise power from the I-f amplifier is doubled. The noise figure is this change in temperature expressed in units of 300°K. It is implicitly assumed that only a single band of r-f frequencies are converted to the intermediate frequency. If two r-f bands are converted to the intermediate frequency, the noise figure is the change in temperature expressed in units of 150°K. It is evident that the noise figure is a function of the magnitude of the resistance. What is commonly quoted as the noise figure of a receiver is the measured value for the optimum value of resistance.

The receiver will accept signals from two bands separated by twice the intermediate frequency. The thermal radiation picked up by the antenna in these two signal bands is converted to i-f noise in the mixer. Because the receiver

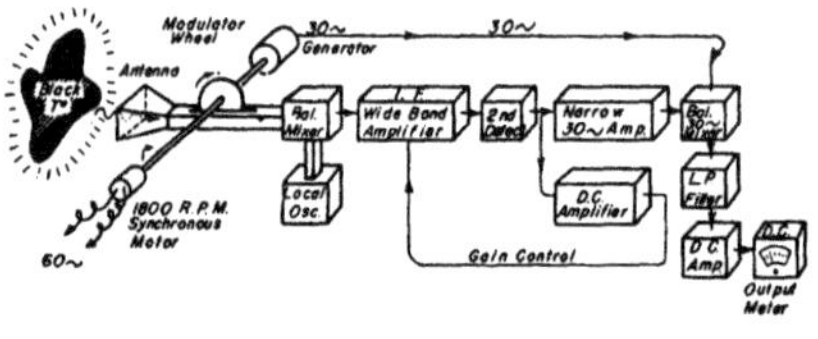

FIG. 3. Microwave radiometer.

is imperfect, noise having its origin in the receiver is added to this converted noise. For the
3 noise figure of 20 assumed above, $\frac{1}{8}$★of the average power entering the second detector has its origin in the 300°K antenna load.

The precision with which thermal radiation can be measured is limited by the statistical fluctuation in the output meter. The order of magnitude of these fluctuations can be obtained from the following argument. With an i-f band-width of 10^7 cycles/sec., about 10^7 rectified noise
4 pulses leave the second detector each second.★ Because of the statistical nature of noise, some of these pulses are large and some are small, but the probability of a given pulse being large or small is completely independent of the sizes of preceding pulses. This situation can be simplified somewhat without changing the basic ideas. It will be assumed that these noise pulses appear like beads on a string and are of two sizes, large and small, in essentially equal numbers. It will be assumed that the small pulses are so small that they do not contribute to the rectified current registered by the d.c. meter assumed to have a time constant of one second. The d.c. meter reading is a number which is essentially the number of big pulses in the preceding second. The problem is now equivalent to flipping pennies a large number of times, always counting the number of heads which appear. It will be remembered that in $2n$ tosses of a coin, the number of heads which appear is likely to differ from n by a number of the order of $(n)^{\frac{1}{2}}$. The implication is fairly obvious. The fluctuation in the number of pulses counted by the d.c. meter in each second is about 10^7,★and the fluctuation in the d.c. meter reading is about 5
$1/(10^7)^{\frac{1}{2}} = 3 \cdot 10^{-4}$ of the meter reading. Since for the noise figure assumed above the radiation from a 300°K load produces about $\frac{1}{10}$ of the d.c. output, the fluctuation in the output meter is about

$$300° \times 10 \times 3 \cdot 10^{-4} \sim 1°\text{C}.$$

Thus it should be possible to measure the temperature of an antenna termination to an accuracy of about 1°C. It should be noted that this fluctuation temperature varies inversely as the square root of the band width, and hence, the wider the i-f band the better the radiometer. The fluctuation temperature also varies directly as the square root of the band width of the d.c. output meter. The more sluggish the meter, the smaller the fluctuations but also, unfortunately, the more time required to make a measurement.

The above discussion was, of course, oversimplified but an exact calculation gives numbers of the same order of magnitude. There is, however, an important source of fluctuation which was intentionally omitted. If the gain of the receiver varies in a random way, the output meter will fluctuate, and this type of fluctuation is likely to be larger than the other. The more obvious sources of these fluctuations are, of course, line voltage variations, and ambient temperature drifts. However, even if power supplies are carefully regulated and temperatures stabilized there are fluctuations having their origins in the tubes themselves. Also fluctuations in the conversion gain and intrinsic noise of a

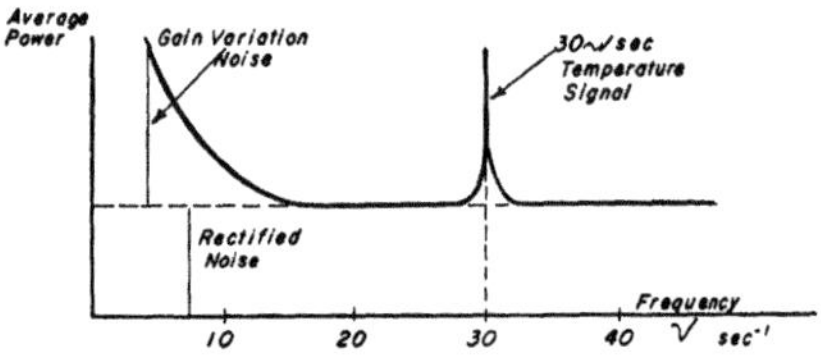

FIG. 4. Second detector output.

crystal mixer must be included. Since a variation of the over-all gain of only 0.1 percent produces a change of 3°C on the output meter, it can be seen that this type of fluctuation is serious. Even though some of these sources of added fluctuation do not have their origin in gain variations, it will be helpful to call all low frequency fluctuations over and above the regular rectified noise "gain variation noise." The regular calculable low frequency noise from the second detector will be called "rectified noise."

THE MICROWAVE RADIOMETER

In order to eliminate "gain variation noise," the radiometer shown in block diagram in Fig. 3 was designed.

The device shown in Fig. 3 consists of an antenna (shown as a horn) connected to a crystal balanced mixer (see Fig. 5 for details of mixer) by a rectangular wave guide. The connecting wave guide contains a slotted section in which an absorbing wheel is caused to rotate. The absorbing wheel is driven by a motor at 30 cycles/sec. and is so shaped that it produces a nearly square modulation with about equal times in and out of the guide. It is assumed that both the wheel and antenna are non-reflecting. Under these conditions the effect of inserting the wheel is one of disconnecting the antenna and connecting an equivalent resistance (represented by the absorbing wheel). If the amount of radiation (or noise power) produced by the wheel while it is in the guide is the same as that being picked up by the antenna, this commutation will produce no change in the output of the second detector. However, if the radiation from the wheel is either more or less than that from the antenna, the output of the second detector will modulate up and down at 30 cycles/sec.

The 30-cycle component of this modulation is amplified by the 30-cycle amplifier and then mixed with 30 cycles from the coherent beat frequency generator to produce d.c. The output from the 30-cycle balanced mixer is fed through a very narrow band low pass filter into a d.c. amplifier which drives a d.c. meter. This d.c. meter measures the difference between the antenna temperature and the modulating wheel's temperature. The meter needle swings either to the left or the right depending upon the sign of this difference.

The device of Fig. 3 has the advantage over that of Fig. 2 of eliminating the "gain variation noise" from the output meter. The reason for the elimination of this type of fluctuation can be understood, at least qualitatively from the following arguments: in the radiometer of Fig. 3 the radiation from the antenna is compared with that generated by a standard reference resistor (the modulator wheel). This comparison is made

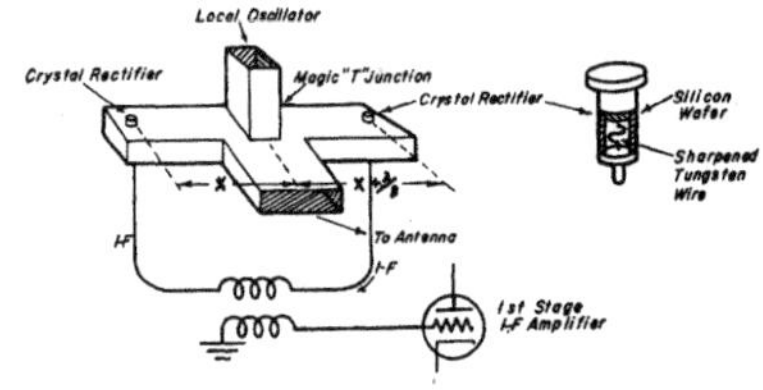

FIG. 5. Microwave balanced mixer with wave guide circuits.

30 times a second and in the short time interval of 1/30 sec. the amplifier gain does not change greatly. The output meter with its long time constant is effectively measuring the average of a great number of individual measurements, each requiring 1/30 of a second to perform.

In Fig. 4 there is plotted the average power against frequency for the output of the second detector of Fig. 3. It is to be noted that the signal, appearing as it does at 30 cycles/sec., avoids the "gain variation noise."

It has been assumed in the description of the operation of Fig. 3 that the antenna and modulating wheel presented equal impedances to the receiver. This equality can, of course, never be perfectly realized and it is necessary to examine the effects introduced by small mismatches in these components. For the purpose of this discussion, it will be assumed that the wheel is a perfect match but that the antenna is slightly mismatched.

There are several possible sources of error due to antenna mismatch but the most important of these seems to be the i-f impedance effect. The i-f impedance of a single crystal mixer depends to a marked degree upon the r-f circuit impedances presented to the mixer. Radiofrequency impedances are transformed through the mixer into i-f impedances, and a 30-cycle/sec. variation in the r-f impedances is transformed into a 30-cycle/sec. variation in the i-f impedance presented to the i-f amplifier. As the gain and noise figure of the i-f amplifier depend upon the i-f impedance of the input load, there is generated a variation in the output of the second detector as a result of this effect.

By employing two crystals in the proper r-f and i-f circuits, a mixer can be constructed which does not convert from i-f to r-f and hence for which changes in the r-f impedances are not reflected into the i-f circuit. Figure 5 is a diagram of such a microwave balanced mixer employing wave guide r-f circuits. In general, it is desirable to make one of the wave guides feeding a crystal rectifier of adjustable length and to include tuners for the two crystal rectifiers.

In addition to minimizing effects arising from impedance transformation to the i-f circuit, the balanced mixer eliminates converted local oscillator noise and minimizes the local oscillator power that enters the antenna. It is desirable to keep local oscillator power out of the antenna line as the load is strictly speaking not in thermal equilibrium with an external source of power being dissipated in it. Effects have been observed which had their origin in poor contacts modulating the local oscillator power to produce noise side bands.

As was pointed out previously, the larger the band width of the i-f amplifier the smaller the fluctuation in the output meter. However, if this band width is too great the noise figure of the i-f amplifier suffers, and there is a certain optimum band width for crystals of particular noise temperature. With present techniques in the microwave region, this optimum is probably in the neighborhood of 10 mc. There is presumably also an optimum band shape for the i-f amplifier, but this question is obscured by a number of unknown factors. If the noise figure of the amplifier is constant over the pass band of the amplifier, a square band pass is the best.

The band width and shape of the low pass filter circuit of Fig. 3 is important in determining the performances of the system. As was mentioned before, the narrower this pass band, the smaller the fluctuation in the output meter. However, this narrowing can only be accomplished at the expense of a sluggish meter with resulting long observation time.

The optimum band shape for the low pass filter is that which gives the most rapid meter response for a given band width. Instead of con-

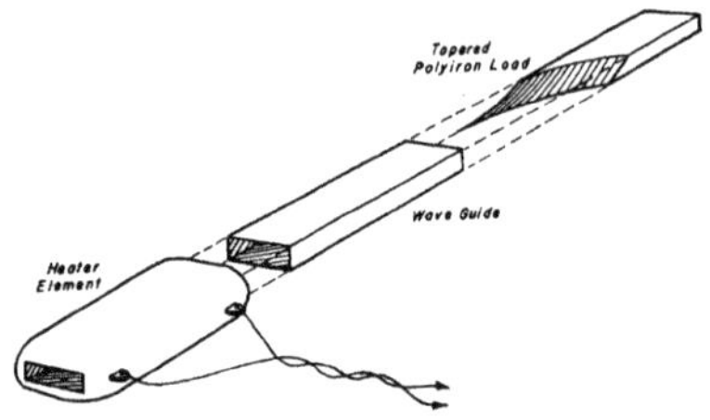

FIG. 6. Exploded view of calibrating load (thermometer not shown).

sidering the most general problem the following simplified calculation has been made: How must a galvanometer be damped to make its response to a step function input as rapid as possible keeping the noise fluctuation constant? The result, as might be expected, is that the critically damped galvanometer is best.

The low pass filter of Fig. 3 is so designed that the d.c. meter has all the characteristics of a long time constant, critically damped galvanometer—all this without broken suspensions.

In Appendix 1, the root mean square temperature fluctuation of the output meter is calculated assuming that all "gain variation noise" is absent.[3] The result for a typical set of system parameters is

$$[\langle(\Delta T)^2\rangle_{\mathrm{Av}}]^{\frac{1}{2}} = 0.46°\mathrm{C}.$$

The system parameters corresponding to this fluctuation are:

Modulator wheel—square wave
Receiver noise figure—25
I-f band width—8 mc/sec.
I-f band shape—square
Second detector—linear
Time constant of low pass filter—2.5 sec.
Band shape of low pass filter—critically damped.

The actual root mean square temperature fluctuations measured for a system having approximately the above parameters was 0.4°.

The power sensitivity of the radiometer with the above parameters is of some interest. The device will detect $\frac{1}{2}$°C and with the total r-f band width of 16 mc/sec. the device has a minimum detectable power of

$$P = kT\Delta\nu = 10^{-16} \text{ watt.}$$

[3] The writer is greatly indebted to Dr. G. E. Uhlenbeck and Miss M. Wang for valuable aid in connection with this problem.

RADIOMETER CALIBRATION

A microwave radiometer can be calibrated by using an r-f absorbing load, the temperature of which can be varied. The absorbing load is substituted for the horn antenna in the calibrating measurement. The calibrating load used in a series of experiments consisted of a section of wave guide containing a tapered "polyiron"[4] absorbing load. A heater element, consisting of resistance wire wrapped around the guide, was used to heat the load. A bi-metallic thermometer was used to measure the temperature of the wave guide. (See Fig. 6 for an exploded view of the calibrating load.) In practice, the temperature of the guide was raised about 60°C and the deflection in the output meter noted.

It is evident that there are many uses to which a microwave radiometer may be put. As an example, an r-f noise source may be calibrated referring to thermal noise as a standard. A noise source so calibrated may be used to make absolute noise figure measurements.

A radiometer operating in a region slightly less than 1 cm^{-1} has been used by R. Beringer and the author to make some observations on radiation from the sun and moon. These results
6 will be reported elsewhere.★

Microwave radiometers have also been used to measure atmospheric absorption at several radio frequencies. An account of these experiments will be given in another paper by R. Beringer, R. L. Kyle, A. B. Vane, and the
7 author.★

ACKNOWLEDGMENTS

The size of the Radiation Laboratory being what it was, it is impossible to acknowledge all the people who have aided in this development. However, special acknowledgment is due the authors colleagues, Dr. R. Beringer, Mr. R. L. Kyle, and Mr. A. B. Vane, for their active participation in the development of the radiometer. Special thanks are also due Mr. C. F. Nawrocki for developing a special oscillator to generate the beat frequency for one of the radiometers. Finally, the author wishes to express his appreciation to Dr. E. M. Purcell for his continued interest, support, and advice.

[4] "Polyiron" is finely divided iron held together with a plastic binder.

APPENDIX 1

Statistical Fluctuations of Output Meter

Assume a unit of time, τ seconds, such that $2\pi\tau$ seconds is much greater than any correlation time of apparatus. All noise voltages will be expanded in a Fourier series of this period.

Let the frequency response function of the i-f amplifier be

$$F(\omega), \tag{1}$$

where $\omega = 2\pi \times$frequency in cycles/τ sec.

An alternative representation for $F(\omega)$ in terms of a Fourier series is

$$F_n = F(n). \tag{2}$$

n is an integer.

Let the response function of the low pass filter be

$$S(\omega) \text{ or } S_n = S(n). \tag{3}$$

Case I

Assume that the modulating wheel and the source are at the same temperature so that there is no temperature signal present.

Let $F(t)$ be the output of the i-f amplifier for some $2\pi\tau$ second period where t is again in units of $2\pi\tau$ seconds.

$$F(t) = \sum_n f_n e^{int}. \tag{4}$$

Assume for the time being that the second detector is square law. The output from the second detector is

$$g(t) = F^2(t) = \sum_n G_n e^{int},$$
$$= \sum_{n,m} f_n f_m e^{i(n+m)t}. \tag{5}$$

Imagine the above function obtained for a large number of periods each of $2\pi\tau$ seconds. An ensemble average will be an average of corresponding quantities from each of these periods.

If the input r-f noise spectrum is essentially flat over the i-f pass band, the ensemble averages:

$$\langle f_n f_{-n} \rangle_{Av} = P F_n F_{-n} = P |F_n|^2, \tag{6}$$
$$\langle f_n f_{-m} \rangle_{Av} = 0 \quad n \neq m,$$

where P is a constant and

$$\langle G \rangle_{Av} = P \sum_n |F_n|^2; \tag{7}$$

also,

$$\langle |G_n|^2\rangle_{Av} = 2P^2 \sum_m |F_m|^2 |F_{m+n}|^2. \quad (8)$$

Let the output from the second mixer after passing through the low pass filter be

$$S(t) = \sum_n S_n(G_{n+k} + G_{n-k})e^{int}, \quad (9)$$

where k is the modulation frequency. Form $S^2(t)$ and take the time average over the period

$$\langle S^2(t)\rangle_{Av} = \sum_n |S_n|^2 \cdot |G_{n+k} + G_{n-k}|^2. \quad (10)$$

Form the ensemble average of (10)

$$\begin{aligned}\langle\langle S^2(t)\rangle_{Av}\rangle_{Av} = \langle S^2(t)\rangle_{Av} &= \sum_n |S_n|^2 \\ &\quad \cdot \langle |G_{n+k}|^2\rangle_{Av} + \langle |G_{n-k}|^2\rangle_{Av}, \\ &= 2P^2 \sum_n |S_n|^2 \cdot [\sum_m |F_m|^2 \\ &\quad \cdot |F_{m+n+k}|^2 + \sum_m |F_m|^2 \\ &\quad \cdot |F_{m+n-k}|^2]. \end{aligned} \quad (11)$$

For small n

$$|F_m| \cong |F_{m+n}|,$$

$$\langle S^2(t)\rangle_{Av} = 4P^2 \sum_{n,m} |S_n|^2 \cdot |F_m|^4. \quad (12)$$

Case II

Let the antenna load temperature differ from the wheel temperature T by ΔT. Assume a noise figure of N for the receiver. The noise power into the second detector varies with the time as the modulating wheel rotates. It will be assumed that this modulation is square wave and that the power into the second detector is

$$P = P_0 + p(t), \quad (13)$$

where $p(t)$ is a square wave function of the time t as shown in Fig. 7 and

$$p_0 = \frac{\Delta T}{T}\frac{P_0}{N}. \quad (14)$$

Equation (13) may be expanded in a Fourier series and becomes

$$P = \sum_n p_n e^{int},$$

$$p_n = 2p_0/\pi|n|. \quad (15)$$

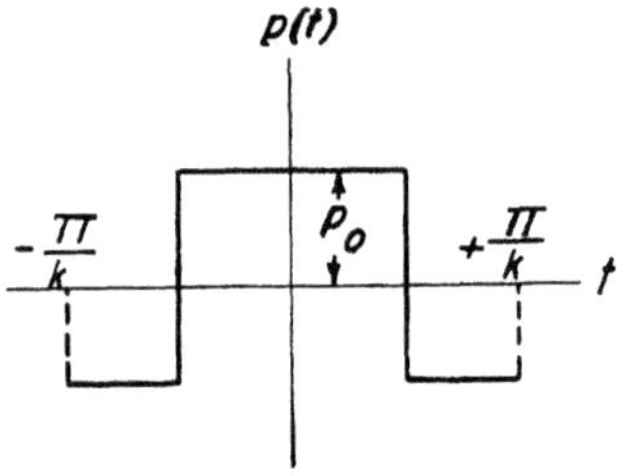

FIG. 7. $p(t)$ square wave function of t.

With the temperature signal present, $n \neq 0$ Eq. (9) becomes modified as follows:

$$S(t) = \sum_n S_n(G_{n+k} + G_{n-k})e^{int} + 2S_0 \cdot G_0 \cdot \frac{P_1}{P_0}. \quad (16)$$

The last term on the right of (16) is the contribution caused by signal and its square is

$$(\text{Signal})^2 = 4S_0^2 G_0^2 \frac{P_1^2}{P_0^2}. \quad (17)$$

Substituting (7) in (17)

$$(\text{Signal})^2 = 4S_0^2 p_1^2 (\sum_n |F_n|^2)^2, \quad (18)$$

since $P = P_0$ for no temperature signal. Equation (12) is the noise fluctuation and (18) divided by (12) becomes

$$\frac{(\text{Signal})^2}{(\text{Noise})^2} = \frac{P_1^2}{P_0^2} \frac{S_0^2(\sum_n |F_n|^2)^2}{(\sum |S_n|^2)\cdot(\sum_m |F_m|^4)}. \quad (19)$$

Substituting from (14) and (15) and writing the sums in the form of integrals

$$\frac{(\text{Signal})^2}{(\text{Noise})^2} = \frac{4}{\pi^2}\left(\frac{\Delta T}{T}\right)^2 \frac{1}{N^2} \times \frac{[\int |F(\omega)|^2 d\omega]^2 \cdot S^2(0)}{[\int |F(\omega)|^4 d\omega] \cdot [\int |S(\omega)|^2 d\omega]}. \quad (20)$$

For a signal/noise = 1, (20) becomes

$$\frac{\Delta T}{T} = \frac{\pi N}{2} \frac{\left\{\left[\int_{-\infty}^{\infty} |F(\omega)|^4 d\omega\right] \cdot \left[\int_{-\infty}^{\infty} |S(\omega)|^2 d\omega\right]\right\}^{\frac{1}{2}}}{S(0)\int_{-\infty}^{\infty} |F(\omega)|^2 d\omega}. \quad (21)$$

To state (21) in words, ΔT is the root mean square temperature fluctuation of the output meter in terms of the system parameters.

Dr. O. E. Uhlenbeck[5] has shown that for an ideal linear detector, the output fluctuation is just $\frac{1}{2}$ that of the square law detector. For a linear detector (21) becomes

$$\frac{\Delta T}{T}=\frac{\pi N}{4}\,\frac{\left\{\left[\int_{-\infty}^{\infty}|F(\omega)|^4 d\omega\right]\cdot\left[\int_{-\infty}^{\infty}|S(\omega)|^2 d\omega\right]\right\}^{\frac{1}{2}}}{S(0)\int_{-\infty}^{\infty}|F(\omega)|^2 d\omega}. \tag{22}$$

Equations (21) and (22) are dimensionless and are valid for any unit of time measure.

Example: Assume

Receiver noise figure—$N=25$
I-f band shape—square
8 I-f band width—$\Delta\omega=5.10^7$ sec.$^{-1}$ ★
Second detector—linear
Band shape of low pass filter—critically damped
8 Band width of low pass filter—.4 sec.$^{-1}$ ★

The response function of a critically damped filter is of the form

$$S(\omega)=-\frac{a^2}{(\omega-ia)^2}, \tag{23}$$

where in this case $a=.4$ sec.$^{-1}$, the integral

$$\int_{-\infty}^{\infty}|S(\omega)|^2 d\omega=\frac{\pi a}{2}, \tag{24}$$

and

$$S(0)=1. \tag{25}$$

Also for a square i-f band of width $\Delta\omega$

$$\frac{\left(\int_{-\infty}^{\infty}|F(\omega)|^4 d\omega\right)^{\frac{1}{2}}}{\int_{-\infty}^{\infty}|F(\omega)|^2 d\omega}=\frac{1}{(2\Delta\omega)^{\frac{1}{2}}}. \tag{26}$$

Substituting (24)–(26) in (22)

$$\frac{\Delta T}{T}=\frac{\pi^{\frac{1}{2}}N}{8}\left(\frac{a}{\Delta\omega}\right)^{\frac{1}{2}}. \tag{27}$$

Substituting numerical values, assuming that $T=300°$K

$$\Delta T=0.46°\text{C}. \tag{28}$$

[5] Private communication.

EDITOR'S NOTES AND AMENDMENTS

1. c.w. = continuous wave; r–f = radio frequency.
2. This corresponds to a double-sideband system temperature of 2700 K.
3. Should be "1/10".
4. It would seem that the intended number here is 2 × 10^7 rectified noise pulses per second.
5. Should be "$\sqrt{10^7}$"
6. Paper 22.
7. Dicke *et al.* (1946).
8. Units are "rad sec^{-1}".

SECTION F

THE PHASE-SWITCHED INTERFEROMETER

One of the achievements for which Sir Martin Ryle was awarded the 1974 Nobel Prize in Physics was his continued development of the technique of radio interferometry. In Paper 12, Ryle describes the principles and practice of the phase-switched interferometer as developed by him in the years before 1952. In the process of this description, he virtually creates a manual of interferometry as he passes on advice concerning many of the tricks and pitfalls of the trade, as well as giving numerous examples of actual interferometric records.

The galactic background radiation and the solar disk emission are so strong at meter wavelengths that measurement of any additional weak radiation present in the antenna lobes is extremely difficult. The primary advantage of phase-switching is shown to be the resultant insensitivity of the interferometer to this extended radiation, allowing for more accurate measurements of the intensities, positions, sizes, and polarization of either discrete "radio stars" or solar bursts. Phase-switching can, in fact, be described as Dicke-switching between two sets of antenna lobes; as in Paper 11, one modulates the desired signal in such a manner that large fluctuations are removed, allowing small residuals to be accurately measured. Ryle makes detailed comparisons of his method with total-power interferometers, sea-cliff interferometers and "pencil-beam" antennas. In particular, he demonstrates the superiority of the phase-switching technique in its relative insensitivity to interference and to any asymmetries in the characteristics (e.g., cable losses) of the two arms of the interferometer. Also discussed are the differing effects of being sensitivity-limited vs. confusion-limited, a topic which was to be pivotal a few years later in the debate over the meaning of the 2C and MSH source counts (Ryle and Scheuer, 1955; Mills and Slee, 1957).

At about this same time in Australia, B. Y. Mills developed a type of antenna which soon came to be known as the "Mills Cross". It also employs a type of phase-switching, but rather than the phase connection between two spaced elements being alternated, it is that between two intersecting line arrays. This principle, also mentioned in section 3(g) of Ryle's paper, is described in detail by Mills and A. G. Little in a 1953 paper. The Mills Cross achieves a narrow pencil-beam with a small collecting area; its design enables a large reduction in the effects of confusion while still maintaining adequate sensitivity. This is particularly advantageous for surveys of radio sources at meter wavelengths where one typically enjoys high signal-to-noise ratios, while at the same time being strongly limited by confusion.

A new radio interferometer and its application to the observation of weak radio stars

BY M. RYLE

(*Communicated by Sir Lawrence Bragg, F.R.S.—Received* 19 *June* 1951—
Revised 10 *October* 1951)

A new type of radio interferometer has been developed which has a number of important advantages over earlier systems. Its use enables the radiation from a weak 'point' source such as a radio star to be recorded independently of the radiation of much greater intensity from an extended source. It is therefore possible to use a very much greater recorder sensitivity than with earlier methods. It is, in addition, possible to use pre-amplifiers at the aerials, and the resolving power which may be used is therefore not restricted by attenuation in the aerial cables.

Besides improved sensitivity, the new system has a number of other advantages, particularly for the accurate determination of the position of a radio source. Unlike earlier systems the accuracy of position finding is not seriously affected by rapid variations in the intensity of the radiation. It also has important applications to the measurement of the angular diameter and polarization of a weak source of radiation.

The new system has been used on wave-lengths of 1·4, 3·7, 6·7 and 8 m for the detection and accurate location of radio stars, and for the investigation of the scintillation of radio stars. It has also been used in a number of special experiments on the radiation from the sun. The results which have been obtained in these experiments have confirmed the advantages predicted analytically.

1. INTRODUCTION

One of the most important experimental problems in radio astronomy is concerned with the detection and location of weak radio stars. An improvement in the accuracy with which their positions may be determined might allow identification with faint visual bodies, or might enable their annual parallax or proper motion to be detected.

Reprinted, with permission, from the 'Proceedings of the Royal Society' **A211**, 351–375 (1952).

The observation of a large number of radio stars would make it possible to carry out a detailed analysis of their distribution in space. It has already been shown that one of the most satisfactory methods for observing radio stars makes use of a radio interferometer arrangement associated with a special type of receiver (Ryle 1950). Observations with this type of apparatus have enabled a number of radio stars to be located, but the extension of these methods to allow the observation of weaker stars presents considerable difficulty. At the same time it is unlikely that the accuracy of the early methods of position finding can be greatly improved without the use of systems of considerably greater aperture.

It is the purpose of this paper to describe a new type of interferometer which has considerable advantages both for the detection and for the location of weak radio stars, and to compare its performance with other systems which have been used. Consideration is also given to the factors which limit the detection of radio stars by present-day methods.

The results of observations carried out with the new system are not discussed in detail in the present paper; some of the earlier results have already been described (Ryle, Smith & Elsmore 1950; Smith 1951). The various methods of determining the position of a radio star by measurements with interferometer systems are discussed in a separate paper (Smith, in preparation).

2. The new phase-switching system

(a) *General description*

Previous interferometer aerial systems used at Cambridge for the observation of the sun and radio stars have made use of two aerials spaced on an east-west line (Ryle 1950). The new phase-switching system also makes use of two spaced aerials to produce an interference pattern, but a switch is arranged so that an additional half-wave-length* of cable may be introduced into one of the aerial cables; by this means it is possible to displace the interference pattern so that the new maxima correspond to the minima of the original pattern. The two reception patterns are indicated in figure 1, where the envelope of the interference patterns is determined by the reception pattern $A(\theta)$ of the individual aerials. ($A(\theta)$ is a function which will be used to represent the effective area of each aerial for reception of radiation incident from a direction θ; it therefore takes account of any directivity in the axial plane, the plane normal to the line joining the two aerials.)

The power receptivity of this system when the aerials are in phase is given by

$$A(\theta)\,[1+\cos\{(2\pi d/\lambda)\sin\theta\}], \tag{1}$$

where θ is the angle between the source and the axial plane, d is the spacing between the aerials and λ is the wave-length.

* It is important that the phase-switching be carried out by the insertion of a half-wave-length into one of the aerial cables, and not by the more symmetrical method of transferring a quarter-wave-length from one aerial cable to the other. In the latter case the parallel impedance of the two aerials will not be the same in the two positions of the switch unless the impedances of the two aerials are exactly equal. A variation of the parallel impedance may modify the internal noise power of the receiver, and as will be shown later this variation will produce a false deflexion of the recorder. This condition cannot arise if 'half-wave' switching is used.

With the aerials connected in anti-phase, the receptivity is

$$A(\theta)\,[1-\cos\{(2\pi\,d/\lambda)\sin\theta\}]. \tag{2}$$

If the analysis is restricted to angles near the axial plane of the interferometer, these expressions may be simplified to

$$A(\theta)\,[1\pm\cos(2\pi\,d\theta/\lambda)]. \tag{3}$$

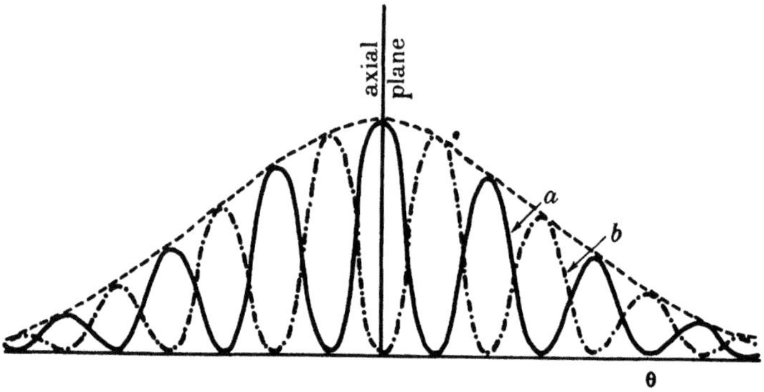

FIGURE 1. Reception pattern of two spaced aerials: (a) connected in phase and (b) connected in anti-phase.

Suppose now that a point source of radiation which produces a power flux p at the aerial is situated in a direction θ; the powers intercepted by the aerial system in the two conditions are given by

$$pA(\theta)\,[1+\cos(2\pi\,d\theta/\lambda)] \quad \text{and} \quad pA(\theta)\,[1-\cos(2\pi\,d\theta/\lambda)].$$

If now the system is switched rapidly between the two conditions, the aerial power will contain an alternating component whose magnitude is the difference between the powers intercepted in the two switch conditions

$$2pA(\theta)\cos(2\pi\,d\theta/\lambda). \tag{4}$$

If the aerial system is connected to a receiver having a square-law detector* the output voltage will contain a steady component which includes the noise power generated in the receiver itself, together with a square-wave alternating component whose amplitude is

$$2GpA(\theta)\cos(2\pi\,d\theta/\lambda),$$

where G is a constant which includes the amplification of the receiver and the constant of the detector. Thus, by the addition of an amplifier which responds to the alternating component from the detector, but not to the steady component, it is possible to obtain an alternating output voltage related to the power received from the point source. By combining this voltage with a reference signal derived from

* A more detailed analysis shows that for any type of detection the same result is obtained when the noise power from the source represents a small fraction of the total noise power in the input circuit of the receiver.

the aerial phase-changing switch, it is then possible to produce a direct current whose magnitude and sign depend on the intensity and direction of the point source. A block diagram of the complete system is shown in figure 2.

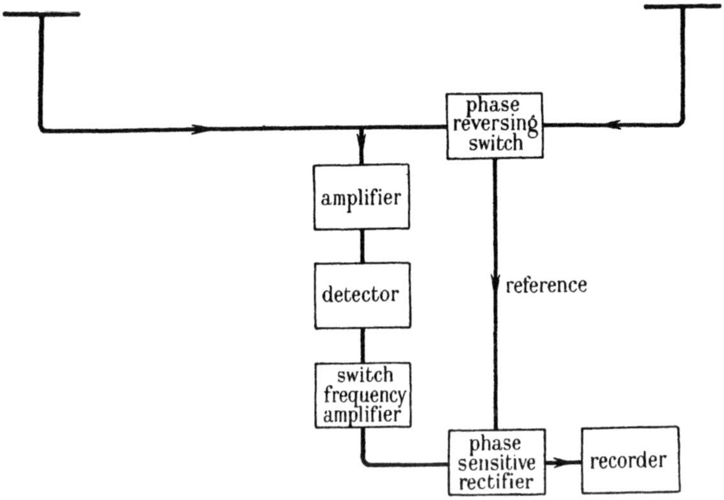

FIGURE 2. Block diagram of the phase-switching system.

Now consider the effect of a slow variation of θ, which might be caused by the rotation of the earth; the relative powers from the aerial system in the two positions of the phase-changing switch will alter, and the amplitude of the square-wave component and hence of the direct-current output from the phase-sensitive rectifier will undergo a periodic variation in time proportional to $2GpA(\theta)\cos(2\pi d\theta/\lambda)$.

If the output current is used to operate a recording milliammeter, a trace having the form shown in figure 3 will be produced.

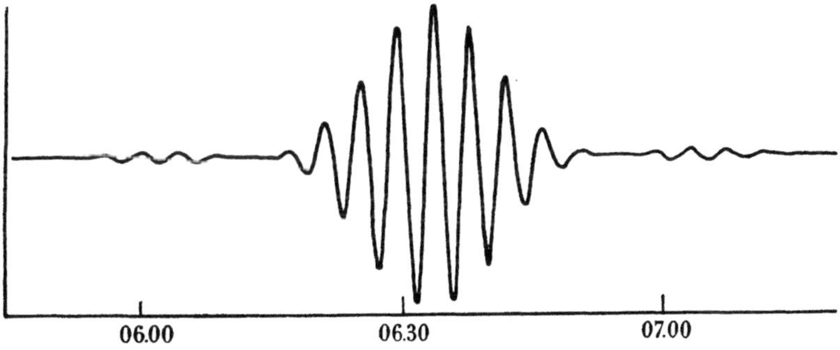

FIGURE 3. Record of the intense radio star in Cassiopeia obtained with the phase-switching system on a wave-length of 3·7 m.

So far it has been assumed that the incident radiation is due to a single point source; it is now necessary to consider the effect of radiation from an extended source, which subtends an angle which is not small compared with the separation (λ/d) of successive maxima of the interference pattern. Let the intensity emitted by a strip bounded by the directions θ and $(\theta + \mathrm{d}\theta)$ be represented by $P(\theta)\,\mathrm{d}\theta$. Then the power intercepted when the aerials are connected in phase is

$$\int P(\theta)\,A(\theta)\left[1+\cos\left(\frac{2\pi d\theta}{\lambda}\right)\right]\mathrm{d}\theta,$$

and with the aerials in anti-phase is

$$\int P(\theta)\,A(\theta)\left[1-\cos\left(\frac{2\pi d\theta}{\lambda}\right)\right]\mathrm{d}\theta.$$

If the variation across the source of the receptivity of each aerial individually, $A(\theta)$, is neglected, the powers intercepted may be written

$$A(\theta)\int P(\theta)\left[1\pm\cos\left(\frac{2\pi d\theta}{\lambda}\right)\right]\mathrm{d}\theta. \tag{5}$$

For a source of uniform 'brightness' and angular width $2\Delta\theta$, the quantity $P(\theta)$ may be replaced by $p'/2\Delta\theta$, where p' represents the total intensity of the radiation emitted by the source. The powers intercepted by the aerial when the centre of the source is in a direction θ_1 are then

$$\frac{p'A(\theta)}{2\Delta\theta}\int_{\theta_1-\Delta\theta}^{\theta_1+\Delta\theta}\left[1\pm\cos\left(\frac{2\pi d\theta}{\lambda}\right)\right]\mathrm{d}\theta = \frac{p'A(\theta)}{2\Delta\theta}\left[2\Delta\theta\pm\frac{\lambda}{\pi d}\cos\left(\frac{2\pi d\theta_1}{\lambda}\right)\sin\left(\frac{2\pi d\Delta\theta}{\lambda}\right)\right]. \tag{6}$$

An extended source of radiation therefore produces a steady component, $Gp'A(\theta)$, at the output of the receiver, together with an alternating component of magnitude

$$\frac{Gp'A(\theta)}{\Delta\theta}\frac{\lambda}{\pi d}\cos\left(\frac{2\pi d\theta_1}{\lambda}\right)\sin\left(\frac{2\pi d\Delta\theta}{\lambda}\right).$$

The latter component will be amplified and will produce, from the phase-sensitive rectifier, a proportional direct current which may be used to operate the recorder. The steady component will have no effect on the phase-sensitive rectifier and will therefore produce no deflexion of the recorder; as will be seen later, however, it is of importance when considering automatic control of the receiver gain.

As the earth rotates, θ_1 will increase and the recorder deflexion will vary periodically; the phase of the variations will be the same as would be produced by a point source situated at the centre of the extended source, but the amplitude will be smaller than that from a point source giving rise to the same total flux by a factor

$$\frac{\sin\left(\frac{2\pi d}{\lambda}\Delta\theta\right)}{\frac{2\pi d}{\lambda}\Delta\theta}.$$

By measuring the amplitudes of the traces obtained with two different values of d, it is therefore possible to deduce $\Delta\theta$, the half-width of the source; a comparison between this method of measuring the angular diameter of a source and methods which have been described previously in which the total aerial power was measured is given in § 3 (d).

From equation (5) it is apparent that the amplitude of the trace is in fact proportional to one term of the Fourier transform of the angular distribution of intensity across the source*; the angular periodicity of the term is determined by the spacing between the aerials and is given by d/λ. A similar analysis, based on the use of a 'total power' recording system, has already been applied to the measurement of the distribution of radio 'brightness' across the undisturbed solar disk (Stanier 1950; Ryle 1950). The same experiment, on a longer wave-length, has recently been carried out by Machin (1951) using the phase-switching system.

(b) The elimination of the background radiation due to the galaxy

It is now possible to apply these results to find the effect of the background radiation from the galaxy on measurements made with the phase-switching system. Suppose that the average intensity of the background radiation is deduced for a number of strips running parallel to the axial plane, and extending over the reception pattern of the primary aerials. Let $f(\theta)$ represent the variation of this average intensity with θ. (For the case of an east-west interferometer, the strips will be parallel to the meridian, and will extend in declination through an angle determined by the north-south coverage of each aerial.) In the absence of any point sources of radiation the deflexion of a recorder used with a phase-switching system will be proportional to the amplitude of the component of the Fourier transform of $f(\theta)$ having a periodicity in angle of d/λ. If the background radiation varies in angle in a continuous manner, the residual periodic variation of the trace can be made indefinitely smaller by increasing the separation (d) between the aerials.

In any practical system it is clear that the function $f(\theta)$ may contain small terms having angular periodicities comparable with d/λ, and that confusion may therefore arise between a genuine radio star and the small irregularities in the background radiation. The importance of this possible limitation depends not only on d/λ but on the area and shape of the individual aerials of the interferometer. The use of a second alternative value of the aerial spacing (d), however, always makes it possible to discriminate between irregularities of this nature and genuine sources of small angular diameter, and no trouble has been encountered with this limitation in practice.

The effectiveness of discriminating against the background radiation in this way may be compared with that obtained with a simple 'pencil-beam' aerial. In the latter case any irregularities having an angular structure comparable with the angular width of the reception pattern will be difficult to distinguish from a radio

* The relation between the magnitude of the varying component of power intercepted by an interferometer and the Fourier transform of the distribution across the source was first pointed out by McCready, Pawsey & Payne-Scott (1947).

star. A 'pencil-beam' aerial system would therefore have to have an aperture of about twice the *separation* between the aerials of an interference system ($2d$) to have the same discrimination.

It is clear that because the background radiation does not produce a deflexion on the record, it is possible to use a very much greater recording sensitivity than is possible with a 'total-power' recording system. As the sensitivity is increased one of two factors may prevent the detection of further radio stars:

(i) The random fluctuations on the record may mask the traces due to radio stars whose flux is less than a certain value.

(ii) The traces from adjacent radio stars may overlap to produce a resultant trace whose amplitude and phase are related to the intensities and positions of two or more radio stars. Under these conditions no accurate position or intensity can be determined for the individual stars, yet they cannot be regarded as an irregularity in the unresolved background radiation, since in principle they could be resolved by observations with a number of different aerial spacings.

The relative importance of these two factors which determine the ultimate limit of detection will be discussed in greater detail in §3 (*a*). Some further properties of the phase-switching system will first be discussed.

(*c*) *The general relationships for the phase-switching system*

It is now important to consider the behaviour of a phase-switching system in the general case where the two aerials are not identical; these differences might arise if the two aerials had different reception patterns, or if the attenuation factors of the two cables were unequal.

Suppose that a point source produces a power flux p_s at the aerials, and that a mean flux p_g per unit solid angle is incident from the background radiation. If A and A' represent the effective areas of the two aerials for radiation incident from the direction of the point source, then the powers available from the two aerials are

$$Ap_s + p_g\lambda^2, \quad A'p_s + p_g\lambda^2.$$

As will be seen later, there are certain applications of the phase-switching system in which the two aerials have very different reception patterns; under these conditions it may be necessary to consider the different area of sky covered by each aerial. If the effective values of the incident flux from the background are represented by p_g and p'_g, the aerial powers may then be written

$$Ap_s + p_g\lambda^2, \quad A'p_s + p'_g\lambda^2.$$

If now the aerials are connected to the receiving system with cables having attenuation constants α and α', the power available at the input of the receiver with the aerials in-phase and with the source on the axial plane of the interferometer is

$$\tfrac{1}{2}\{\sqrt{(\alpha A p_s)} + \sqrt{(\alpha' A' p_s)}\}^2 + \tfrac{1}{2}\lambda^2(\alpha p_g + \alpha' p'_g).$$

When the aerials are connected in anti-phase the power will be

$$\tfrac{1}{2}\{\sqrt{(\alpha A p_s)} - \sqrt{(\alpha' A' p_s)}\}^2 + \tfrac{1}{2}\lambda^2(\alpha p_g + \alpha' p'_g).$$

If the receiving amplifier generates a noise power P_N, the total power at the input of the receiver in the two positions of the phase-changing switch will be

$$\tfrac{1}{2}p_s\{\sqrt{(\alpha A)} \pm \sqrt{(\alpha' A')}\}^2 + \tfrac{1}{2}\lambda^2(\alpha p_g + \alpha' p_g') + P_N. \tag{7}$$

This result may be interpreted in terms of a steady component of noise power

$$\tfrac{1}{2}p_s(\alpha A + \alpha' A') + \tfrac{1}{2}\lambda^2(\alpha p_g + \alpha' p_g') + P_N, \tag{8}$$

and an alternating component

$$2p_s\sqrt{(\alpha\alpha' AA')}. \tag{9}$$

The latter component may be used to produce an output current which will vary sinusoidally as the position of the point source changes due to the rotation of the earth. The steady component of noise power determines the amplitude of the statistical fluctuations on the record, and therefore determines the limiting sensitivity of the system.

Some applications of this result will now be examined.

(d) *The use of pre-amplifiers*

In order to obtain good discrimination against the background radiation, and to determine the positions of radio stars as accurately as possible, it is important to use the maximum possible separation between the aerials. Apart from observations with the 'cliff' type of interferometer (McCready, Pawsey & Payne-Scott 1947), the use of a large separation introduces considerable attenuation (α) in the cables from the aerials. At the shorter wave-lengths, where the total noise power at the input of the receiver is mainly due to the internal noise P_N, an increase of α would seriously reduce the overall sensitivity of the system, and accurate observations of the weaker radio stars would therefore be impossible.

It is clear that if an amplifier were installed at each aerial, having a gain somewhat greater than $1/\alpha$ and a noise power comparable with that of the main receiver, then the limiting sensitivity would be the same as if there were no attenuation. In the earlier experiments, in which the total aerial power was measured, the use of pre-amplifiers would not have been possible unless their gains and noise powers remained extremely constant. It will now be shown that these difficulties do not occur with the phase-switching system; the system therefore enables weaker radio stars to be detected, and also permits an indefinite increase of the resolving power which may be realized.

Suppose that the two pre-amplifiers have gains β and β', and that they generate noise powers P_N and P_N'. If a point source situated in the axial plane of the interferometer produces a flux p_s, and if, as before, the background radiation produces an effective flux per unit solid angle of p_g and p_g', then the total noise power available at the input of the main receiver in the two positions of the phase-changing switch is

$$\tfrac{1}{2}p_s\{\sqrt{(\alpha\beta A)} \pm \sqrt{(\alpha'\beta' A')}\}^2 + \tfrac{1}{2}\lambda^2(\alpha\beta p_g + \alpha'\beta' p_g') + \tfrac{1}{2}\alpha\beta P_N + \tfrac{1}{2}\alpha'\beta' P_N'. \tag{10}$$

(If $\alpha\beta \gg 1$ the noise power generated in the main receiver may be neglected.)

This result may be interpreted as a steady component of noise power

$$\tfrac{1}{2}p_s(\alpha\beta A + \alpha'\beta' A') + \tfrac{1}{2}\lambda^2(\alpha\beta p_g + \alpha'\beta' p_g') + \tfrac{1}{2}\alpha\beta P_n + \tfrac{1}{2}\alpha'\beta' P_n', \tag{11}$$

and an alternating component

$$2p_s\sqrt{(\alpha\alpha'\beta\beta' AA')}. \tag{12}$$

Since only the latter component is effective in producing a deflexion of the recorder, it may be seen that in the absence of a point source no deflexion is produced even if the aerials, cable attenuation, amplifier gains or noise powers are unequal or vary with time. The amplitude of the trace produced by a point source is a function of the geometric mean of the overall gain of the two halves of the system, and as will be seen in the next section it is unnecessary to adopt any elaborate arrangement for equalizing the gains or noise powers of the pre-amplifiers.

(e) The absolute measurement of power flux with the phase-switching system

It is clear that the amplitude of the trace produced by a radio star is a function of the overall gain of the system whether pre-amplifiers are used or not, and it is therefore important to examine the problem of obtaining absolute measurements of the intensity of a radio star over long periods of time.

When measuring the total aerial power, the automatic balancing system which is mentioned in § 3 (*a*) enabled absolute measurements to be made without the necessity for accurate stabilization of the gain of the receiver; a similar result can be obtained with the phase-switching system by incorporating an automatic-gain control circuit having a response time which is long compared with the switching period. There is little difficulty in designing such a circuit to maintain the mean noise power at the detector constant to within 1 %.

First consider the application to a phase-switching system without pre-amplifiers. By reference to equation (8) it can be seen that the gain G of the amplifier will be continuously adjusted so that

$$G[\tfrac{1}{2}p_s(\alpha A+\alpha' A')+\tfrac{1}{2}\lambda^2(\alpha p_g+\alpha' p_g')+P_N] = C,$$

where C is a constant determined by the setting of the gain control. The amplitude of the trace produced by the passage of the radio star will then be

$$2Ggp_s\sqrt{(\alpha\alpha' AA')} = \frac{2gp_s\sqrt{(\alpha\alpha' AA')}\,C}{\tfrac{1}{2}p_s(\alpha A+\alpha' A')+\tfrac{1}{2}\lambda^2(\alpha p_g+\alpha' p_g')+P_N},$$

where g represents the amplification of the remaining circuits following the detector (i.e. the switch-frequency amplifier, the phase-sensitive rectifier and recorder circuits). In practice $g \ll G$, and there is little difficulty in design to ensure that g is sufficiently stable.

Except for observations of the most intense radio stars with very large aerial systems, $p_s A \ll p_g\lambda^2$, and the amplitude on the record may be written

$$\frac{2gp_s\sqrt{(\alpha\alpha' AA')}\,C}{\tfrac{1}{2}\lambda^2(ap_g+\alpha' p_g')+P_N}.$$

Since A, A' are constants depending only on the dimensions of the aerial structure, and since α, α' represent the attenuation constants of the cables, it can be seen that for a given gain setting C, the deflexion sensitivity depends only on $\lambda^2 p_g$ and P_N, and is not dependent on the maintenance of constant gain of the main amplifier. The two limiting cases are of interest.

(i) $\alpha\lambda^2 p_g \gg P_N$

For wave-lengths greater than about 3 m, the aerial power available at the input of the receiving amplifier may be appreciably greater than the noise power of the receiver itself. (In the absence of pre-amplifiers, this condition may only arise for comparatively small resolving powers, where the cable attenuation (α) is small; the use of pre-amplifiers will be discussed later.) If $\alpha\lambda^2 p_g \gg P_N$, it can be seen that the recorder sensitivity is independent of the noise power generated in the receiver, and depends (apart from the characteristics of the aerials and cables) only on the average intensity of the background radiation within the solid angles of the two aerials. The flux from the radio star is, in effect, compared with that from the galactic background, and when the aerial characteristics are known an absolute measurement is available without knowledge of the characteristics of the receiving amplifier itself. For absolute measurements of the intensity of a number of radio stars it is, of course, also necessary to determine p_g in different directions; this may conveniently be done by making recordings of the total aerial power.

(ii) $P_N \gg \alpha\lambda^2 p_g$

At the shorter wave-lengths (< 2 or 3 m), the noise power generated in the receiver itself (P_N) normally exceeds that from the aerial. Under these conditions the recorder sensitivity does not depend on the direction of observation, but is a function of P_N. Fortunately, it is very much easier to maintain the input noise power of an amplifier within small limits than it is to maintain its gain, and there is little difficulty in practice in maintaining the noise power within 5 to 10 % despite large variations in the supply voltage.

(f) *The absolute measurement of power flux when pre-amplifiers are used*

The use of pre-amplifiers not only allows a very considerable extension of the limiting sensitivity and resolving power which may be used, but also increases the range of wave-lengths over which the recorder sensitivity is determined by the galactic background radiation. Previously it was shown that the sensitivity would be controlled by the galactic background as long as $\alpha\lambda^2 p_g > P_N$. With pre-amplifiers the limit is given by $\lambda^2 p_g > P_N$. For example, observations with an aerial separation of 100λ would only be controlled by the background on wave-lengths greater than about 8 m (the precise figure depends, of course, on the type of aerial cable used). If pre-amplifiers are used, the sensitivity will be controlled by the background with any aerial separation and with low-grade cables, at wave-lengths as short as about 3 m. It is, however, now necessary to examine the effects of inequality in the gains or noise powers of the two pre-amplifiers.

(i) $\lambda^2 p_g \gg P_N$

The noise power at the input to the main receiving amplifier will be

$$\tfrac{1}{2}\lambda^2 p_g(\alpha\beta + \alpha'\beta'),$$

where $\beta\beta'$ are the gains of the two pre-amplifiers. (The noise power generated in the main receiver may be neglected provided $\alpha\beta \gg 1$.)

The alternating component of noise power due to the radio star is given by

$$2p_s\sqrt{(\alpha\alpha'\beta\beta'AA')},$$

and the deflexion of the recorder will be

$$\frac{4gp_s\sqrt{(\alpha\alpha'\beta\beta'AA')}\,C}{\lambda^2 p_g(\alpha\beta+\alpha'\beta')}.$$

It can thus be seen that, providing the overall gains ($\alpha\beta A$) of the two halves are equal, the deflexion sensitivity is independent of the gains of the pre-amplifiers.

Now suppose that one-half has a gain equal to half that of the other (e.g. $\beta = \frac{1}{2}\beta'$). The deflexion will be reduced by a factor $\frac{2}{3}\sqrt{2} = 0{\cdot}94$. Thus a deterioration of 2 : 1 in the gain of one pre-amplifier will only produce a 6 % change in the overall sensitivity of the system. In practice the total gain of the pre-amplifiers need only be sufficient to overcome the attenuation constant of the cables ($\alpha\beta \gg 1$), and the maintenance of a gain stability considerably better than 2 : 1 is very simple.

(ii) $P_N \gg \lambda^2 p_g$

At the shorter wave-lengths the gain of the main receiving amplifier is determined by the total noise power from the two pre-amplifiers $\frac{1}{2}(\alpha\beta P_N + \alpha'\beta' P'_N)$, and the deflexion of the recorder due to a radio star of intensity p_s is given by

$$\frac{4gp_s\sqrt{(\alpha\alpha'\beta\beta'AA')}\,C}{(\alpha\beta P_N+\alpha'\beta' P'_N)}.$$

If the gains and noise powers of the two pre-amplifiers are equal, the sensitivity depends only on the value of this noise power. If the gain of one pre-amplifier falls by a factor of 2, the overall sensitivity decreases by 6 %; a similar change would be produced by an increase of the noise power of one pre-amplifier by 12 %. As has already been stated, the maintenance of such figures of stability presents no great difficulty.

3. Comparison of the phase-switching system and other systems for observing radio stars

(a) *The detection of radio stars*

The problem of detecting a weak 'point' source of radio waves has already been considered in some detail (Ryle 1950); it was there shown that a lower limit may be set to the intensity of the detectable radio stars by any one of the three following experimental difficulties.

(i) The difficulty of measuring the small radio-frequency power produced in the aerial by the radio star owing to the random fluctuations on the record. The amplitude of these fluctuations depends first on the total noise power at the input of the receiver, which includes that due to the galactic background and that generated in the receiving amplifier itself, and secondly on the ratio of the bandwidths of the receiving amplifier before and after the detector.

(ii) The difficulty of discriminating between the aerial power due to the radio star and the very much greater power due to the galactic background.

(iii) When sensitive receiving apparatus is used in conjunction with an aerial system of large resolving power an additional limitation may arise, which is due to the confusion between the traces produced by adjacent radio stars.

It was shown that the use of an interferometer aerial system provided a powerful method of overcoming the second difficulty, and allowed a discrimination which was comparable with that of a simple 'pencil-beam' aerial system having an aperture of twice the *spacing* between the aerials of the interferometer.

Two main types of interferometer have been used: (i) those in which the interference pattern is produced by two spaced aerials (Ryle & Vonberg 1946; Ryle 1950) and (ii) those based on the use of a single aerial mounted on a high cliff to produce interference by reflexion at the surface of the sea (McCready *et al.* 1947; Bolton & Stanley 1949; Stanley & Slee 1950). The latter type has the advantage of simplicity, and allows a large resolving power to be obtained without the use of two long and equal lengths of high-quality cable from the aerials, but it suffers from the disadvantage that observations are necessarily made at small angles of elevation, where refraction effects become serious. Observations with the spaced-aerial interferometer, on the other hand, are made when the radio star is near the meridian, and refraction effects are a minimum. The cable attenuation introduced when large resolving powers are used does not affect the limiting sensitivity when the phase-switching system is used, since pre-amplifiers may be mounted at the aerials.

The problem of measuring a small random radio-frequency power in the presence of a much larger noise power (P_N) generated in the receiving amplifier has been examined in an earlier paper (Ryle & Vonberg 1948), where it was shown that the additional power should be detectable if it exceeded $P_N/\sqrt{(Bt)}$, where B is the band-width of the receiving amplifier and t is the time-constant of the recorder circuits following the detector. The product Bt is commonly of the order of 10^5 to 10^6. In practice the necessity for very accurate stabilization of the gain of the amplifier makes it difficult to achieve a sensitivity as good as this, and in earlier observations at Cambridge of the radiation from the sun and radio stars an improved system of recording was used in which the output of a local source of noise power was continuously and automatically balanced against the aerial power (Ryle & Vonberg 1948). This system has been found to have a limiting sensitivity close to that expected theoretically,* and since the aerial power is measured absolutely, in a manner which is independent of the characteristics of the receiving amplifier, it has proved a most satisfactory way of measuring the total aerial power.

It is now important to examine the extension of these ideas to the detection of the much smaller power from weak radio stars. The aerial power due to the galactic background must now be considered with the receiver noise power as an unwanted background, which not only increases the statistical fluctuations on the record $(P_N+\lambda^2 p_g)/\sqrt{(Bt)}$, but may also introduce other difficulties which are discussed

* Since the aerial is connected to the receiver for only half the time in this system, the smallest detectable power is increased by a factor of $\sqrt{2}$, as compared with the theoretical limit; this loss does not occur with the phase-switching system, where the aerial system is connected all the time.

below. It should first be noted that the theoretical limit of detection set by the statistical fluctuations on the record may be very much smaller than was considered in earlier work on solar radiation. It is often possible to use a much longer recording time-constant t, determined only by the rate of movement of the interference pattern due to the rotation of the earth, and the corresponding statistical fluctuations on the record may represent less than $\frac{1}{1000}$th of the total aerial power. Even with the automatic balancing system mentioned above, other difficulties may arise when attempting to measure additional aerial powers of this magnitude.

A typical record obtained with a spaced-aerial interferometer and the total-power recording system mentioned above is shown in figure 4. The periodically varying component of aerial power due to the passage of the two most intense radio stars can be seen superimposed on the slowly varying component due to the galactic background. It is clear that records of this type are adequate for observing radio stars which produce an aerial power comparable with that due to the background radiation, but it would be difficult to make accurate observations of a radio star

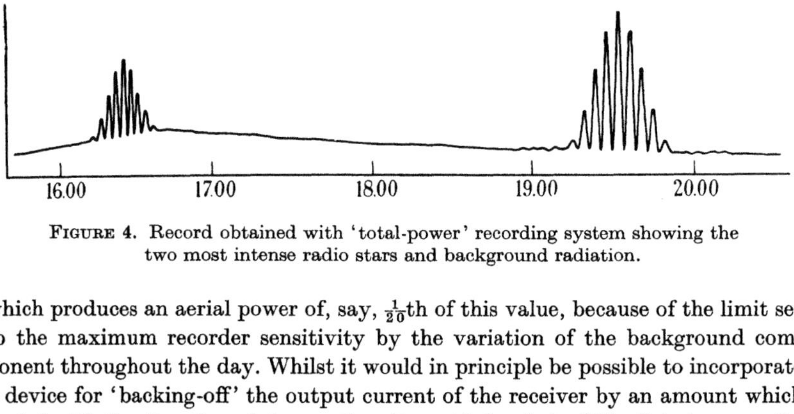

FIGURE 4. Record obtained with 'total-power' recording system showing the two most intense radio stars and background radiation.

which produces an aerial power of, say, $\frac{1}{20}$th of this value, because of the limit set to the maximum recorder sensitivity by the variation of the background component throughout the day. Whilst it would in principle be possible to incorporate a device for 'backing-off' the output current of the receiver by an amount which varied with the direction of observation, it would clearly be difficult to increase the sensitivity sufficiently to enable the observation of a radio star which produced an aerial power only $\frac{1}{1000}$th of that due to the background radiation. In the phase-switching system, no deflexion of the recorder is produced by the background radiation, and, as has already been shown, a very much greater recorder sensitivity can therefore be used.

Apart from purely practical difficulties of this type, there are, however, more fundamental problems involved in detecting weak radio stars by 'total power' recording methods. No attempt will here be made to analyze these problems in detail, but two specific difficulties will be mentioned.

(i) As has already been shown (Ryle & Vonberg 1948), small variations of the aerial impedance will alter the effective load resistance in parallel with the input circuit and will thereby modify the internal noise power (P_N) generated in the amplifier. It was shown that a change of 10 % of the aerial impedance might displace the recorder by an amount corresponding to 5 % of the receiver noise power P_N. If observations of a radio star which produces an aerial power of only $P_N/1000$ are attempted, it is clear that small variations of the aerial impedance, such as are likely to occur in wet weather, will produce irregular variations on the recorder which may mask the small periodic trace due to the radio star. In the phase-switching system the impedance appearing across the input circuit of the receiver is the same in both positions of the phase-changing switch (provided that 'half-wave' switching is used); no component of the output noise power at the switching frequency will therefore be produced, even if one or both of the aerials is badly mismatched, and no deflexion of the recorder will therefore occur.

(ii) Even in the absence of difficulties associated with the noise power generated in the receiver, similar difficulties may arise when the aerial power produced by the radio star is only about $\frac{1}{1000}$th of that due to the background radiation. Small variations of the attenuation or reflexion loss in one of the cables from the aerials may produce sufficient variation of the total power from the galactic background to mask the periodic variations due to the movement of the radio star through the interference pattern. No deflexion of the trace of the phase-switching system is produced by the background radiation, and small variations of the cable loss will therefore not affect the record.

It has been shown that many of the difficulties encountered in the detection of weak radio stars by methods based on the measurement of the total aerial power have been overcome by the use of the phase-switching system. The limiting sensitivity determined by the random fluctuations on the record has been found to be in close agreement with that predicted theoretically for a total noise power equal to the sum of the aerial and pre-amplifier powers, and this limiting sensitivity is not affected by attenuation in the aerial cables. The use of aerial spacings of 100 to 200λ has allowed virtually complete elimination of the background radiation and inequalities or variations of the impedances of the aerials or the attenuation of the cables have no effect. The limit of detection with the phase-switching system is therefore determined either by the theoretical random fluctuations on the record, or by the confusion of the traces produced by adjacent radio stars. The relative importance of these two limitations depends not only on the intensity and distribution of radio stars and of the background radiation, but on the dimensions of the aerial system and on the wave-length used. It is outside the scope of the present paper to discuss the effect of these two limiting factors on the choice of aerial design, or of their corresponding effects on measurements of the positions of radio stars, but it is clear that at the longer wave-lengths, the solid angle for reception is likely to be large, whilst the random fluctuations on the record depend only on the background radiation from the galaxy; under these conditions the limitation set by confusion is likely to be the more serious. At short wave-lengths, on the other hand, the resolving power is likely to be greater, whilst the random fluctuations will be mainly due to the pre-amplifier noise power and are more likely to be the limiting factor.

In addition to improved sensitivity and available resolving power the phase-switching system has a number of other advantages over 'total power' recording systems, and these will now be discussed.

(b) *The determination of the position of a radio star*

Accurate measurements of the positions of radio stars have all been based on observations of the movement of the radio star through the reception pattern of an interference system. A discussion of the methods which may be used is given in another paper (Smith, in preparation). In observations with 'total power' recording systems, such measurements are best made by determining the times of successive minima of the recorded power. It is, however, clear that a marked improvement in the accuracy should be possible if measurements were made at a time when the *rate of change* of the recorded power was a maximum rather than a minimum.

In the phase-switching system, zero deflexion of the recorder occurs whenever the two signals from the aerials are in quadrature; at this time the rate of change of the recorder deflexion is a maximum and very much greater timing accuracy is therefore possible. Furthermore, since the addition of any two voltages in quadrature gives a resultant whose magnitude is unaltered if the phase of one voltage is reversed, the time at which zero deflexion of the recorder occurs is not altered if the two signals are of unequal magnitude; the 'time of cross-over' on the record is therefore not affected by inequalities in the effective area of the aerials, or by differences in the gains of the pre-amplifiers or in the attenuation constants of the cables.

Since the time of cross-over corresponds to zero amplitude of the switch-frequency component at the detector, the accuracy is also unaffected by non-linearity in the succeeding amplifiers or by changes in the intensity of the incident radiation. It therefore becomes possible to use a very great deflexion sensitivity, such that the maxima of the trace would correspond to perhaps 50 times full-scale deflexion of the recorder, and complete overload of the switch-frequency amplifier and subsequent circuits. If a very great deflexion sensitivity is used in this way, the time during which the recorder deflexion is changing is very small (perhaps only 0·01 of the total period of the trace), and it becomes possible to determine the cross-over time with an accuracy which corresponds to only 10^{-3} of the period of the pattern. This accuracy corresponds to a determination of the quadrature condition of the voltages from the two aerials to a fraction of a degree; with aerial separations of 100 to 200λ this error is equivalent to an error in the apparent right ascension of the source of a few seconds of arc.* Experimental records of this type are shown in figures 5 and 6.

(c) *Determination of the position of a source of variable intensity*

The ability to observe the time of cross-over, and hence the position of the source, in a manner which is independent of the intensity of the radiation is particularly important for investigating the diffraction effects caused by irregularities in the terrestrial ionosphere, and for observing the intense sources of radiation associated with sunspots. In both cases large and rapid fluctuations of intensity occur, and it is clear that serious errors will be introduced if attempts are made to estimate the apparent position of the source with a 'total power' recording system. Even with

* The determination of the *absolute* position of a source also requires a knowledge of the refraction effects in the terrestrial atmosphere, and of the positions of the electrical centres of the two aerials of the system.

very large and rapid variations of intensity the timing error involved in the phase-switching system can only be of the order of the time-constant of the recording system. In many cases the observed variations of intensity are considerably slower than this quantity, and the errors are correspondingly reduced. Phase-switching systems have been used in this way to determine the relation between the variation of apparent position and the variation of intensity of a radio star due to ionospheric diffraction (Ryle & Hewish 1950), and to investigate the angular movement of the effective centre of sources of sunspot radiation.

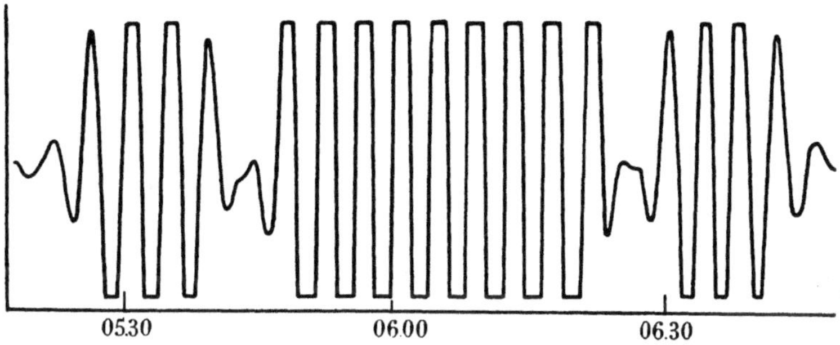

FIGURE 5. Typical record of intense radio star using increased recording sensitivity.

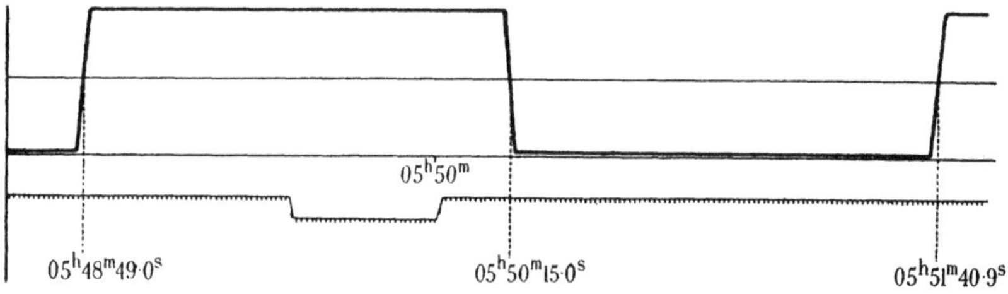

FIGURE 6. Typical high-speed record used for accurate determination of the position of a radio star.

The method described allows accurate observations of the position to be made at each half-period of the interference pattern; an extension of this method to allow a continuous record to be obtained of a source whose intensity and apparent position are both varying has also been used (Hewish, in preparation).* 1

(d) The measurement of the angular diameter of a source

It has been shown that when an extended source is observed with an interferometer having an aperture d, the amplitude of the trace produced by a phase-switching system, and the amplitude of the *variable component* of the trace produced by a total power recording system, are proportional to one term of the Fourier transform of the intensity distribution; the angular frequency of the term is given by d/λ. If the source has a uniform distribution of intensity, and a width $2\Delta\theta$

(e.g. a square source), the amplitude of the trace relative to that produced by a point source emitting the same total power is given by

$$\frac{\sin \frac{2\pi d}{\lambda}\Delta\theta}{\frac{2\pi d}{\lambda}\Delta\theta} = \gamma.*$$

Previous measurements with total power recording systems have taken advantage of the fact that provided observations are made of the received power at both maxima and minima of the interference pattern, their ratio $(1+\gamma)/(1-\gamma)$ enables a determination of γ without a knowledge of the total intensity of the source. It is therefore possible to deduce the angular diameter with a single value of the aerial spacing d. Observations with the phase-switching system give a trace whose amplitude is proportional to γ, and it is therefore necessary to compare the amplitudes obtained with two different aerial spacings before the angular diameter can be deduced.

It is now important to compare the practical difficulties of the two methods, with particular reference to the determination of very small angular diameters, and those of sources of small intensity. It is first useful to show (table 1) how the quantities γ and $(1+\gamma)/(1-\gamma)$ vary with $2\Delta\theta/(\lambda/d)$.

TABLE 1

$2\Delta\theta/(\lambda/d)$	γ	$(1+\gamma)/(1-\gamma)$
0	1·0	∞
0·01	0·9999	12200
0·02	0·9993	3040
0·05	0·996	485
0·1	0·984	121
0·2	0·936	31
0·3	0·86	14·3
0·4	0·76	7·3
0·5	0·64	4·1
0·6	0·51	3·0
0·8	0·23	1·6
1·0	0	1·0

The resolving power of either system is determined by the largest value of γ or $(1+\gamma)/(1-\gamma)$ that can be measured. In the phase-switching system this limit is set by the accuracy with which the overall gains of the systems at the two different spacings are known. The total-power system is limited by the accuracy with which the minimum power may be found, and is therefore affected by inequalities in the gains of the two aerials (or the losses in their cables) and by uncertainty concerning the contribution of the galactic background radiation.

Early observations of the angular diameter of the intense sources of radiation associated with sunspots were made with interferometers having apertures of 100 to

* A circular source which is small compared with λ/d will produce a trace equal to that from a 'square' source whose width is $0{\cdot}82 \times$ its angular diameter.

200λ. Since the diameters of these sources are of the order of 3 to 10 min. of arc, it was only necessary to work with values of $2\Delta\theta/(\lambda/d)$ of 0·1 to 0·5, and no special difficulties were encountered in equalizing the gains or cable losses of the two aerials. Furthermore, since the intensity of the radiation emitted by these sources was normally considerably greater than that from the galactic background, there was little difficulty in estimating the effective zero level on the record when determining the minimum power from the sun.

If attempts are made to extend such observations to measure the angular diameter of radio stars, two serious limitations are encountered with total-power recording systems:

(i) The angular diameters are known to be very much smaller than those of sunspots, and present estimates of the maximum diameters using interferometers of aperture 100 to 200λ have involved maximum/minimum ratios of the interference pattern of about a hundred. An increase of the aerial spacing would of course allow the same resolving power to be obtained with a smaller ratio, but the additional cable attenuation would decrease the ratio of signal to random fluctuations on the trace, and therefore again introduce a limitation to the resolving power; pre-amplifiers could only be used if it were possible to ensure that their gains remained equal to within 1 %.

Although the problem of cable attenuation does not arise in 'cliff' type interferometers, other difficulties are encountered in this method, both because of the effects of imperfect reflexion at the surface of the sea (Stanley & Slee 1950) and because of the geographical difficulties of obtaining a sufficiently high cliff site.

(ii) Both 'cliff' and spaced aerial interferometers using total-power recording suffer from an even more serious limitation when observing sources of small intensity; the uncertainty of determining the deflexion produced by the background radiation then makes it impossible to obtain an accurate measure of the minimum power received from the source. For this reason it is likely to be difficult with existing systems to make any estimate of the diameter of a source which produces an aerial power less than about 1 % of that due to the background radiation, whilst a source which produces a power of 10 % could probably only be resolved if its angular diameter exceeded 0·25 (λ/d).

Both of these difficulties are largely overcome by the use of the phase-switching system. Since no deflexion is produced by the background radiation, the intensity of the source does not present a limitation provided that it can be observed satisfactorily in the presence of the random fluctuations on the record. The limiting resolving power is then determined solely by the accuracy with which the overall gain of the system is known at the two aerial spacings; if the gain remains constant to within 5 % when the aerial spacing is changed, it is possible to resolve a source whose diameter is about 0·15 (λ/d). Increased resolution is possible if the overall gain can be determined directly by observing another source (such as a more remote radio star) whose angular diameter is less than that of the source under investigation. In addition, the use of simple pre-amplifiers makes it possible to operate with a very much greater aerial spacing than was possible with a total-power system.

(e) *The measurement of the polarization of the radiation from a radio star*

The measurement of the polarization of the radiation from a weak source, in the presence of a large unpolarized background radiation, has already been discussed in some detail (Ryle & Vonberg 1948, Ryle 1950). It was shown that the use of an interferometer aerial system in which the planes of polarization of the two aerials could be altered, allowed a complete analysis of the polarization of the radiation from a source of small angular diameter, in the presence of intense unpolarized radiation from the galactic background. The application of the phase-switching system to this method has similar advantages to those considered in the detection of weak radio stars, and makes it possible to detect a very small degree of polarization. If the aerials are arranged to respond to a circularly polarized component, no deflexion of the recorder will be produced either by the background radiation or by randomly polarized radiation from the radio star; it therefore becomes possible to use a very much greater recording sensitivity.

In another application which has been used to study the instantaneous polarization of the radiation from sunspots, a phase-switching system has been used in conjunction with a conventional crossed-aerial polarization system. No deflexion is produced by the background radiation or by randomly polarized radiation from the sun, but the occurrence of a circularly polarized component produces a proportional deflexion whose direction depends on the sense of the polarization. In this way a continuous record can be obtained of the instantaneous changes of polarization.

(f) *Freedom from man-made interference*

The incidence of man-made interference, particularly the radiation from unsuppressed ignition systems of cars and aircraft, presents one of the most serious practical limitations to the detection sensitivity of 'total power' systems.

From equation (9) it can be seen that the output deflexion in the phase-switching system is a function of the product of the powers intercepted by the two aerials. An interfering signal which is received by only one aerial will not therefore deflect the recorder. This property enables a very considerable immunity from local sources of interference to be obtained with the large aerial systems used for detecting radio stars. As an example the present system which has been used at Cambridge on a wave-length of 3·7 m consists of two aerials having an aperture of 20λ and separated by a distance of 0·5 km; the reception pattern of each aerial therefore extends over about $\pm 1\frac{1}{2}^\circ$ in an east-west plane, and a source of interference cannot lie in both reception patterns unless it is at a distance greater than about 10 km. Sources much nearer than this, which would cause serious interference with a total-power system whenever they passed through the reception pattern of either aerial, do not therefore affect the phase-switching system. At distances greater than 10 km the intensity of the interference is of course greatly reduced, particularly for sources at ground level, for which the received power falls off as the fourth power of the distance.

(g) *Control of the primary reception pattern of the phase-switching system*

It was shown in equation (12) that the recorder deflexion was proportional to $\sqrt{(\alpha\alpha'\beta\beta'AA')}$. The envelope of the interference pattern is thus determined by the geometric mean of the reception patterns of the individual aerials. Where the two halves of the interferometer are identical, the envelope of the pattern on the record will therefore correspond to the power reception pattern of each aerial. The ability to control the overall reception pattern by independent adjustments to either aerial has some important possibilities both for improving the effective resolving power of the system, or for reducing side lobes. One application of this feature is related to the detection of radio stars at the longer wave-lengths, where the limit of detection is usually determined by confusion between adjacent radio stars and not by the limiting sensitivity of the system. Under these conditions a decrease of the solid angle for reception without necessarily increasing the total area of the aerials would allow an increase in the number of sources which could be detected. By using aerials of equal area, one of which has a large aperture in the north-south plane, whilst the other has a large aperture in the east-west plane, it becomes possible to reduce the effective solid angle for reception without increasing the total area of the system.

4. Some illustrative practical examples of the phase-switching system

In this section an attempt will be made to illustrate some of the results, which have been demonstrated analytically in earlier sections, by experimental records obtained during observations at Cambridge. These observations have been made on wave-lengths of 1·4, 3·7, 6·7 and 8 m with a number of different types of aerial system. Most of the observations on 3·7 m have been made with an interferometer which covers a narrow strip near the meridian extending from the North pole to the equator (Ryle *et al.* 1950). This system allows daily observations of a considerable number of radio stars in the northern hemisphere to be made. In another series of observations on 1·4 m two parabolic mirrors of diameter 6·5λ and spacing 205λ have been used to provide a system capable of more accurate position finding (Smith 1951). Observations on a wave-length of 8 m have been made with an interferometer having an aerial spacing of 60λ. In addition, a number of experiments have been made to investigate the scintillation of radio stars on all four wave-lengths.

(a) *Elimination of the background radiation*

Examples of records obtained with the phase-switching system have already been given in figures 3, 5 and 6. Examination of these records and of that shown in figure 9 (in which full-scale deflexion of the recorder corresponds to an aerial power of about 5 % of that due to the galactic background) shows the absence of any slow variations due to the background radiation which are visible in records such as figure 4 obtained with the same aerial system and a 'total power' recording system. In one of the series of observations made to investigate fluctuations of intensity caused by the diffraction effects in the terrestrial ionosphere, it was necessary to make continuous observations of the intense radio star in Cassiopeia (Ryle & Hewish 1950). To avoid complications of rotatable aerial structures, a phase-switching system was used with an interferometer in which each aerial had a sufficiently large solid angle for reception to observe the radio star in Cassiopeia continuously (power gain $\simeq 4$). A section of the record obtained with this system is shown in figure 7; full-scale deflexion corresponds to about 3 % of the aerial power due to the galactic background.

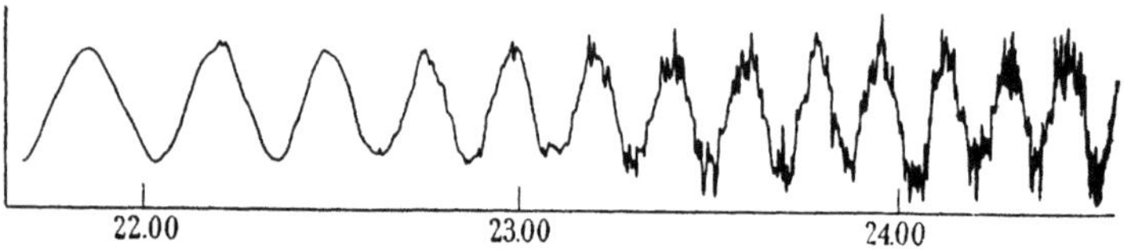

FIGURE 7. Record obtained with phase-switching system using very simple aerials (gain $\simeq 4$) to observe the radio star in Cassiopeia continuously.

In order to obtain a more exacting test of the effect of a comparatively localized area of increased background radiation, some special tests were carried out. Since, apart from the weak radiation from four of the extra-galactic nebulae (Ryle *et al.* 1950) no such localized areas are known, a series of observations was made of the sun in the presence of very weak sunspot radiation. Observations were made with the 1·4 m interferometer (aerial spacing 205λ), both with the 'total-power' recording system and with the phase-switching system. The results of some of these observations are shown in figure 8. During the first half of the time during which the sun was in the reception pattern of the aerials, the total-power system was connected, and the record shows both the galactic background radiation and the unresolved increase due to the radiation from the undisturbed solar disk. A small periodic component due to the small sunspot can also be seen. During the latter half of the record, the phase-switching system was used with a sensitivity approximately ten times as great; no deflexion associated with the galactic background or the undisturbed sun is visible, but the periodic component due to the sunspot is correspondingly greater.

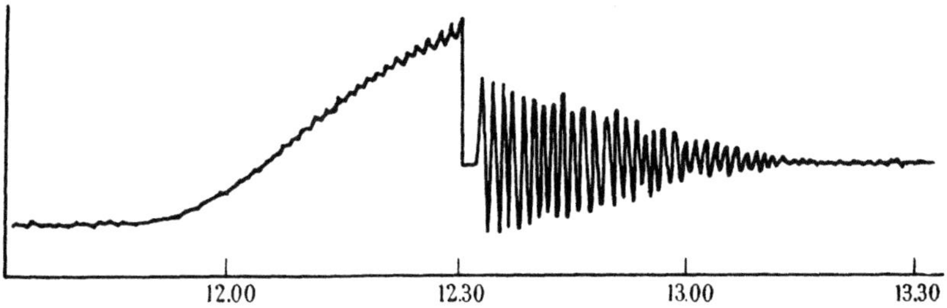

FIGURE 8. Observations showing radiation from undisturbed sun, and very small variable component from a minor sunspot: (*a*) 11.30–12.30 'total power' recording, (*b*) 12.30–13.30 phase-switching system.

From the results of such observations on the sun it can be seen that the use of an interferometer with the phase-switching system provides a satisfactory method of eliminating any likely distribution of the background radiation from the galaxy.

(*b*) *The limit of detection*

It was shown in § 3 (*a*) that the intensity of the weakest detectable radio star might be limited either by the random fluctuations on the record, or by confusion between the traces produced by adjacent radio stars. In particular, the relative importance of these two limitations depends on the size of the aerials, and on the wave-length.

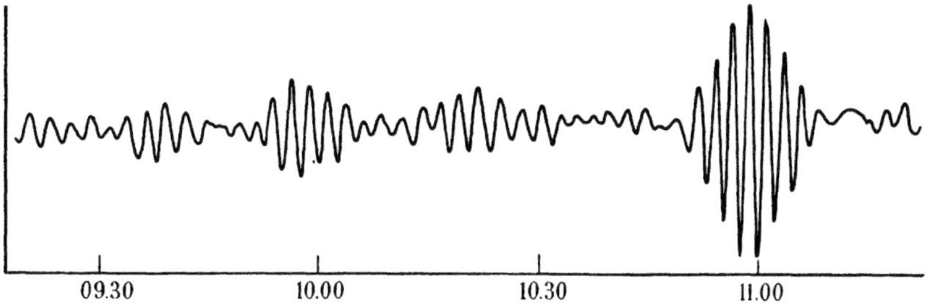

FIGURE 9. Section of record obtained with 3·7 m interferometer, showing a number of weak radio stars.

A typical section of record obtained with the 3·7 m interferometer is shown in figure 9. It can be seen that the random fluctuations are considerably smaller than the amplitude of the smallest trace which can be reliably identified. It is, in fact, found that many of the unidentified variations repeat from day to day and are therefore not due to random effects, but to the superimposition of the traces from a number of weak radio stars. Observations with this system are therefore limited entirely by confusion.

Observations with the 1·4 m interferometer, on the other hand, show that there is no confusion between any traces which are visible above the random fluctuations.

In § 2 (*e*) and (*f*) it was shown that absolute intensity measurements with the phase-switching system did not involve accurate stabilization of the gain of the receiving amplifier, and that at the longer wave-lengths the sensitivity was also independent of the noise power of the pre-amplifiers. In a series of observations on 3·7 m (where the background radiation produces a power approximately equal to that of the pre-amplifier) it was found that the deflexion sensitivity remained constant to within 5 % over a period of a year, although no special precautions were taken to stabilize the gains of the pre-amplifiers.

(*c*) *The determination of the position of a source*

An example of the type of record used for accurate determination of the position of a radio star was shown in figure 6. In the absence of ionospheric diffraction the accuracy of such observations is limited by the stability of the differential phase-angle of the aerials and pre-amplifiers and of the differential length of the aerial cables. Experiments have shown that extremely good stability can be achieved. In one particular series of observations with the 3·7 m interferometer over a period of 8 months no variation of the apparent position of the two intense radio stars could be detected greater than the limit set by the residual ionospheric diffraction. These results indicated a stability of phase of the aerials, pre-amplifiers and cables of better than 8° (corresponding to a variation in the apparent right ascension of the radio star of 40 sec. of arc).

In another series of experiments on 1·4 m (where the ionospheric diffraction effects are smaller), a phase stability of 5° was observed, corresponding to a variation in the apparent right ascension of 15 sec. of arc.

It was shown in § 3 (*c*) that the phase-switching system allowed accurate observations to be made of the position of a source of variable intensity. Observations of the variation of the apparent position of a radio star due to ionospheric diffraction have already been described (Ryle & Hewish 1950). Similar observations have also been made to investigate the movement of the effective centre of the intense sources of radiation associated with sunspots. An example of such a series of observations is shown in figure 10, in which the variation of intensity and apparent position of the source of 3·7 m solar radiation is shown.

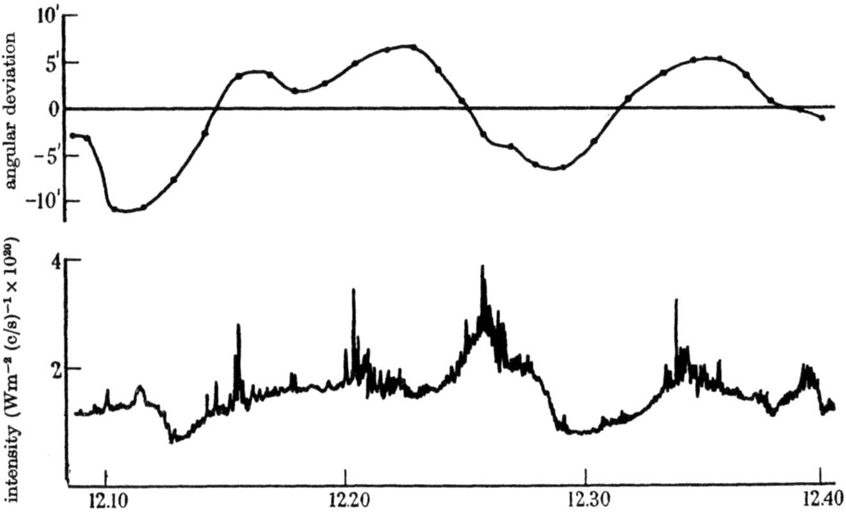

FIGURE 10. Record showing variation of intensity and deviation of apparent position of source of 3·7 m radiation associated with sunspot, 23 January 1951.

(*d*) *The reduction of man-made interference*

Figure 11 shows part of a record obtained with the 3·7 m interferometer in the presence of serious local interference generated in the reception pattern of one of the aerials. The first trace shows the record obtained with the 'total-power' recording system; the second trace shows that obtained with the phase-switching system, whose gain has been adjusted so that the deflexion sensitivity is the same as that of the 'total-power' system.

(e) *Control of the primary reception pattern*

It was shown in § 3 (*g*) that by the use of two aerials having different primary reception patterns the effective reception pattern of the interferometer could be improved. An example of this property of the phase-switching system is given in

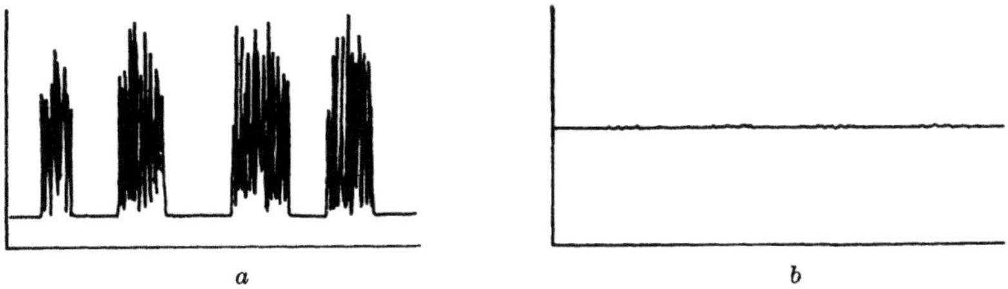

FIGURE 11. Records showing the effect of local interference on observations with (*a*) the 'total power' system and (*b*) the phase-switching system.

figure 12, which shows the observed envelope pattern of an interferometer in which each aerial consisted of six full-wave dipoles specially arranged to produce a restricted east-west coverage. For comparison the calculated reception pattern which would have been obtained with six aerials arranged as a conventional broadside array at each end of the interferometer is also shown.

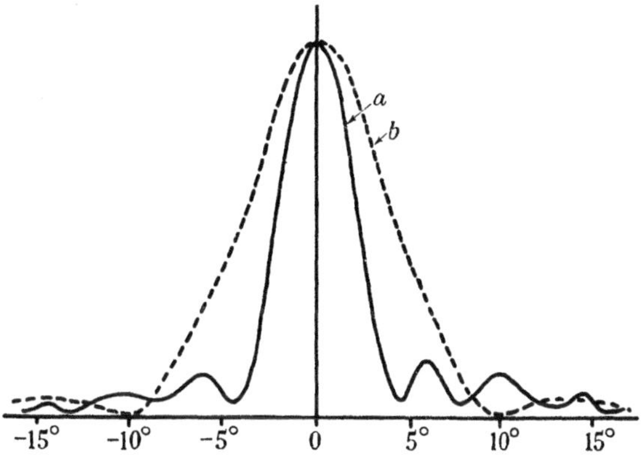

FIGURE 12. Envelope of reception pattern of interferometer in which six full-wave dipoles were used at each aerial: (*a*) special spacing of aerials; (*b*) uniform aperture.

This work was carried out as part of a programme of radio research supported at the Cavendish Laboratory by the Department of Scientific and Industrial Research. The author is indebted to his colleagues, especially to Messrs K. E. Machin and F. G. Smith, for their assistance during the development of the phase-switching system, and to Mr B. Elsmore for the preparation of the diagrams.

References

Bolton, J. G. & Stanley, G. J. 1949 *Aust. J. Sci. Res.* A, **2**, 139.
Machin, K. E. 1951 *Nature, Lond.*, **167**, 889.
McCready, L. L., Pawsey, J. L. & Payne-Scott, R. 1947 *Proc. Roy. Soc.* A, **190**, 357.
Ryle, M. 1950 *Rep. Progr. Phys.* **13**, 184.
Ryle, M. & Hewish, A. 1950 *Mon. Not. R. Astr. Soc.* **110**, 381.
Ryle, M., Smith, F. G. & Elsmore, B. 1950 *Mon. Not. R. Astr. Soc.* **110**, 508.
Ryle, M. & Vonberg, D. D. 1946 *Nature, Lond.*, **158**, 339.
Ryle, M. & Vonberg, D. D. 1948 *Proc. Roy. Soc.* A, **193**, 98.
Smith, F. G. 1951 *Nature, Lond.*, **168**, 555.
Stanier, H. M. 1950 *Nature, Lond.*, **165**, 354.
Stanley, G. J. & Slee, O. B. 1950 *Aust. J. Sci. Res.* A, **3**, 234.

EDITOR'S NOTES AND AMENDMENTS

1. *Proc. Roy. Soc.* **A209**, 81 (1951).

PART III

THE SOLAR SYSTEM

The Sun was not only the first discrete source of extraterrestrial radio emission to be detected, but it was also the object of several unsuccessful experiments long before the time of Jansky. In the 15 years following the laboratory experiments of Heinrich Hertz (1857–1894), attempts to detect solar radio waves were made in several countries. In fact, as early as 1890 Thomas A. Edison and his associate A. E. Kennelly are known to have considered such a possibility, using an antenna consisting of a cable draped on poles around a mass of iron ore (Shane, 1958). Although it is not known if this experiment was ever carried out, even its very contemplation reveals yet another facet of the remarkable Mr Edison.

The next known attempt was in England, sometime during the period 1897 to 1900, by Sir Oliver J. Lodge (1851–1940), an early leader in the development of electrical techniques and sometime maverick in many another field. The entire available documentation reads as follows (Lodge, 1900):

> I [hoped] to try for long-wave radiation from the sun, filtering out the ordinary well-known waves by a blackboard or other sufficiently opaque substance. I did not succeed in this, for a sensitive coherer in an outside shed unprotected by the thick walls of a substantial building cannot be kept quiet for long. I found its spot of light liable to frequent weak and occasionally violent excursions, and I could not trace any of these to the influence of the sun. There were evidently too many terrestrial sources of disturbance in a city like Liverpool to make the experiment feasible. I don't know that it might not possibly be successful in some isolated country place; but clearly the arrangement must be highly sensitive in order to succeed.
>
> Also I exhibit [Figure 4] a small complete detector made by my assistant, Mr. Davies, which is quite portable and easily set up. The essentials (battery, galvanometer, and coherer) are all in a copper cylinder, A, three inches by two. A bit of wire, B, a few inches long, pegged into it, helps it to collect waves.

It is not certain that the apparatus shown in Figure 4 was exactly the same as that used in this try for the Sun, but it was probably quite similar and sensitive to centimeter wavelengths. The *coherer* or radio-conductor was an early detector separately developed by Branley and Lodge. It typically consisted of a tube of metal filings or other such substance which in its

Fig. 4. An early radio detector and antenna developed by Lodge. This probably was similar to one employed in his attempt to detect solar radio waves (see text) in 1897.

normal state provided poor electrical contacts and, therefore, a high resistance. An incident radio wave was detected by the fact that it caused the resistance to decrease greatly, an effect easily measurable.

Johannes Wilsing (1856–1943) and Julius Scheiner (1858–1913), well-known astrophysicists whose careers were primarily concerned with photographic and spectroscopic work on a variety of objects, carried out the most quantitative and detailed of the early experiments at Potsdam in 1896. They met with no success on any of several days of observations, and finally concluded that they were being foiled by atmospheric absorption of the solar rays (Paper 13). This conclusion in turn prompted Charles Nordmann (1881–1940) to make an attempt from a research station high in the French Alps. Despite what today appears to be an antenna and receiver of adequate sensitivity for solar bursts, his one day of observations also was to yield no signals (Papers 14 and 15).

Following a long and somewhat puzzling gap in further attempts, during World War II, Hey (February, 1942; Paper 16), Southworth (June, 1942; Paper 17), and Reber (November, 1943; Paper 6) all made independent and quantitative measurements of the solar radio radiation.[1] As mentioned previously, Jansky would undoubtedly have also detected the Sun a decade before if the early 1930s had not been a period of solar minimum. Later, near the maximum of that same solar cycle in the late 1930s, there were, in fact, numerous reports by both amateurs and professionals of intense short-wave noise associated with radio fade-outs and other ionospheric effects known to be related to solar activity (Hey, 1973; Appleton, 1945). None of these solar radio bursts, however, were recognized for what they were.

Solar radio astronomy developed extremely rapidly in the post-war years in many countries, primarily Australia, England, and France. This exciting new field yielded discovery after discovery concerning the intense bursts as well as the quiescent solar corona. When better angular resolution was found to be necessary, the first radio observers extended radar techniques in a manner familiar to optical astronomers. High angular resolution was achieved either through the use of an eclipse or, more handily, an interferometer. Thus great efforts were made in these years to travel to the sites of solar eclipses. At the same time, groups in Cambridge and Sydney developed the radio analogues to the Michelson interferometer and to Lloyd's mirror. The following papers are primarily limited to the major solar studies and discoveries which occurred during or immediately following the war. If there were room to follow the subject to a later date, also included in particular would be the paper by J. P. Wild and L. L. McCready (1950) on the development and use of the dynamic solar spectrograph, a study which led to a much improved understanding of solar bursts.

NOTE

[1] German radar antennas in Denmark in 1939–1940 also detected the Sun at 125 MHz and, to some extent, even studied its emission (Schott, 1947). Furthermore, other military units around the world also detected the solar radiation, but their findings were never made public.

SECTION G

EARLY SOLAR SEARCHES

The first properly published experiment in radio astronomy is Paper 13 by Wilsing and Scheiner (1896). Despite careful construction and calibration of their apparatus (shown schematically in Figure 5), they are unable to record

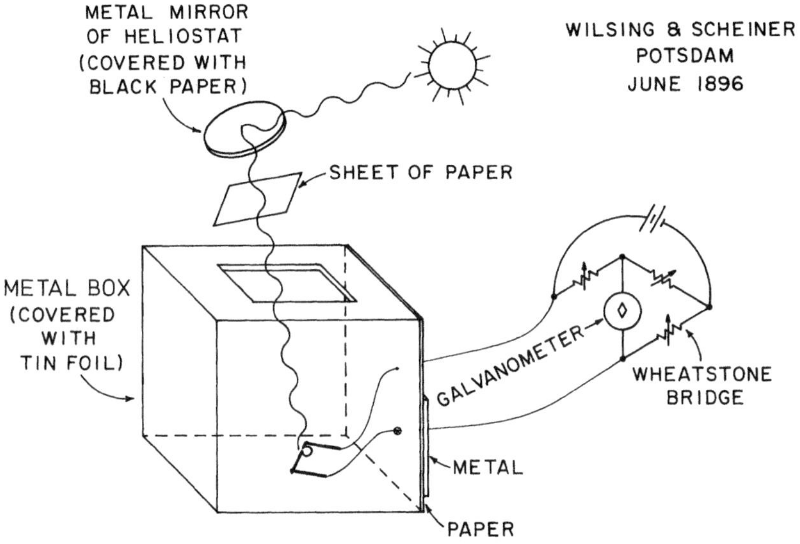

Fig. 5. A schematic representation of the set-up employed by Wilsing and Scheiner (1896) in an attempt to detect solar radio waves (see the detailed description within Paper 13).

a significant deflection on their galvanometer on any of eight days of observation. This negative result apparently is not totally unexpected, for they cite the work of Langley on infrared absorption in the Earth's atmosphere. Then, making a natural extrapolation, they conclude that "electrodynamic waves" are also probably screened by the atmosphere.

In September, 1901, Nordmann, a graduate student at the University of Paris, conducted a remarkable experiment which is recounted in Paper 14. He apparently agreed with Wilsing and Scheiner that their experiment failed due to atmospheric absorption, and so he made his observations under the difficult conditions found on a glacier at 3100 m altitude near Mount Blanc. The addition of a long wire for an antenna meant that he not only achieved a greater effective collecting area, but also that his sensitivity was primarily to radiation of longer wavelengths. These factors would have helped him in detecting solar bursts, but Nordmann was unlucky. If it had not been a time of solar minimum or if he had been persistent enough to observe for more than one day, he might well have succeeded and thereby drastically changed the history of astronomy.[1]

Paper 15, read to the *Académie des Sciences* only one month later by Henri Deslandres (1853–1948) and Louis Décombe (1873–?), is a theoretical follow-up to Nordmann's negative results. It remains a remarkable paper despite various conceptual errors. The main value is in the prediction that intense solar Hertzian radiation should be associated with solar prominences and flares. In addition, a most sensible proposal is put forth: the establishment of a network of radio monitoring stations distributed over the Earth so as to distinguish global effects from local interference. The proposed task of these stations is of course remarkably reminiscent of the work actually carried out by Jansky thirty years later. As part of this network, they even suggest that the Eiffel Tower might just make a fine antenna for radio astronomy! It seems certain that if such a network had, in fact, been established, solar bursts would have been eventually recognized.

NOTE

[1] The Sun was not the only possible radio source in Nordmann's mind. He also, at this time, wrote several other papers (summarized in a review (1902b)) in which he attempted to use Hertzian waves to explain such diverse phenomena as the solar corona, comets, bright nebulae, terrestrial magnetic storms, aurorae, and even a recent nova!

ON AN ATTEMPT TO DETECT ELECTRODYNAMIC SOLAR RADIATION AND ON THE CHANGE IN CONTACT RESISTANCE WHEN ILLUMINATING TWO CONDUCTORS BY ELECTRIC RADIATION*

(Ueber einen Versuch, eine electrodynamische Sonnenstrahlung nachzuweisen, und über die Aenderung des Uebergangswiderstandes bei Berührung zweier Leiter durch electrische Bestrahlung)

J. WILSING and J. SCHEINER

Proceeding from the theory established by Stoney and von Helmholtz concerning the valence charge of atoms, Herren Ebert [1] and Richarz [2] have interpreted the oscillations of opposite charges as [resulting in] Hertzian waves from "oscillators of molecular dimensions". Herr Richarz finds that the length of the period of the oscillations of hydrogen [molecules] at 0° is $T = 23 \times 10^{-15}$ sec, while the period of the longest red waves is 25×10^{-16} sec. The gas thus emits electric oscillations identical with infrared heat radiation. Herr P. Lebedev [3] succeeded in experimentally detecting with a thermopile the effect of electrodynamic oscillations of 3 mm wavelength. Herr Langley recognized infrared heat radiation in the solar spectrum at 3 μ with a bolometer; substantially longer waves of 5.3 and 5.7 μ, and even longer at 30 μ in terrestrial heat sources, were measured using mineral-salt prisms by Herren Langley and Rubens [4]. All of this radiation is therefore separated by only a small gap from the electrodynamic oscillations. The excitation of light, heat or electrodynamic waves appears to depend only on the nature of the oscillators. If a continuity of the transition from the radiation occurring as light and heat to the electric oscillations were demonstrated, then one would have to admit the possibility of the occurrence of such long-period oscillations in connection with the emission of light rays. It is therefore natural to think that electric rays exist in the complex of rays sent to us by the Sun. One complication which makes the detection of electric solar radiation difficult, however, should be pointed out. It is namely that, since the radiation which is open to observation will be affected by absorption in the atmosphere, the results of any investigation will depend on the transparency of the atmosphere. Herr Langley contends that the spectrum before

* Modified version of the translation from the German by H. Byhre (1974) from the original article published in *Annalen der Physik und Chemie* 59, 782–792 (1896). (Volume 59 refers to the series of this journal under the editorship of Wiedemann ("*Wied. Ann.*"); all such references in this paper are also in this series.)

absorption by the upper levels of the atmosphere extends much further into the infrared than at the surface of the Earth [5]. One could compare this atmospheric screening effect, which would absorb the electric radiation, with that of a continuous, conducting covering which is impenetrable to long-period oscillations. One must, however, also expect [optical] emission from the rarefied gases in the upper atmospheric layers, as the experiments of Herren Ebert and Wiedemann [6] have shown that these gases only screen when the oscillations have caused them to glow. The usual atmospheric absorption, which Herr Langley says decreases with increasing wavelength, would substantially differ from this screening effect, as would the absorption by carbon dioxide [7], which is limited to certain parts of the spectrum. When assuming an electrical screening by the atmosphere, we should not consider it absolute; rather, the waves gradually lose the ability to penetrate the screen with increasing wavelength. The edge of the spectrum thus significantly depends on the sensitivity of the observational method. Therefore, when choosing a method for the detection of electric solar radiation, the highest possible sensitivity was important. We considered as particularly suitable the method which we have given in the *Proceedings* [*of the Prussian Academy of Science*] [8]. This is based on the changes in galvanic resistance, discovered by Herr Lodge [9], which are initiated by electric oscillations incident on two metals loosely in contact. A considerable advantage of this method appears to be that these resistance changes do not directly cause temperature changes, as does the effect of electric oscillations on a bolometer or a thermopile, but rather are an effect peculiar to the longer period oscillations.

In the reference quoted above we reported experiments which show that the changes in galvanic resistance (which were at the same time studied by Herr von Lang [10]) consist exclusively of a change in the contact resistance at the point of contact of the two conductors, and that they are little influenced by the shape or the size of the conductor itself. For these experi-
1 ments we inserted into the circuit of a cell both a multiplier,* whose pair of 6-cm long needles had an oscillation time constant of 10 sec, and a "bridge" sensitive to electric oscillations. The bridge consisted of a steel wire, a few millimeters thick and several centimeters long, which had been loosely laid over two other steel wires of similar dimensions, thus closing the circuit. The needle [of the galvanometer] showed no noticeable deflection for the amount of contact resistance, normally more than 1000 Ohms, at the point where the wires touched. But as soon as a spark crossed between the gaps of a small 10 cm long induction coil, or between the poles of an induction
2 machine,* the resistance at the points of contact of the bridge wires dropped

to a few Ohms. The needle thus received a strong impulse due to the sudden increase in current of about 0.01 amp. The decrease in resistance remained constant so long as the point of contact of the wires was not changed. The surface of the wires had to be carefully cleaned in order to achieve a high sensitivity. A high sensitivity also required that the pressure at the contact points and the intensity of the current had to be adjusted such that the slightest vibration of the bridge would cause the contact resistance to disappear. Furthermore, we proved that the decrease in resistance not only demonstrated the existence of electrodynamic oscillations in the vicinity of the bridge, but also that the amount of decrease was determined by the energy of these oscillations. As we increased the distance between the bridge and the primary conductor, we recorded the following needle deflections.

TABLE 1

Distance (m)	Deflection (deg)
1.75	51
3.5	44
5.8	35
10.8	24
15.8	13
20.8	7
24.8	8

Because of the sensitivity of the bridge, any kind of screening [attempted by us] was masked by reflection of the waves from the walls of the room. We therefore could only achieve a decrease of the energy of the oscillations by enlarging the distance. We achieved a complete isolation against electric waves only when the bridge, the galvanometer, the cell, and the conducting wires were all enclosed in a box covered with tin foil [11].

In order to be able to use this method for measuring purposes, we had to keep the waves away from the contact points of the wires without enclosing the galvanometer and the cell in the box. After many attempts, we achieved this in the following manner. On the upper side of a cube-shaped sheet metal box we cut an opening 100 cm^2 in area through which we could bring the bridge into the box. The opening was then closed again by means of a tight fitting metal lid. From the bridge a conducting wire led to the inside wall of the box at which point on the outside a wire was soldered which led to the cell. The second wire, which connected the bridge with the cell, led from the cell first to the center of a metal plate 25 cm^2 in size and then was insulated as it went to the bridge through a small opening in the side of the box. That side of the box was covered with a thin layer of paper which prevented the passage of current from the above mentioned, tightly fitting metal plate to the box. The oscillations which were received by the lead-in wires and conducted to the outer wall of the box were unable to get through to the inside of the box as they would have had to cross through the space between the two conducting areas separated by the layer of paper [see Figure 5 in Introduction].

We then managed to detect quantitative differences in the effect of the reflected oscillations on the resistance of the bridge. When the lid was removed, a decrease in resistance occurred as soon as the oscillations reached

TABLE 2

w_1 (Ohms)	w_2 (Ohms)	w_3 (Ohms)
First run		
16 700	46	14
17 500	100	22
14 600	43	15
11 500	33	16
9 900	96	26
[mean]	64	19
Second run		
8 100	150	(42)
8 400	129	27
15 600	156	15
13 300	151	18
9 600	133	26
[mean]	144	22
Third run		
11 500	550	22
26 700	525	29
9 000	372	(10)
17 500	443	24
22 800	(1570)	34
[mean]	473	27

the bridge. Since the direct line of sight from the bridge to the spark gap went through the metal side of the box, the size of the decrease depended only on the energy of those oscillations which were reflected from the walls of the room. When the energy of the oscillations in the vicinity of the bridge was increased by holding a metal plate with sides of 16 cm over the opening of the box, we found a further decrease in the resistance.

In the first column of Table 2 we give the initial resistance w_1, in the second the resistance after opening the case, and in the third we list the resistance after reflection of the oscillations from the metal plate.

Between each series of measurements we varied the type of oscillations from the primary conductor through small changes in the size of the spark gap. The influence of these changes was especially evident in the effect exerted by those unavoidable oscillations that reached the bridge after reflection off the walls of the room.

We now give a few measurements, the first of which shows how much the change of resistance depends on the amount of the initial resistance w_1. We used two spheres of 4.0 cm diameter in connection with the small induction coil and a Leyden jar to excite the oscillations in the above mentioned (as well as in all following) experiments. The distance between the spheres and the bridge was about 3 m. The oscillations were reflected upon the encased bridge by means of a 24 cm^2 metal plate.

TABLE 3

w_1 (Ohms)	w_2 (Ohms)	w_1 (Ohms)	w_2 (Ohms)
50 000	11.1	1470	16.3
24 000	10.6	890	56.3
13 290	8.5	786	85.3
11 500	10.9	316	104.1
1 550	18.5	266	96.2

The resistance w_2 stays constant as soon as the initially existing resistance w_1 becomes larger than several thousand Ohms. At smaller resistances (w_1 less than 1000 Ohms), however, a marked increase in the resistance w_2 occurs.

The bridge was set up free from and outside the box in the experiments reported in Table 4. These experiments show the effect of the individual sparks. The first column contains the initial resistance w_1, the second the resistance after releasing the first spark, etc., the last the resistance after the

TABLE 4

w_1 (Ohms)	1 spark (Ohms)	2 sparks (Ohms)	n sparks (Ohms)
2 070	3.8		
6 690	3.1		2.3
8 270	3.6		3.4
15 700	5.1		3.5
16 200	3.9		2.7
17 500			2.6
10 900	3.6	3.2	3.2
10 100	3.3	2.7	2.7
10 600	5.2	4.7	4.0
mean	3.9	3.5	3.0

w_1	spark: 1	2	3	4	5	n
19000	6.0	4.9	4.4	3.9	3.6	3.1

stream of sparks had been crossing over between the two spheres for about 1 sec.

Since the resistance of the section of the circuit which contained the bridge was 0.50 Ohms, [Table 4 indicates that] only 2.5 Ohms of the initial resistance remained at the point of contact of the two wires.

After inserting a metal screen of about 1/10 m^2 area into the path of the rays, we continued the experiments [Table 5].

Inasmuch as the resistance increases by 2.3 Ohms as compared to not using the screen, a definite screening effect is present. The fact that this screening effect is relatively small stems from the reflection of the oscillations from the

TABLE 5

w_1 (Ohms)	1 spark (Ohms)
12 500	7.5
11 500	6.9
11 800	4.7
19 800	5.4
10 400	6.4
25 000	6.1
[mean]	6.2

walls of the room. Moreover, a portion of the oscillations are collected by the unscreened lead wires and conducted to the bridge.

For the following experiments [Table 6] the spheres were taken off, so that the spark jumped directly between the ends of the induction coil, which ended in 0.1 mm thick wires.

TABLE 6

w_1 (Ohms)	w_2 (Ohms)
10 400	16.0
7 300	12.1
9 400	13.1
6 700	14.9

From this sequence we clearly see a decrease in the effect induced by electrodynamic oscillations of lesser energy. The effect of the oscillations was in all cases still so strong, however, that the resistance of the bridge was depressed to almost nil. Differences in the effects of the oscillations, therefore, were only discernible to a small extent. Considerable attenuation, achieved by screening the bridge with the metal box, was necessary to investigate the changes in resistance from very weak electrodynamic oscillations. The opening of the box was covered by two independently rotatable gratings composed of parallel, 2 mm wide tin foil strips separated from each other by 1 cm. Attached above the box was a metal plate which reflected the oscillations toward the grating. By positioning the grating in different orientations after excitation of the oscillations, we found the following resistances w_2:

TABLE 7
Orientation of gratings

Perpendicular		Parallel		Inclined 45° [or 30° (*)]	
w_1	w_2	w_1	w_2	w_1	w_2
9 640Ohms	7 200	(off scale)	9.6	19 000	1240
35 000	35 000	10 000	26.6	23 000	1240
4 900	3 900	8 000	47	3 900	240
3 200	3 200	9 500	65	5 100	265
(off scale)		(off scale)	49	(off scale)*	366*
6 000	$= w_1$	6 000	26	6 000*	453*

The lack of agreement of these measurements for highly attenuated oscillations came mostly from the fact that the closure of the box opening by the grating was not perfect enough to completely prevent the penetration of unwanted oscillations. The energy of the oscillations polarized by the grating was also probably subject to variations. These tests with strongly attenuated oscillations, however, show how very sensitively the resistance reacts to the smallest changes in the state of oscillation. The results of all these experiments show that the resistance decreases, for constant initial resistance, by an amount depending on the energy of the oscillations. Variation in the resistance thus not only shows the existence of electrodynamic oscillations, but also can be used for quantitative measurements.

For the present purpose of investigating the solar radiation, it was sufficient to demonstrate that the closed box totally screened the oscillations. Furthermore we showed that the resistance, after opening the box and suitably reflecting the oscillations, was brought down to nearly the same amount as when the bridge was under the effect of direct radiation while outside.

Now, in exactly the same manner as with the oscillations emitted from the primary conductor, we directed solar rays reflected from the metal mirror of a heliostat towards the box. With the lid of the box removed, the rays then struck the bridge. When the heliostat mirror was covered with black paper, a highly sensitive thermopile at the same location as the bridge exhibited only a small heating effect. This effect could be made entirely imperceptible by inserting a paper screen [between the mirror and the thermopile]. Since the paper is transparent to electrodynamic oscillations, however, we would have to find, [for example] by observing the resistance changes in the bridge and making measurements with the thermopile, different values for the screening effect of the paper on electrodynamic oscillations [as opposed to heat radiation].

First of all, a strong effect, in the sense of a decrease in resistance, was exhibited with the mirror uncovered. Since these changes continued for a long time after the radiation was stopped, we thought the principal cause of this to be elastic deformations which resulted from thermal expansion of the bridge and which increased the pressure at the contact points of the wires. But the bridge showed almost no sensitivity [to thermal effects] if used in the following manner. The [upper] wire, which had previously been loosely lying on both lead-in wires, was firmly attached at one end by means of a thin, 10 cm long wire bent into a curve; the other end of the wire made loose contact, so that a large resistance existed only at that spot. In this case the thermal expansion essentially caused only a bending in the elastic wire without a change in pressure at the contact point.* 3

In order to measure the resistance changes, we used a Wheatstone bridge connected with a Siemens galvanometer whose bell-shaped magnets had been replaced with a lighter system [See Fig. 5 in introduction]. The movement of the reflected image of the scale, 2 m distant from the galvanometer, was monitored with a telescope. After determining the resistance of the bridge, we measured the sensitivity, i.e., the value of the scale unit in Ohms, by inserting a known small resistance into the same section [of the circuit]. The box was then opened and the effect of the radiation on the resistance observed. Finally, we checked the sensitivity of the bridge by excitation of the spark gap.

The second column of Table 8 contains the initial resistance w_1 before the radiation; the third column the value of a scale unit at the zero point; the fourth column, the upper limit, determined by the uncertainty of the readings, to any resistance change resulting from [solar] radiation; and finally, the fifth column, the change in resistance observed when sparks were excited.

These experiments led to no positive results. If we separate the domain of the investigated oscillations from that of the heat radiation by requiring that the oscillations have the ability to an appreciable extent to penetrate a nonconductor, then it has not been possible to measure in these experiments the amount of the energy from any such solar radiation. Due to the possible screening effect of our atmosphere (as emphasized in the introduction), however, this does not mean that we can deduce the absence [of electrodynamic oscillations] in the original complex of rays emitted by the Sun.

The extraordinary sensitivity to heat radiation which the bridge exhibits in certain circumstances appears, as already mentioned, to be essentially of a mechanical nature. Our experiments nevertheless do not rule out the possibility that the heat radiation also exerts a direct effect on the change in resistance.

Potsdam, Royal Observatory August, 1896.

TABLE 8

Date	w_1 (Ohms)	Sensitivity (Ohms)	Radiation (Ohms)	Spark (Ohms)	Remarks
1896 23 June 11^h 0^m	75	0.0083	0.05	50	overcast sky, mirror not covered.
	2000	1.0	5.0	1950	overcast sky.
24 June 11 30	87	0.0012	0.005		overcast sky.
		0.017	0.02		mirror not covered.
	1780	0.40	4	1740	
„ 6 15	56	0.00077	0.02	40	overcast sky, mirror not covered.
25 June 11 45	83	0.0014	0.07	74	hazy sky, needle jittery, mirror covered. When mirror uncovered, a resistance decrease of more than 0.5 Ohm occurred.
	5020	1.4	0.0	5011	hazy sky, mirror covered. When mirror uncovered, a resistance decrease of about 60 Ohms occurred.
„ 1 0	128	0.033	0.1	113	hazy sky, clouds.
„ 5 15	175	0.0033	0.03	150	hazy sky, mirror covered.
27 June 12 0	100	0.00071	0.01		very clear skies between clouds, mirror covered.
„ 6 0	66	0.00050	0.01	45	hazy sky.
	200	0.0025	0.05	175	mirror covered.
8 July 11 30	120	0.0050	0.05		very clear, mirror covered. When mirror uncovered, a resistance decrease of more than 3 Ohms occurred. After covering, the resistance slowly rose again.
9 July 12 0	260	0.17	1.0		thin cloud cover, needle very jittery.
	230	0.01	0.1		mirror covered.
	390	0.042	0.2		
10 July 10 35	6140	500	400	6121	very clear, mirror covered.
	210	0.0083	0.03		When mirror uncovered, a considerable decrease in resistance occurred.
11 July 11 15	320	0.020	0.2		very clear, mirror covered. When mirror uncovered, a considerable decrease in resistance occurred.

REFERENCES

1. Ebert, H., 1893, *Wiedemann's Ann.* **49**, 651.
2. Richarz, F., 1894, *Sitzungsber. Niederrh. Ges. Bonn* **47**; *Wiedemann's Ann.* **52**, 385.
3. Lebedev, P., 1895, *Wiedemann's Ann.* **56**, 11.
4. Rubens, H. 1892, *Wiedemann's Ann.* **45**, 254.
5. Langley, S. P., 1883, *Wiedemann's Ann.* **19**, 400.
6. Ebert, H. and Wiedemann, E., 1893, *Wiedemann's Ann.* **49**, 49.
7. Ångström, K., 1890, *Wiedemann's Ann.* **39**, 279.
8. Wilsing, J. and Scheiner, J., 1895, *Sitzungsber. der K. Preuss. Akad. d. Wiss.*, December.
9. Lodge, O., 1894, *The Work of Hertz and Some of his Successors.*
10. Von Lang, V., 1895, *Sitzungsber. der K. Akad. der Wissensch. in Wien*, July; *Wiedemann's Ann.* **57**, 34 (1896).
11. Aschkinass, 1894, *Sitzung der Physik. Gesellschaft zu Berlin*, 30 May.

EDITOR'S NOTES AND AMENDMENTS

1. It is not certain whether the term *Multiplicator* refers to this entire circuit (*cf. amplifier*) or to only the static voltmeter with its pair of charged 'needles'.
2. *Influenzmaschine* was any of a variety of machines which set up oppositely charged surfaces, usually for the purpose of creating a spark discharge.
3. The arrangement of the curved and straight portions was apparently such that thermal expansion would be largely compensated and thus cause little movement at the contact point.

A SEARCH FOR HERTZIAN WAVES EMANATING FROM THE SUN*

(Recherche des ondes hertziennes émanées du Soleil)

CHARLES NORDMANN
(Introduced by M. Janssen at the meeting of 3 February 1902)

As part of the project at Mount Blanc Observatory, which M. Janssen was willing to assign to me for the month of last September, I have conducted various experiments with the object of determining whether the Sun emits electromagnetic waves. Indeed it seems natural to assume that this star, an intense source of luminous and caloric radiations, should likewise emit electric undulations, these being of an identical nature to light and the caloric radiation.

The choice of a high station for this research was definitely indicated since it eliminated, to the largest extent possible, the absorption by the atmosphere, especially that due to water vapor, of the hypothesized waves which were to be detected.

These experiments were carried out at the Grands-Mulets station at 3100 m altitude. Bad weather prevented me from repeating them in good conditions at the very summit of Mont Blanc[1].

The receiver for the waves consisted of a horizontal antenna 175 m in length which I laid out on the Glacier des Bossons, on insulating wooden posts, in such a manner that at about noon the solar rays fell normally on it. I placed this antenna so that it was sheltered from any disturbances originating in the high-voltage electrical line from Chamonix.

The choice of a glacier as a base for the antenna was of very great importance. As follows in particular from the telegraphic experiments conducted during the past years at Mont Blanc under the direction of M. Janssen, the glacier can in fact be considered as a nearly perfect insulator; as such, it is transparent to Hertzian waves. On the other hand, the thickness of the glacier at the spot where the antenna was situated can be estimated to be at least 25 m, based on the depth of the crevasses which one finds in the vicinity. This, coupled with the fact that at the time of these experiments (the autumnal equinox) the solar rays were very inclined to the vertical,

* *Comptes Rendus Acad. Sci. (Paris)* **134**, 273–275 (1902); translated from the French by W. T. Sullivan, III (1974).

means that one serious cause of possible error, arising from interference between the direct solar waves and those reflected from the conducting ground, was completely eliminated.

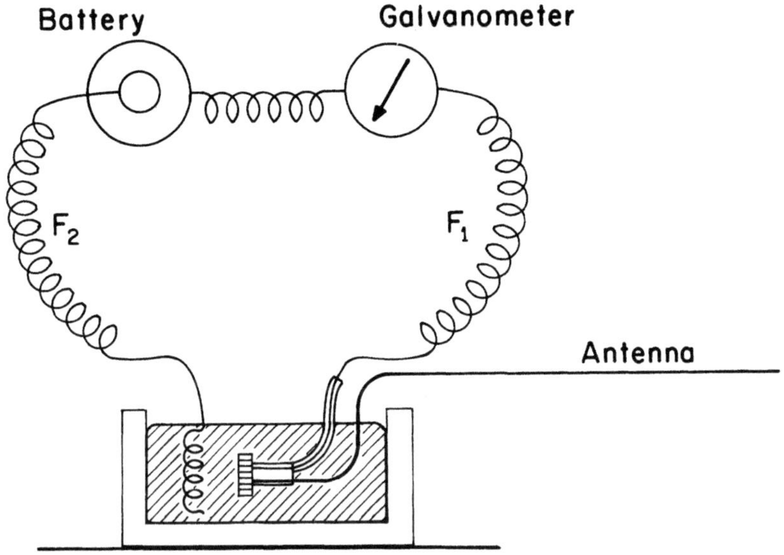

Fig. 1

The wave detector which was used was a radio-conductor immersed and held fixed in the midst of a vessel containing mercury [Figure 1]. This constituted an opaque enclosure to the exterior Hertzian waves. One of the poles of the radio-conductor was in contact with the mercury; the other was insulated and connected to a similarly insulated wire F_1 which passed through the mercury. The circuit was closed outside by a Leclanché battery and a galvanometer which in turn were connected with the mercury in the vat by a non-insulated wire F_2. The wire F_1 was itself enclosed in a conducting sheath, thus creating an electromagnetic screen such that the radio-conductor could be considered completely protected from outside interference. The antenna, not insulated, was connected through the mercury with one of the radioconductor leads; in this state the antenna could not transmit any perturbations to the radio-conductor.

For these experiments I employed two adjustable radio-conductors. I effected the adjustments in the very interior of the enclosure by immersing my fingers in the mercury. One of these radio-conductors was composed of nickel filings. It was of such a sensitivity that simply closing the circuit for an ordinary bell, powered by two Leclanché cells at a distance of more than 4 m, was sufficient to cause the luminous spot of the galvanometer to deviate more than 100 divisions on a scale placed at a distance of 1 m. The other radio-conductor, equally sensitive and adjustable, consisted of 30 small steel balls in mutual contact in a crystal tube of the same diameter as the balls. The galvonometer was of the ordinary Deprez-d'Arsonval type.

Once the apparatus under the mercury was adjusted, the mercury was gently drawn off through a stopcock so as to expose the radio-conductor, on which the antenna could then act.

Experiments with this equipment were repeated throughout the whole day of 19 September, a day of beautiful weather without a cloud in the sky. I alternately used the two radio-conductors and checked their adjustment before and after each experiment. All the experiments led me to the same negative result, i.e., I was not able to record with any certainty the least deviation of the luminous spot on the calibrated scale.

It follows that the sun does not emit electric radiation which can propagate the length of the wires and which is capable of exciting the radioconductors. If the sun does emit such radiation, it is completely absorbed by the solar atmosphere and the upper sections of the terrestrial atmosphere.

This result is consistent with that which one could have predicted: the very rarefied gases indeed do drastically absorb the Hertzian waves. The object of my experiments was to find out if a part of the electric oscillations which emanated from the Sun somehow escapes absorption in the upper rarefied atmospheres of the Sun and Earth[2].

NOTES

[1] Messieurs Wilsing and Scheiner in 1896 established that when solar rays are reflected by a heliostat and caused to fall on a steel wire resting on two other parallel wires, the resistance of the system does not vary. This experiment needs to be repeated using a more sensitive and precise method.

[2] I have been assisted in these experiments by M. F. Haberkorn, electrical engineer.

ON THE SEARCH FOR HERTZIAN RADIATION EMANATING FROM THE SUN*

(Sur la recherche d'un rayonnement hertzien émané du Soleil)

H. DESLANDRES AND L. DÉCOMBE
(Introduced by M. Janssen at the meeting of 3 March 1902)

The remarkable discoveries by Hertz in 1889 concerning the propagation and nature of electromagnetic induction have presented the following question to astronomers: Is the new radiation, which is of the same nature as the caloric and luminous radiations, emitted by the Sun and received by the Earth? The answer to this interesting question has been provided almost directly by research, pursued on many fronts since 1895, on wireless telegraphy with the Hertzian waves over longer and longer ranges of transmission. The receiving instrument, whose sensitivity has been steadily increasing, comprises a Branly radio-conductor and a long wire called an *antenna* (more often than not vertical or inclined 40° to the vertical), which collects the waves. Consider the fact that during the day such an antenna continuously receives solar radiation, and yet the receiving instrument does not exhibit a continuous signal. Therefore, by reason of the large number of observations done under a great variety of conditions, and although the angle between the solar rays and the antenna is both variable and not always the most favorable, one is led to the following conclusion: *the Earth does not continuously receive detectable Hertzian waves of length similar to those used in wireless telegraphy (between 10 m and 1000 m)*[1]. Moreover Messieurs Scheiner and Wilsing have especially investigated the solar Hertzian radiation, at first in 1896 and then in 1899,[*] 1
with a horizontal antenna, and recently M. Nordmann has resumed the same study on the slopes of Mont Blanc under conditions favorable in certain regards.

This negative result is less surprising if one notes that on the Earth incandescent substances, which emit heat and light, do not normally emit Hertzian waves. The Sun nevertheless presents two very distinct sources of radiation, its surface and its atmosphere. The surface, which supplies the greatest part of the solar heat and light, is composed of incandescent particles; it is similar to terrestrial sources and very probably does not emit

* *Comptes Rendus Acad. Sci. (Paris)* **134**, 527–530; translated from the French by W. T. Sullivan, III (1974).

electromagnetic waves under normal conditions. But the probabilities are just the opposite for the solar atmosphere, whose illumination must have an electric origin (at least if one considers its gaseous part, namely the chromosphere, prominences, and gas bound up in the coronal particles). One of us has developed these ideas at great length in a dissertation[2] some time ago.

It is also probable that the upper solar atmosphere emits cathode rays. With great fanfare M. Arrhénius has recently developed this hypothesis to explain all of the details of comets, the aurora borealis, and terrestrial magnetic storms. One of us has already presented the same hypothesis, with all its consequences, in several papers from 1896 to 1898[2].

We can, therefore, conclude that the chromosphere and the prominences emit Hertzian waves similar to the electric discharges of our own atmosphere. But do these waves reach the Earth? The upper layers of low pressure in the solar and terrestrial atmospheres (from 100 mm to 0.01 mm [pressure] on the earth) strongly absorb these special waves, and under normal conditions it is not likely that a measurable amount of this energy reaches the surface of the earth. But is it the same for the large *eruptive* prominences? These occasionally rise to an upper height of one-third of the solar diameter (such as the one which one of us photographed in Paris in 1895[3]) and, for a very short time, exhibit an intense illumination from enormous volumes of material well beyond the boundaries of the chromosphere. The electromagnetic waves emitted in these extraordinary cases must be of a long wavelength, longer than the waves studied on the Earth. On account of their wavelength and intensity, they are able to penetrate the barrier of our atmosphere. The waves then act on the wireless telegraphy receivers, or other suitable receivers, but become blended with the waves ascribed to terrestrial storms.

The distinction between the terrestrial and cosmic waves is assuredly difficult. In order to tell them apart, the most reliable method will be to distribute receivers or automatic wave recorders over several points on the earth, and then to investigate the signals which are simultaneous and thereby attributable to a single source for the entire Earth. It is similar to the situation where one has been able to ascribe certain disturbances of the magnetic needle to the Sun and its spots.

The wave recording instruments will naturally be placed in astrophysical observatories beside apparatus devoted to the study of the Sun and its atmosphere. The observers in that way should be better able to understand the expected relations between the two classes of phenomena[4].

In organizing the recorders, one can draw inspiration from the former studies of Popoff, Boggio Lara, Tommasini, and Fenyi on recording waves emitted in thunderstorms, but our search will favor the reception of very long solar waves. Hence, one can replace the vertical antenna by one parallel to the Earth's axis; this antenna will be equipped with supplementary Slaby wires terminated at their end by a very large capacitance and inductance[5]. A complete solution requires three antennas (each perpendicular to the plane containing the others) capable of receiving waves of any polarization. On the other hand, a workable wave detector must be compatible with an inexpensive recorder, continuous and easy to look after.

In summary, a long series of observations will be necessary to decide in the end if the surface of the Earth receives waves from the Sun similar to the Hertzian waves.

NOTES

[1] One of us has directly measured the length of waves emitted by similar apparatus (*Comptes Rendus* **126**, 518 [1898], paper by M. Décombe).

[2] Deslandres, *Observations de l'éclipse totale de 1893, au Sénégal*; Gauthier-Villars, 1896, pp. 62–70, and *Comptes Rendus* **124**, 687 and 945, and **126**, 1323. Moreover, Zenger and several others have attributed the same phenomena to induction or to Hertzian radiation from the Sun.

[3] *Comptes Rendus* **124**, 171.

[4] It will [also] be necessary, in order to permit a more precise distinction between the diverse causes to which the phenomenon can be attributed, to bring the results of the wave recorders together with the results provided by the recording devices for telluric currents and magnetic variations.

[5] Along this same line, the Eiffel Tower, equipped with a very large inductance in its upper section, would undoubtedly make a good receiver. One could also use the Blondel oscillographs to test for waves of period greater than 1/100 second.

EDITOR'S NOTES AND AMENDMENTS

1. There is no known record of this second experiment by Scheiner and Wilsing.

SECTION H

DETECTION OF THE SUN

James S. Hey holds a prominent place in the history of radio astronomy in that either he or groups which he led during the period 1942–1946 made three fundamental discoveries in widely different areas: (1) the Sun emits radio waves (Paper 16), (2) other non-solar discrete sources of radio emission exist (Paper 23), and (3) radar is a powerful technique to study meteors (Hey and Stewart, 1946). The reader is referred to Hey's own book (1973) for detailed accounts of these achievements as well as other work on mapping the galactic background. The detection of the Sun was not by scientific intent, but happened accidentally when British Army radar sets picked up excessive interference in February, 1942. Fearing a new jamming capability by the Germans, the Army ordered an investigation by Hey, its chief troubleshooter for such matters. He quickly deduced that, in fact, it was the *Sun* which was causing this "jamming", and that the enhanced radio emision was probably associated with an exceptionally active sunspot group then on the central meridian of the Sun. A full report was of course circulated, but these exciting results necessarily remained unknown to most of the scientific community at large for almost four years. Follow-up work on the solar radiation can be found in a paper by Appleton and Hey (1946b).

George C. Southworth (1890–1972) was also working in a war-time environment, at the Bell Telephone Laboratories, conducting research primarily concerned with waveguide techniques for microwave radar. After constructing a sensitive receiver at 9.4 GHz, he mounted it on a 5 ft paraboloid and, in June, 1942, pointed the antenna at the Sun and immediately detected an excess in the total noise output. He thus became the first person to detect "normal", i.e., thermal radiation from the quiescent Sun. Further experiments over the next year are described in his paper. Despite the fact that some of his data and interpretations seem incorrect today, his measured fluxes at two of the three wavelengths turn out to be reasonably accurate when compared with modern values. The paper is also interesting from a philosophical point of view because Southworth considers himself to be working in the far-far-infrared region more than at radio wavelengths. Rather than a qualitatively new result, he sees his work as a continuation of the 50 years of solar infrared work which had preceded him.

Southworth was not as patient as Hey and, for more than a year, made vigorous efforts to have his results released by the military for general publication. The results finally appeared in April, 1945, but only under the stipulation that no actual frequencies could be mentioned or details of the receiving system given. Southworth was determined, however, to enhance the scientific value of his article by cleverly including much of this information while still satisfying the censor. Thus, the operating frequencies can be directly measured to about 10% accuracy from Figure 2, the dish diameter is easily deduced from Figure 4, and Figures 3 and 4 give the reader a good idea of the sensitivity of all three receivers!

Solar Radiations in the 4–6 Metre Radio Wave-Length Band

THE solar radiation spectrum does not normally extend into the 5-metre wave-length region with sufficient intensity to be detectable on radio receiving equipments in commercial or Service use. It is now possible to disclose that, on one occasion during the War, Army equipments observed solar radiations of the order of 10^5 times the power expected from the sun, assuming that the sun behaves as a perfect black-body radiator at a temperature of 6,000° K.

This abnormally high intensity of solar radiation occurred on February 27 and 28, 1942, when Army radar receiving equipments, working at various wave-lengths in the 4–6 metre band, noticed strong directional radiations similar in character to the random fluctuations of internal receiver noise (thermal and valve noise). The radiation was first detected in the afternoon on February 26, 1942, and was almost continuous, with some variations of intensity, between dawn and sunset on February 27 and 28, 1942. It extended over the whole receiver tuning range of about 4–6 metres. It was not observed by any site at night, and there has been no recurrence since February 28, 1942.

The main evidence that the disturbance was caused by electromagnetic radiations of solar origin was obtained by the bearings and elevations measured independently by the receiving sets, sited in widely separated parts of Great Britain (for example, Hull, Bristol, Southampton, Yarmouth). The operators determined the bearings according to the normal practice for finding the direction of a source of interference. It was found that the bearings moved throughout the day and were always within a few degrees of that of the sun. The most striking results came from two sites, about 150 miles apart, where the elevation was also measured. These sites were able to follow the source continuously in bearing and elevation, and observation through the equipment

Reprinted, with permission, from 'Nature' 157, 47–48 (1946).

telescope revealed that they were looking directly at the sun.

Precise measurements of the intensity of the radiation were not made, but all reports indicate that its magnitude on the display cathode ray tubes was several times normal noise-level. From the known receiver noise and aerial characteristics, and by making an allowance for cosmic noise, it can therefore be shown that the noise-power received from the sun on this occasion was of the order of 10^{-13} watts per square metre per megacycle band-width. This unusual intensity, of the order of 10^5 times that corresponding to the calculated black-body radiation, appears to have been associated with the occurrence of a big solar flare reported to be in a central position on February 28, 1942.

J. S. HEY.

Ministry of Supply,
London. Oct. 17.

MICROWAVE RADIATION FROM THE SUN.

BY

G. C. SOUTHWORTH,

Research Engineer, Bell Telephone Laboratories, New York.

INTRODUCTION.

During the summer months of both 1942 and 1943, the writer and certain of his associates observed a small but measurable amount of microwave radiation coming from the sun. This appeared as random noise in the output of a conventional double-detection radio receiver designed to work at centimeter wave lengths. Other observations made at the time showed that the noise output of such a receiver was sensibly less when it was pointed at open sky than when pointed at nearby objects. A high noise level also prevailed when the receiver was located in a closed room. All of these results are in general accord with ordinary black-body radiation theory. In this connection, it is convenient to regard the room as a black-body enclosure in which the receiver is in thermal equilibrium. When the receiver is pointed at the sun, equilibrium conditions are such that more energy is absorbed than is radiated and when it is pointed at interstellar space, conditions are reversed and more energy is radiated than is absorbed.* In the latter case, the result is almost as though a sensitive pyrometer calibrated to work at high temperatures had been turned suddenly on a block of ice.

NATURE OF THERMAL RADIATION.

Before considering further the experiments and their results it may be well to review briefly certain well-known facts about thermal radiation and follow with an account of some of the more important steps that have led to our present knowledge of the sun's spectrum.

According to the accepted views of radiation, most solids and liquids and also a few of the gases radiate energy when heated. In order to distinguish this form of radiation from that associated with fluorescence and light from ionized gases it is generally known as temperature radiation. The relationship between the energy radiated and temperature is generally rather complicated except for a certain class of surfaces, known as black bodies. The latter are not characterized so much by their apparent color as by their ability to absorb completely a wide range of wave lengths.

* While these simple statements serve a useful purpose as regards orientation of ideas, it is quite possible that they do not apply strictly to the apparatus actually used. We can be certain only that as the apparatus was pointed alternately at the sun and at interstellar space, there was a net differential in favor of energy coming from the sun.

Reprinted, with permission, from the 'Journal of the Franklin Institute' **239**, 285–297 (1945).

It is almost an axiom that good radiators are also good absorbers. Two similar bodies at the same temperature placed in close proximity receive from each other as much energy as each respectively radiates to the other and the two are said to be in temperature equilibrium. If they are black bodies and are not in temperature equilibrium, the transfer takes place at a rate proportional to the difference of the fourth power of their respective absolute temperatures. The latter relations hold for the *total* energy being emitted on *all* wave lengths. If we are interested in some small part of the total spectrum then the rate of emission is given by the Planck or Rayleigh-Jeans formulas to be discussed later.

Measurements made on hot bodies, as for example the sun, show that the energy is small for the shorter wave lengths, rising to a maximum at some intermediate wave length and diminishing again for long waves more or less as shown in Figs. 1 and 2. As the temperature of a

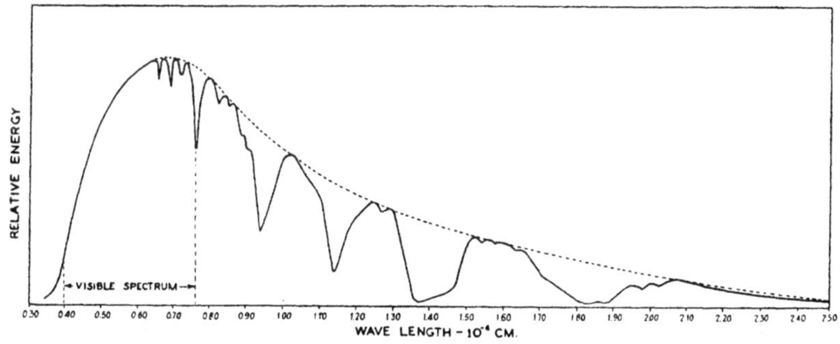

FIG. 1. Tracing of Langley's original graph showing distribution of energy in the solar spectrum.

radiating surface is increased, all of the ordinates of the characteristic increase, and the maximum shifts toward the shorter wave lengths. This shift of the optimum accounts, in part, for the fact that as the temperature of a metallic object is raised, it first appears red, then yellowish, then white and finally blue-white. This shift was first expressed quantitatively by Wien in the formula that follows:

$$\lambda_m T = 0.294 \text{ cm. degree,} \qquad (1)$$

where λ_m is the wave length in centimeters at which the energy E_m is at a maximum and T is the absolute temperature.

Wien also devised a formula for the distribution of energy in a black-body spectrum but it seemed to work well only at the shorter wave lengths. Lord Rayleigh proceeding along somewhat different lines also arrived at a formula but his formula fitted only at high temperatures and relatively long waves. After some modification by

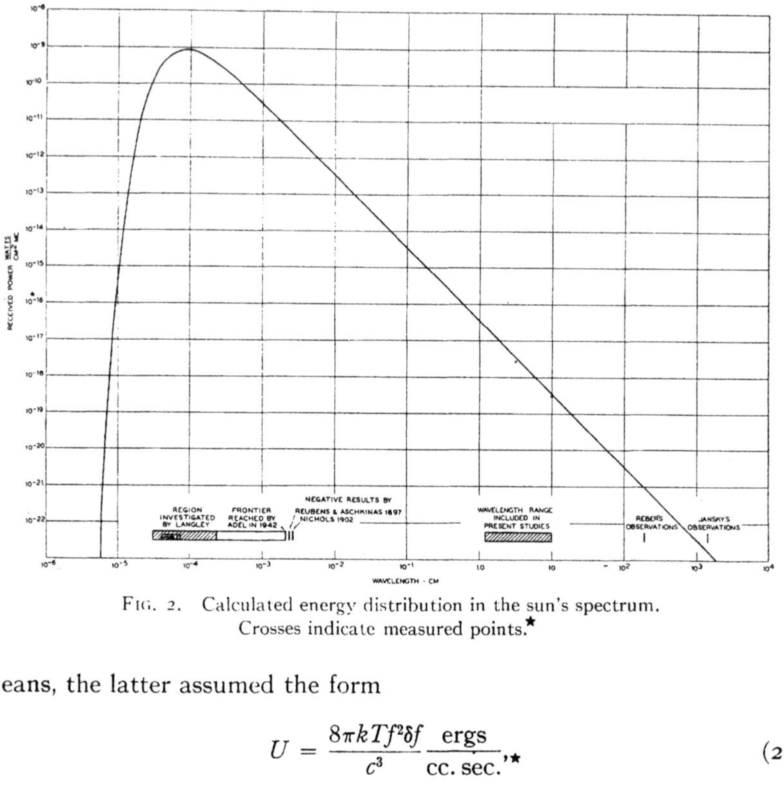

FIG. 2. Calculated energy distribution in the sun's spectrum.
1 Crosses indicate measured points.★

Jeans, the latter assumed the form

1 $$U = \frac{8\pi kTf^2\delta f}{c^3} \frac{\text{ergs}}{\text{cc. sec.}},\star \qquad (2)$$

where f is frequency, c is the velocity of light in free space $\left(\text{i.e., } 3 \times 10^{10} \frac{\text{cm.}}{\text{sec.}}\right)$, T is the absolute temperature and k is the Boltzmann constant = 1.37×10^{-16} ergs per degree. δf is the frequency band under consideration. A more complete formula, derived by Planck from quantum considerations, fits well throughout a large part of the spectrum. It may be put in the form

1 $$U = \frac{8\pi hf^3\delta f}{c^3\left(\epsilon^{\frac{hf}{kT}} - 1\right)} \frac{\text{ergs}}{\text{cc. sec.}},\star \qquad (3)$$

1 where h is the Planck constant = 6.5×10^{-27} ergs per second.★

At high temperatures and long wave lengths, a condition in which we are particularly interested in the case of radio waves from the sun, the Planck relation reduces to the Rayleigh-Jeans' formula. At short waves it reduces to the Wien formula.

Assuming that the sun's temperature is 6,000° K. and its volume is 1.41×10^{33} cc.,* we may calculate the power radiated per second over 2
an element of band width δf. From this we may compute the radiation per square centimeter falling on the surface of the earth one hundred fifty million kilometers away. The result is in general accord with experiment throughout a considerable part of the visible and infra-red portions of the sun's spectrum as already mentioned. The experiments to be reported below show that it also fits well in parts of the centimeter-wave region. At first sight the latter may appear rather surprising for we are applying the formula to a range of frequencies far beyond that for which it was originally conceived.

EARLY WORK.

More than fifty years ago Professor S. P. Langley, one-time Secretary of the Smithsonian Institution and pioneer in aviation as well as solar physics, measured the energy in the sun's spectrum throughout the visible region and to a considerable distance either side. After correcting for absorption by the earth's atmosphere his results agreed well with radiation theory assuming the sun to be a black-body having a temperature of 6000° K. and a volume of 1.41×10^{33} cc.* His results 2
provided the basis for a new and much more accurate value for the so-called *solar constant*.

Figure 1 attached was traced from one of Langley's composite curves published some time after the experimental work itself had been completed.[1] This shows the distribution of solar power as observed at sea level and also after being corrected for absorption by the earth's atmosphere. Although the graph stops at a wave length of 0.00025 cm. it is said that Langley's observations were carried to a wave length of 0.00054 cm. Beyond this point the available energy presumably became too small for the methods then available. In the years that followed, slow but consistent progress toward the long-wave part of the spectrum was made. By 1917 Fowle,[2] and his colleagues, Abbot and Aldrich, had extended the solar spectrum to 0.0014 cm. and by 1942 a new frontier of 0.0024 cm. had been reached as was reported by Arthur Adel [3] of the Lowell Observatory.

Of the attempts to carry Langley's work to longer waves, mention should be made of a lecture by Sir Oliver Lodge [4] before the Royal Institution in 1894 in which he expressed a hope "to try for long wave radiation from the sun." The author has found no record that such an experiment was actually carried out. In this case, the wave lengths under consideration were presumably a few centimeters and therefore comparable to those of the experiments reported below.* 3

In 1898, Reubens and Aschkinass [5] reported that they were unable to find measurable amounts of power in the sun's spectrum at wave lengths of about 0.0025 cm. It will be noted that this is near the region later investigated by Adel.[3] The failure was attributed to carbon dioxide and water vapor in the earth's atmosphere. In fact, laboratory experiments were performed that seemed to justify these conclusions. In 1907, E. F. Nichols [6] reported similar measurements made at the Mt. Wilson Observatory, elevation 6,000 ft., where the atmosphere is relatively dry and also where the carbon dioxide content is probably low. In this case the wave length was about 0.005 cm. Again nothing was observed. Finally Reber [7] using radio methods, at a wave length of 187 cm., made similar observations in 1940 and also reported negative results. In a paper published in 1942 Reber characterized his results as inconclusive but still more recently (1944) he has reported the actual observation of solar radiation of approxi-
4 mately the same magnitude as that calculated by black-body theory.*
In passing it may be of interest to note that Jansky [8] in 1931 discovered radiation coming from the vicinity of the Milky Way. This radiation was subsequently known as cosmic noise. His experiments, which were carried out at a wave length of 14.6 meters, were performed in such a way as to detect solar radiation had it been present in any considerable magnitude. He informs the writer that none was found. The relative positions in the spectrum of these various observations are shown in Fig. 2.

A somewhat different approach to the general subject of heat radiation came in 1926 when J. B. Johnson [9] reported a particular kind of noise in the first circuits of high-gain amplifiers. Experiments were performed that enabled him to relate the noise power quantitatively to the resistance and the prevailing temperature as well as to the mean frequency and the width of the band of frequencies under observation. A corresponding formula was established. A short time afterward Nyquist [10] derived the Johnson expression from well-known principles of thermodynamics and statistical mechanics. In the years that followed, observations of the Johnson or resistance noise, as it was sometimes called, were extended from a few hundred cycles to the highest of the radio frequencies where it was found that the formula continued to hold. In the meantime it became increasingly evident that the expression for resistance noise was the one-dimensional equivalent of the Rayleigh-Jeans expression. This equivalence, was pointed
5 out by Burgess [11] in 1941.* It was but a step further to speak of the noise associated with the radiation resistance of an antenna and to look for this particular kind of noise when an antenna was pointed at the sun.

Whether one regards solar noise as a form of resistance noise or black-body radiation may be a matter of personal preference. If so, the writer leans toward the second view. In either case there seems to be a difference, superficially at least, between energy that originates in a local circuit at ordinary temperatures and that from an incandescent mass ninety-three million miles away and conveyed to us solely by electromagnetic waves.* A reader born with an eye for large numbers will probably note that, except possibly for cosmic noise where the length of path is not yet known, this represents a record distance for radio transmission. Presumably, the time of travel from sun to earth is more than eight minutes.

EXPERIMENTAL RESULTS.

The apparatus used in this work consisted of a more-or-less conventional double detection receiver designed to work at wave lengths in the centimeter region. Associated with it was a parabola of fairly high directivity on an improvised altazimuth mounting. Previous calibrations made it possible to measure noise power in terms of second detector plate current in accordance with methods generally used in the past. Observations were carried out at three rather widely spaced wave lengths in the region between one centimeter and ten centimeters. These three frequencies will be referred to below as the high frequency, intermediate frequency, and low frequency, respectively. In each case, the band over which noise was measured was roughly one megacycle.
It is expected that further details of the method will be published later.* 6

Measurements of received power made at the longest of the three wave lengths mentioned above agreed well with that calculated by the Rayleigh-Jeans formula. It happened that out of about 125 separate observations made on four different days in August 1943, the mean was
about 3 per cent. *below* that calculated.* There was no perceptible 7
diurnal variation such as prevails at the shorter wave lengths. At sunrise the received power increased rapidly more or less in proportion to the area of the sun's disc appearing above the horizon until the level prevailing during the day was reached. The variation at sunset was very similar but in reverse order. The maximum deviation of any observation from the mean was about 12 per cent. The mean of all the deviations was about $\pm$ 5 per cent.

Similar measurements of solar radiation were made at the intermediate wave length mentioned above. Of these, perhaps eight hundred were made between June and October 1942, and about a hundred fifty during the summer of 1943. In this case, the mean value was about 12 per cent. *below* that calculated. Again there was no very obvious diurnal variation. The maximum deviation was about 13 per cent. and the mean was about $\pm$ 4 per cent.

* By a somewhat attenuated and possibly specious line of reasoning, it may be said that Johnson's original observations, which were conducted at frequencies as low as 296 cycles, were, in effect, measurements of thermal radiation at wave lengths of the order of a million meters.

Measurements made at the shortest wave length also gave sensibly constant levels during the daylight period, but this time the level was roughly one-ninth of that calculated by the Rayleigh-Jeans formula. Also, it was observed that the transitions incidental to sunrise and sunset extended over a period of roughly an hour instead of two or

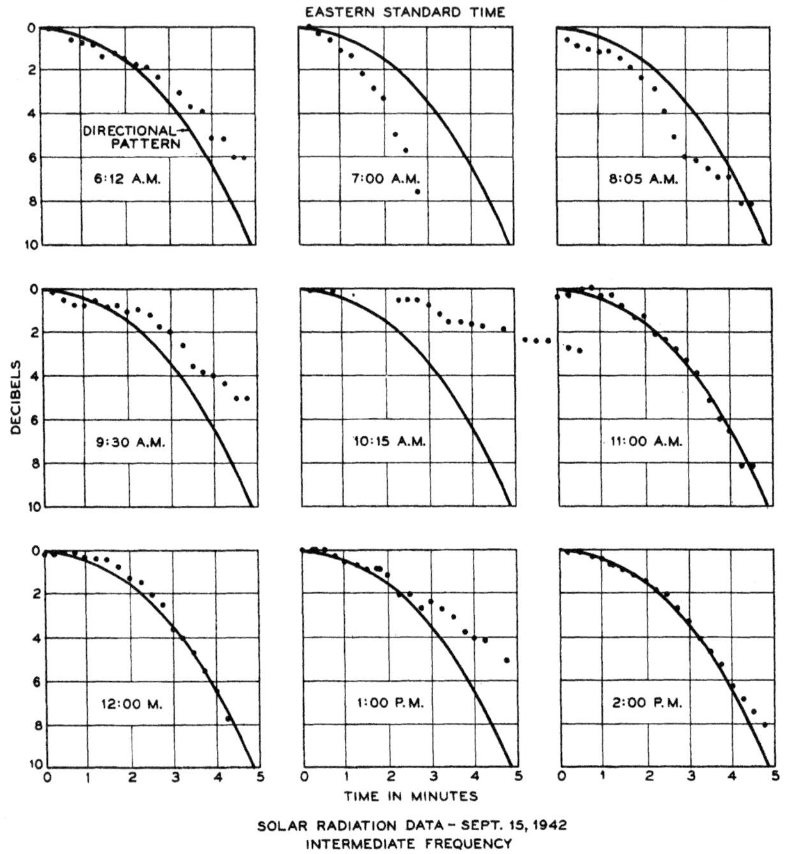

FIG. 3. Representative measurements of the sun's radiation made by centering the receiver on the sun and observing output as the sun passed out of the aperture.

three minutes as in case of the longer waves. The first of these results might, at first sight, suggest attenuation by the earth's atmosphere. However, this is not substantiated by the second result. If the attenuation had been substantial, it might have been expected that, due to increased path lengths through the attenuating medium, the solar signal in the early morning and late afternoon would have been much less accordingly than that in the middle of the day. Because there was no marked diurnal variation such as prevails at the infra-red wave lengths, one is inclined to look elsewhere for the explanation of the discrepancy noted. This matter is being left for further study.* 8

A particularly interesting part of the work consisted of some directional measurements made by centering the receiver on the sun and observing the output as the sun's disc moved out of the aperture of the receiving mirror. Several representative patterns are given in Fig. 3 along with corresponding patterns made with a local source. It will be observed that there is present a random component of considerable magnitude suggesting a variation in the angle of arrival of solar noise.

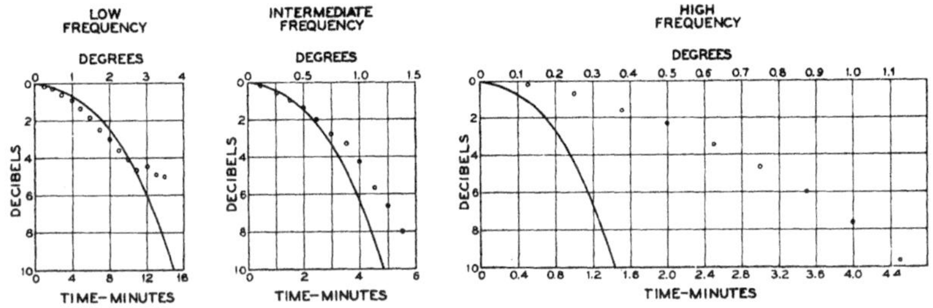

FIG. 4. Measurements like those of Fig. 3 but based on the means of numerous patterns all taken in August 1943 as follows: Low frequency, 16 patterns; Intermediate frequency, 16 patterns; High frequency, 26 patterns.

Attempts were made to obtain a direct measure of the angle of arrival of solar noise hoping that this direction might remain off axis long enough at a time to make such measurements possible. To this end a simple set of sights was arranged on the receiver whereby the angle of arrival of optical waves could be used as a reference. In a few cases, especially during the earlier stages of this work, the direction to the sun as determined optically and electrically seemed to differ by as much as a degree for several minutes at a time. However, this result could not be verified a year later and consequently it was relegated to a position of doubt.

During the period between August 17 and August 28, 1943, numerous directional patterns somewhat more accurate but otherwise similar to those shown in Fig. 3 were taken. For the most part, they showed the same general characteristics observed a year earlier. There seemed to be no very significant difference between patterns taken at various times of the day or on succeeding days so a mean pattern for each wave length was plotted. These are shown in Fig. 4 attached. Each point shown represents the mean of several observations as indicated. Presumably much of the random component has been eliminated by the averaging process, thereby leaving the features that are characteristic of the system as a whole.

Perhaps the most outstanding feature is the general broadening of the solar pattern as compared with the pattern measured locally. This is particularly noticeable at the highest frequency. It appears also in less definite form at the intermediate frequency. The apparent broadening appearing in the low-frequency pattern may be due largely to errors of measurement for in this case the received levels were very small compared with the set noise.

Some of the broadening noted above is no doubt related to the fact that the sun is an apparent disc and not a point source of waves. In

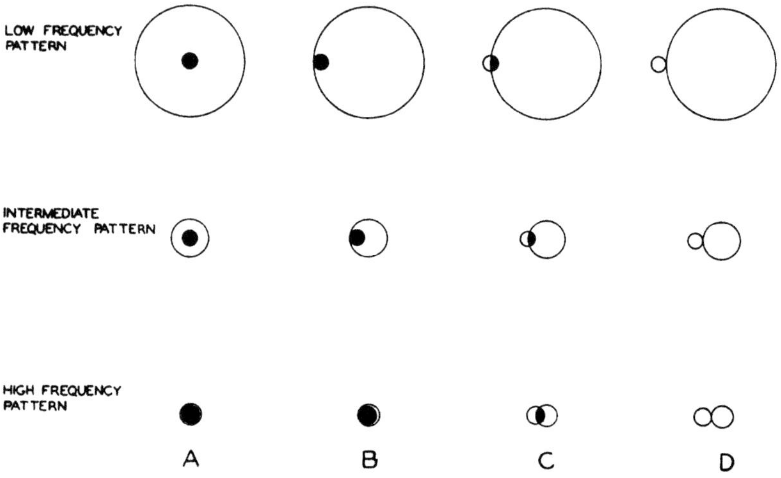

FIG. 5. Relative dimensions of the sun (smaller circle) and the directional pattern of the receiver (larger circle) for each of the three frequencies represented in Fig. 4.

the case of the highest frequency used, the aperture angle of the receiver is comparable with that of the sun. A little reflection will show that as the sun's disc passes the aperture of the receiver, power will continue to be received until the circle corresponding to the aperture angle of the receiver becomes tangent externally to the sun's disc. It may be expected that, to a first degree of approximation, the solar pattern so measured will have a breadth equal to the sum of the aperture angle and the angular diameter of the sun. The relative dimensions involved are made more obvious by Fig. 5 which shows several stages in the transit of the sun across the receiver for each of the three frequencies in question. In each case the larger circle represents the pattern of the receiver while the smaller circle represents the sun. It may be expected that in the transit of the sun, as condition *A* becomes condition *B* the observed change in output will reflect mainly the directional characteristics of the receiver whereas a change from B to D will reflect the effect of a progressively smaller area seen by the receiver. At the lowest frequency the change from A to B occupies a large part of the total transit period while at the highest frequency it is negligible compared with that from B to D. At the intermediate frequency, conditions are, of course, more or less intermediate between those for high and low frequencies.

Reference to Fig. 4 above will show that, at the highest of the three frequencies, the directional pattern may be even wider than that corresponding to the sum of the aperture angle and the angular diameter of the sun.* One reasonable explanation is based on the tentative 9
hypothesis that the angle of arrival of waves from the sun varies rather rapidly over a considerable range. According to this view the sun would appear to the receiver as a shimmering body whose apparent diameter is considerably larger than the true diameter with the result that the pattern is correspondingly broadened. It is quite possible too that this increase in apparent diameter may be due to scattering in the earth's atmosphere. The lack of any very considerable diurnal variation suggests that this scattering may take place in a relatively thin layer.

This increase in apparent diameter may account in part for the discrepancy between measured and calculated energy levels noted above. The comparisons so far made assume that the sun is a point source. This simplification may be warranted in the cases of the two lower frequencies where the directive patterns are only moderately sharp but they may not be warranted in the case of the highest frequency where the angular diameter of the sun is comparable with the angular diameter of the directive pattern. If in addition it turns out that the effective diameter of the sun is perhaps twice that assumed above the error may be very substantial indeed. It is believed that a more accurate calculation should await further measurements of effective diameter.

HORIZON EFFECTS.

While carrying out the above work, it was found that the noise appearing in the output of the receiver was sensibly less when the receiver was pointed at open sky than when pointed at nearby objects or when it was located inside a room. The magnitude of this horizon effect for the longest wave mentioned above was of the order of 10^{-19} watts per sq. cm. per mc. of band width. That for the intermediate frequency was around five times that quantity while that for the highest frequency was barely measurable with the apparatus at hand. More specifically, these figures represent the increase in indicated noise power when the receiver was pointed below the horizon than when pointed above. As might be expected, distant horizons were substantially as definite as those nearby. The quantitative aspects of this problem will be given further attention.*

In the course of the above work, numerous attempts were made to observe cosmic noise coming from the general region of the Milky Way. As already mentioned, noise of this kind had been reported by Jansky who located also its approximate position in space. His observations were made at a frequency of 20.6 mc. Similar noise in the vicinity of 160 mc. was also reported by Reber [7] in 1940. His observations published in 1944 appear to locate the source more definitely in the region of Cygnus and Sagittarius. It was hoped that cosmic noise might also be found in the microwave range and with the highly directive apparatus at hand, its position might be located still more precisely. So far it has not been found.

Although there was no reason to believe, from general radiation theory, that any appreciable power could be picked up from the moon or from any of the planets or from any particular star, a search for such noise was made. Nothing was found coming from any of the more prominent bodies visible at the time.

It may be of interest to note that a measurable amount of centimeter power was found in fluorescent lights. As is well known, the mercury vapor arc which is an important component of the fluorescent light is one of the common sources of wave power in the far infra-red. Similar noise power was also found in mercury vapor arcs of the so-called Lab-Arc variety.

CONCLUSIONS.

The results reported above establish with little doubt the existence, in the sun's spectrum, of components much longer than have been found generally heretofore. Over a considerable portion of the frequency range, the energy appears to be substantially that predicted by black-body radiation theory. A matter of comparable interest is the fact that temperature differences, of a kind, exist between points on a distant landscape and the sky above. The latter effects are readily explicable, qualitatively at least, from the usual equilibrium considerations of black-body theory.

The above experiments also point rather definitely toward an atmosphere surrounding the earth that is fairly transparent to centimeter waves. However, rapid changes in angle of arrival of the solar signal with a consequent shimmering of the sun's image suggest rapid changes in index of refraction. There is also evidence of small amounts of absorption and perhaps also scattering particularly at the shortest of the three wave lengths at which observations were made. Although

* The possibility of circuit noise being radiated into space by an antenna must certainly have been entertained by many radio engineers and physicists in the past. Reber for one pointed out this possibility in one of the articles cited above.

many of the more important results reported above appear firmly established, it must be recognized that certain details must be regarded as tentative and be contingent upon subsequent verification.

Perhaps the most interesting result of this work, from the radio engineer's point of view, is the establishment experimentally of a much closer relationship between black-body radiation and ordinary radio waves. As already mentioned, such a relationship has existed in a rather attenuated way, since the discovery of the so-called Johnson or first-circuit noise, but the recent appearance of such radiation in a source ninety-three million miles away and conveyed solely by radio means would appear to be a much more direct and tangible result. It may also be of interest that no cosmic radiation of the kind observed by Jansky and Reber has been found at these very high frequencies. This tends to support Reber's 1940 view that cosmic radiation varies inversely as the frequency rather than that of Jansky who was inclined to regard cosmic radiation as a form of black-body radiation.

This paper suggests, perhaps more than anything else, a need for further work preferably carried out with more refined apparatus and with better facilities for calibration. The apparatus needed for the next stage should incorporate the best ideas both of the radio engineer and the designer of astronomical equipment with the hope that additional accuracy will bring out features so far overlooked. Such a program is contemplated. Perhaps there lies ahead a new field of astronomical and terrestrial research comparable in interest and scope with that that has prevailed during the last two decades with respect to the ionosphere. The sun being a localized source of radio waves situated entirely outside the earth's atmosphere provides, for the first time, angles of attack on the structure of the earth's atmosphere not previously possible.

REFERENCES.

(1) S. P. Langley, "The solar constant and its related problems," *Astrophys. Jour.*, Vol. 17, pp. 89–99, 1903.

(2) F. E. Fowle, Smithsonian Miscellaneous Collection, *68*, No. 8 (1917).

(3) Arthur Adel, "Extension of the prismatic solar spectrum from 14μ to 24μ through a new atmospheric window in the infrared," *Astrophys. Jour.*, Vol. 96, pp. 239–241, September 1942.

(4) Signalling through space without wires—A lecture by Sir Oliver Lodge before the Royal Institution of Great Britain—June 1, 1894, published as a booklet of the Electrician Series.

(5) H. Reubens and E. Ashkinass, "Infra-red absorption," *Wied. Ann.*, Vol. 64, series 3, pp. 584–610, 1898.

(6) E. F. Nichols, "Absence of very long waves from sun's spectrum," *Astrophys. Jour.*, Vol. 26, p. 46, 1907.

(7) Grote Reber, "Cosmic static," *Proc. IRE*, Vol. 30, pp. 367, February 1940.★ 10
Grote Reber, "Cosmic static," *Proc. IRE*, Vol. 30, pp. 367–378, August 1942.
Grote Reber, "Cosmic static," *Astrophys. Jour.*, Vol. 100, pp. 279–287, November 1944.

(8) K. G. JANSKY, "Directional studies of atmospherics at high frequencies," *Proc. IRE*, Vol. 20, pp. 1920–1932, December 1932.

K. G. JANSKY, "Electrical disturbances of extra terrestrial origin," *Proc. IRE*, Vol. 21, pp. 1387–1938, October 1933.

K. G. JANSKY, "A note on the source of interstellar interference," *Proc. IRE*, Vol. 23, pp. 1158–1163, October 1935.

K. G. JANSKY, "Minimum noise levels obtained on short wave radio receiving stations," *Proc. IRE*, Vol. 25, pp. 1517–1530, December 1937.

(9) J. B. JOHNSON, "Thermal agitation of electricity in conductors," *Physical Review*, Vol. 32, pp. 97–109, July 1928.

(10) H. NYQUIST, "The thermal agitation of electric charge in conductors," *Physical Review*, Vol. 32, pp. 110–113, July 1928.

(11) R. E. BURGESS, "Noise in receiving aerial systems," *Proc. Phys. Soc.*, Vol. 53, pp. 293–304, 1941.

ERRATUM

One of the author's colleagues, Dr. C. H. Townes, has pointed out that the calculated values of the sun's radiation as given in the above-mentioned paper are too high by a factor of about three. Accordingly the graph of Fig. 2 should be lowered by about one-half division. This increases the discrepancy between calculation and experiment for two of the experimental points and reduces it for the third. The newer value being well within the range of experimental errors does not materially alter the outstanding question as to whether or not the radiation noted follows the Rayleigh-Jeans Law. It should be noted too that the units of Equations 2 and 3 should be ergs per cc. and the Planck constant should be written $h = 6.5 \times 10^{27}$ erg sec.

EDITOR'S NOTES AND AMENDMENTS

1. See the Erratum.
2. This total *volume* of the sun is of course not relevant to the problem at hand.
3. The reference here is apparently to an edition of *Signalling Through Space Without Wires* earlier than that in 1900 quoted in the introduction to this section.
4. This statement is incorrect. Reber did not compute from his measurements an equivalent black-body temperature for the solar disk; his values imply, however, a value of 9×10^5 K.
5. Burgess demonstrated the equivalence of the three-dimensional Rayleigh-Jeans radiation law and the Nyquist relation for thermal radiation from a (one-dimensional) resistive element.

6. In a retrospective article published in 1956, Southworth gives some of these details. The three operating frequencies were 3.06 GHz (9.8 cm wavelength), 9.40 GHz (3.2 cm), and 24.0 GHz (1.25 cm). Quoting further: "The antenna system consisted of a 5-foot-diameter paraboloid together with a waveguide pickup that was located at the principal focus. The waveguide led to a box on the side that contained the first detector and beating oscillator. The intermediate-frequency power so obtained (at 60 megacycles per second) was connected to a nearby amplifier and second detector."
7. As noted in the Erratum, the calculated curve of Figure 2 is too high by a factor of ~3 or ~$\frac{1}{2}$ division for blackbody radiation at $T = 6000$ K. The plotted values of intensity at the wavelengths 1.25 cm, 3.2 cm, and 9.8 cm correspond to brightness temperatures (for a disk equal to the size of the optical sun) of 2900 K/η_B (where η_B is the main beam efficiency), 15,000 K/η_B and 16,000 K/η_B, respectively.
8. These effects at 24 GHz were probably due to atmospheric absorption.
9. The evidence in Figure 4 for any significant extra broadening is not very strong.
10. Should be "*Proc. I.R.E.*, Vol. 28, pp. 68–70, February 1940".

SECTION I

FIRST SOLAR INTERFEROMETRY

The value of the next two selections lies not only in their contributions to solar astronomy, but even more in their pioneering techniques of observation. Martin Ryle and Derek Vonberg were the first in radio astronomy to employ an antenna configuration analogous to Michelson's optical interferometer; they soon demonstrated how source sizes may be estimated by measuring the fringe amplitudes associated with various spacings of the receiving elements. In the tradition of the Cavendish Laboratory, they hit upon this scheme as a means to reach the desired scientific objectives in the least expensive and most elegant manner possible. In addition, they developed a unique method for making reliable measurements of signal noise powers as small as 1% of the total receiver noise power. The method, described in detail in a 1948 article by the same authors, involves switching against a reference noise source, the latter being continually adjustable in such a manner that the receiver is always balanced. In practice, this was achieved by varying (and measuring) the anode current in a temperature-limited diode; over a wide range of conditions, the output noise power from such a device is directly proportional to this current. For conditions typical of those at meter wavelengths, where signal noise power is large and intrinsically variable over a wide range, this system is less sensitive to gain fluctuations than the conventional Dicke technique of switching against a fixed load (Paper 11).

Concurrently at the Radiophysics Laboratory in Sydney, Australia, a group led by Joseph L. Pawsey (1908–1962) was attacking the same problems in a different manner. With their wartime experience in "radio direction finding" for locating aircraft, it was only natural that they should apply precisely the same techniques to scientific ends. The largest convenient baselines were secured by using antennas already situated for defense on the edges of seaside cliffs. This arrangement is exactly analogous to what is known in optics as Lloyd's mirror. Appendix 2 of Paper 19 outlines in some detail the major complicating factors which needed to be considered: Earth curvature, heights of tides, refraction, scattering by ocean waves, and receiver bandwidth. For instance, the large difference between radio and optical refraction at the horizon meant that the first fringes were in fact detected several minutes before the (optical) Sun actually rose (see their Figure 5).

Eventually, the more conventional Michelson interferometer, aided by the technique of phase switching (Paper 12) and its greater flexibility in changing baselines and positions of observation on the sky, became the standard in radio astronomy. The sea-cliff interferometer nevertheless was an extremely clever means of gaining high resolution with only one small antenna, and it played an important role for several more years (see Papers 24 and 26). It should also be noted that Paper 19 contains the first explicit statement of the important principle that measurements of a sufficient number of fringe amplitudes and phases at different baselines provide all Fourier components of the sky brightness distribution within the primary beam pattern. It was not until four years later, however, that such a (one-dimensional) synthesis was first carried out, on the Sun, by Stanier (1950).

In Paper 18, Ryle and Vonberg establish that the observed solar bursts originate from a region of size $\lesssim 10$ arcmin in extent. They also measure substantial circular polarization in this enhanced radiation; credit for this discovery, however, must be shared with concurrent independent observations by Appleton and Hey (1946a) and Martyn (1946a).

The importance of the Australian study in Paper 19 is in convincingly showing that not only are the emission regions small, but they also are coincident in position with sunspot groups. Moreover, daily intensity of radio emission correlates well over a long period with total sunspot area.

Three final points about Paper 19 should be made. Firstly, these novice astronomers were somewhat lucky in being able to observe the great sunspot group of February, 1946, one whose main sunspot is amongst the largest ever photographed. Secondly, we can recognize the "slowly varying background" (not to be confused with the so-called "slowly varying component") which was observed on 7 February 1946 as that which we now call Type IV radiation. Thirdly, co-author Ruby Payne-Scott carries the distinction of being the first woman radio astronomer.

Solar Radiation on 175 Mc./s.

Appleton[1] and Hey[2] have directed attention to the fact that radio-frequency energy, with some of the characteristics of random 'noise', is emitted with greatly increased intensity from the sun under the conditions of violent disturbance associated with a large sunspot. These observations were confined mainly to the region of frequencies near 60 Mc./s.

Pawsey, Payne-Scott and McCready[3], who have made observations on 200 Mc./s., suggested that radiation of this type is also observable under less disturbed conditions.

In order to investigate other aspects of this phenomenon, we have constructed a device which automatically records and measures the 'noise' received on 175 Mc./s., and which has a sensitivity such that a power of 3×10^{-15} watts (approximately 1 per cent of the receiver noise power) can be detected. This sensitivity corresponds to a thermal energy temperature of 30° K., and it has been possible to record the 'noise' received from the galaxy on a small broadside aerial consisting of eight half-wave dipoles.

For the purpose of investigating solar radiation under conditions of low solar activity, it is necessary to discriminate against the background of galactic radiation. While this could be achieved by building an aerial to give a sufficiently narrow beam, a very large structure would be required, and observation would be restricted to a short time every day unless arrangements were made for moving the polar diagram of the aerial. An alternative method was therefore used, analogous to Michelson's method for determining stellar diameters. Two aerial systems were used with a horizontal separation of several wave-lengths, and their combined output was fed to the receiving equipment. Such an arrangement produces a polar diagram of the form shown in Fig. 1 where the angle between zeros is governed by the spacing of the two aerials and the envelope is determined by the polar diagram of each individual aerial system. If the angle between minima is sufficiently large compared with the solar angular diameter, then, as the aerial polar diagram is swept past the sun by the earth's rotation, any radiation from the sun should be recorded as an oscillatory trace.

Fig. 2 shows a typical record obtained with an aerial separation of 10 λ, and with only slight solar activity (July 17). The oscillatory contribution due to radiation from the sun can be seen superimposed on the slowly varying background of the galactic radiation. Records of this type enable an estimate to be made of the level of solar radiation even when it is only about one quarter the galactic contribution, and at the present time we have found that the sun is usually sufficiently disturbed to give such records. The power is indicated on the diagram in terms of an 'equivalent aerial temperature', and is the power which has to be fed to an aerial in a black-body enclosure of this temperature, to maintain equilibrium. The temperature of a distant source whose radiation obeys a black-body distribution may be estimated from the observed equivalent aerial temperature by correcting for the ratio of solid angles of source and aerial polar diagram.

Reprinted, with permission, from 'Nature' 158, 339–340 (1946).

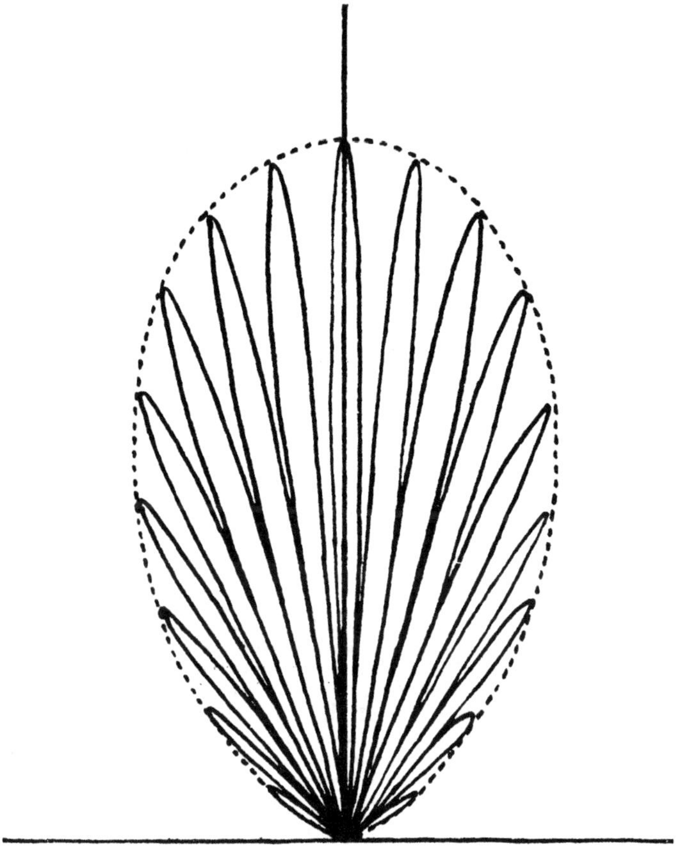

Fig. 1. POLAR DIAGRAM OF TWO 8-ELEMENT AERIAL SYSTEMS WITH SEPARATION OF 10 λ

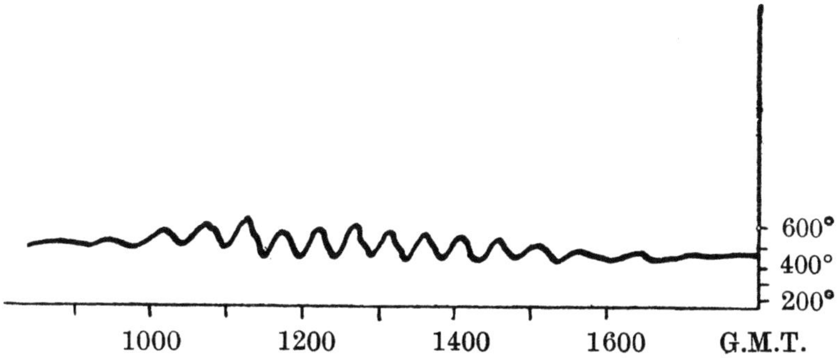

Fig. 2. RECORD OBTAINED WITH 10 λ SEPARATION (JULY 17, 1946)

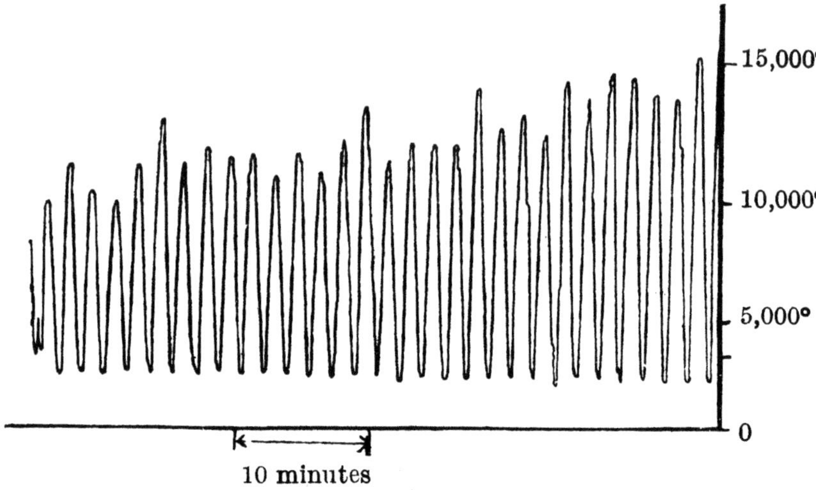

Fig. 3. RECORD OBTAINED WITH 140 λ SEPARATION (JULY 26, 1946)

During the appearance of a large sunspot between July 20 and August 1, the solar radiation was much increased, and the opportunity was taken to use the apparatus to determine the angular diameter of the source, by observing the ratio of maximum to minimum intensity as the polar diagram of the two aerials with a separation of many wave-lengths was swept past the sun. This experiment was carried out with a series of different aerial spacings, the final value being 140 λ, and a sample of the records obtained with this spacing is shown in Fig. 3. The maximum/minimum ratio obtained under these conditions corresponds to a source diameter of 10 minutes of arc. Any inequalities in the two aerial systems would result in an over-estimate of diameter, and this is therefore a maximum value.

Since the value obtained does not greatly exceed the diameter of the visual spot, it is reasonable to relate the source of this radiation with the visual spot itself, or a region closely associated with it.

During the afternoon of July 25 the observed intensity attained a value which would correspond, in the case of black-body radiation from a source of this diameter, to a temperature greater than $2 \cdot 10^{9}$ ° K.

Since the existence of such temperatures in a region from which radiation of this wave-length would escape seems improbable, we considered that the radiation was non-thermal in origin, and the possibility of ordered electron motion was therefore investigated by an examination of the polarization of the radiation. This was carried out by arranging the two aerial systems of the 'Michelson' device to be polarized in planes at right angles to each other. If the radiation were emitted by a completely random 'thermal' source, the two perpendicularly polarized components would not be phase-coherent and no interference effects would be observed. The existence of interference effects would show the presence of phase coherence, and hence prove that the radiation was not of 'thermal' origin. Further, by noting the direction of the sun relative to the aerial systems when an interference maximum was produced, it would be possible to differentiate between plane and right- and left-handed circular polarization.

Using such a system it was found that during periods of intense radiation the polarization was, within the accuracy of measurement, completely circular. (Inequalities in the aerial system limit the accuracy, but at least 90 per cent of the incident energy was circularly polarized.)

Measurements taken over the period July 27–August 3 showed the polarization to be anti-clockwise, viewed along the positive direction of propagation (left-handed). Between August 3 and August 7 the degree of polarization diminished, being virtually completely random on August 7. On August 8, 40 per cent polarization was observed again, but with right-handed polarity—the result, presumably, of increased activity in a subsidiary sunspot.

Any theory of the emission of circularly polarized radiation from sunspots must presumably be given in terms of the magnetic field known to be present in those spots. In considering the mechanism of such a process account must be taken of the magnetic field and electron density not only in the region appropriate to the observed frequency, but also in the overlying layers, where selective absorption of the radiation will occur, in a manner similar to the 'gyro-magnetic' phenomena familiar in the terrestrial ionosphere.

It will be necessary to collect more experimental data before possible theories can profitably be considered in detail.

M. Ryle
D. D. Vonberg

Cavendish Laboratory,
Cambridge.
Aug. 22.

[1] Appleton, *Nature*, **156**, 534 (1945).
[2] Hey, *Nature*, **157**, 47 (1946).
[3] Pawsey, Payne-Scott and McCready, *Nature*, **157**, 158 (1946).

Solar radiation at radio frequencies and its relation to sunspots

By L. L. McCready, J. L. Pawsey and Ruby Payne-Scott
Division of Radiophysics, Council for Scientific and Industrial Research, Commonwealth of Australia

(*Communicated by Sir David Rivett, F.R.S.—Received* 22 *July* 1946)

Experimental studies of solar radiation on a frequency of 200 Mcyc./sec. are described. This radiation has characteristics similar to those of thermal radiation but is always hundreds of times greater than the thermal radiation anticipated from the photosphere and sometimes greater by a factor of 10^4.

The day-to-day intensity variations over a period of 6 months confirm a correlation with sunspots. The received intensity of radiation is subject to rapid fluctuations; sudden increases, or 'bursts', of duration from a fraction of a second to a minute are characteristic. These rapid fluctuations are similar at widely-spaced receiving points, and it is concluded that most of them are extraterrestrial, and presumably solar, in origin.

Directional observations, based on the interference phenomenon as the sun rises over the sea, indicate that the radiation originates not uniformly over the sun's disk but in restricted areas in the immediate vicinity of a sunspot group.

Values of received intensity are at times too great to be accounted for in terms of thermal radiation, so that another mechanism producing radiation must exist. Radiation from gross electrical discharges is suggested.

Introduction

The discovery of radio-frequency radiation, with the characteristics of fluctuation noise, arriving at the earth from the direction of the Milky Way, was announced by Jansky (1933). This discovery is potentially of fundamental importance to astrophysics, since it provides a source of information concerning extraterrestrial phenomena other than that obtained through the use of light. Up to the present, however, the interpretation of such observations has contributed little to astrophysics, and it appears that more complete observational data are necessary.

Jansky's original work on cosmic noise was confirmed and extended by himself and others to cover the frequency range 15 to 160 Mcyc./sec., but, at first, no measurable radiation was observed from the sun. It was therefore suggested that the radiation originates not in the stars but in collision processes in the residual ionized matter in interstellar space.

The development of microwave radar suggested the possibility of detecting at these wave-lengths the black-body radiation from the sun to be expected on Planck's law, assuming the optical temperature of 6000° K. The intensity of this radiation per unit frequency range is proportional to the square of the frequency at radio frequencies. It is too small to be detected in the ordinary short-wave region but should be detectable at centimetre wave-lengths. In 1942 Southworth detected centimetre radiation from the sun (Southworth 1945) and showed that it was of the order to be expected from the Planck formula.

Reprinted, with permission, from the 'Proceedings of the Royal Society' **A190**, 357–375 (1947).

In the period between Southworth's discovery and its delayed publication, Reber (1944) observed solar radiation on 160 Mcyc./sec. Reber did not attempt to explain this radiation, but, had it been of thermal origin, it may be deduced from data given in his paper that a temperature about a hundred times the optical value would be required.

Solar radiation was also observed in 1942 by English radar stations in the frequency range 40 to 200 Mcyc./sec. (Hey 1946). The effect continued for several days and was noted to coincide with the presence of the exceptionally large sunspot of February of that year. Early in 1945 certain New Zealand Air Force radar stations working on 200 Mcyc./sec. observed similar radiation and showed that it varied from day to day (unpublished). Radiation on still longer wave-lengths was reported by Appleton (1945), who described reports on radio noise received in the short-wave communication band at times of marked solar activity during the sunspot maximum about 1936. The observed intensities were greatly in excess of 6000° K black-body radiation, and he concluded that the noise originated in active areas on the sun.

Observations on 200 Mcyc./sec. similar to those of the New Zealand stations were begun by us towards the end of 1945, and the initial results showed that the radiation was associated with sunspots and that it was subject to surprisingly rapid fluctuations (Pawsey, Payne-Scott & McCready 1946). The present paper describes an extension of this work. It includes more detailed observations, almost exclusively on 200 Mcyc./sec., of the variations in intensity of solar radiation, and the use of direction-finding methods to locate its place of origin on the sun.

Observations

Method of observation

In making measurements, a directional aerial is pointed in turn at a 'cold' part of the sky (i.e. a part relatively free from stars) and at the sun. The output of a receiver connected to the aerial is partly due to the noise generated in the receiver itself and partly to radiation received from outside. At the wave-lengths used the external noise is effectively constant over the 'cold' sky but increases appreciably when the aerial is pointed at the sun. The proportional increase in output is taken as a measure of the intensity of the solar radiation.

The intensity of the radiation is specified as the power falling on the earth in watts per square metre normal to the rays per megacycle band width. It is derived from the instrumental constants as in appendix 1. The polarization is assumed random, so that the values quoted are twice those measured on plane-polarized aerials. It is assumed that the intensity of the radiation is constant over the pass band of the receiver (a few megacycles).

The sensitive radar receivers used have output meters measuring second-detector current.* Initially ordinary meters were used, but were later replaced by recording 1
milliammeters. The records obtained from these meters read from right to left and have been reproduced without transposition. The time co-ordinates are curved as

indicated in the figures. Two types of aerial have been used; large horizontally polarized radar aerials capable of rotation only about a vertical axis, and so able to 'see' the sun only near dawn or sunset, and smaller aerials capable of being tilted in any direction. The first type were situated some hundreds of feet above sea-level and had a clear view over the Pacific Ocean towards the rising sun at dawn but an outlook over irregular land to the west.

Table 1 gives the nominal characteristics of the equipments used on 200 Mcyc./sec., the frequency at which most observations were made. Measured values of intensity may have errors up to 40 %.

TABLE 1. LOCATION AND CHARACTERISTICS OF 200 MCYC./SEC. EQUIPMENT

station	location	aerial: effective area (A) (sq.m.)	aerial: tiltable	receiver noise factor (N)
Collaroy	lat. 33° 43′ 45″, long. 10 h. 5 m. 10 s.; hill-top ½ mile inland 400 ft. above mean sea-level	18	no	7
Dover Heights: (1)	lat. 33° 52′ 26″, long. 10 h. 5 m. 8 s.; edge of cliff on coast 278 ft. above mean sea-level	9·5	no	11
(2)	do.	6·5	yes	11
Stromlo	lat. 35° 19′ 30″, long. 9 h. 56 m. 0 s.; inland	9	yes	13
North Head	Sydney	7·5	yes	11

Note. Collaroy is approximately 10 miles north of Dover Heights, and Stromlo approximately 160 miles south-west of Dover Heights.

Characteristics of solar radiation

During the 6 months since observations began solar radiation has been observed whenever an observation has been made. The radiation appears to have the characteristics of receiver fluctuation noise when observed on headphones or on a cathode-ray oscillograph except that the mean level fluctuates continually. Figures 1 a, b, c and d indicate the kinds of variation believed to be typical. Figure 1 a is a tracing of a record obtained when the intensity was very high and large rapid fluctuations were absent. Figure 1 b_1 shows a similar slowly varying signal of much lower amplitude interrupted just before 1614 hr. by a huge burst of radiation lasting about 10 sec. This was the only burst which occurred in several hours of continuous recording if modest ones like that at 1613 hr. are excluded. Figure 1 b_2, obtained simultaneously from a different site, is discussed below. Figure 1 c, an earlier part of the same record, shows a series of intense bursts which are so frequent as to be almost continuous. This condition lasted from the beginning of observations at 1100 to 1400 hr. and then suddenly ceased, the only subsequent large burst on that day being the one shown in figure 1*b*. Figure 1*d* shows a graph of average levels over 5 hr. on a day when the variation of intensity was similar to figure 1*a*, and illustrates a slower type of variation.

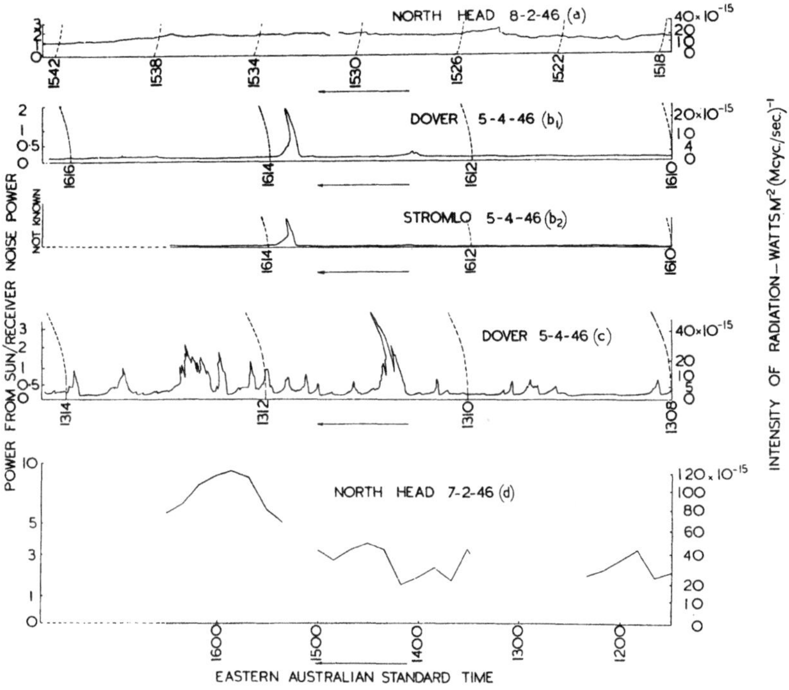

FIGURE 1. Typical records of 200 Mcyc./sec. solar radiation. The upper records show variations over a few minutes, and the lower one over several hours.

Summarizing, there appear to be two distinct types of variation, a relatively slowly varying type with intensity ranging from about $0{\cdot}5 \times 10^{-15}$ to 100×10^{-15} W m.$^{-2}$ (Mcyc./sec.)$^{-1}$, and a type consisting of intense bursts of duration between a fraction of a second and a minute, and of widely varying frequency of occurrence. Their intensity may be tens of times greater than the general level on comparatively quiet days. On days of high general level of intensity, bursts gave off-scale deflexions several times greater than the mean level.

These rapid variations are a very striking feature for a solar phenomenon, and it is desirable to check that they actually originate in the sun and not in some atmospheric effect analogous, perhaps, to the twinkling of stars. No systematic difference in the variations as the sun rose from zero elevation towards the zenith had been detected and the somewhat subjective conclusion had been reached that the variations consisted of *increases* above a datum level. Both observations suggest a solar origin. Further confirmation was obtained by taking simultaneous records at different places and comparing the variations. Figure 2*a* shows, superposed, two records taken 10 miles apart on a north-south line. The two records were taken with different and arbitrary sensitivities and have two different interference patterns, discussed fully in a succeeding section, so that it is only possible to look for the correspondence of sharp dips and peaks. Nearly all agree to within a second, the limit of timing accuracy. Figure 2*b* shows simultaneous recordings at Mt Stromlo and Dover Heights, Stromlo being 160 miles south-west of Dover Heights. The variations are smaller, but again generally agree in timing and shape on the two records. Figures 1b_1 and b_2 show another recording at the latter two stations when an isolated burst was recorded simultaneously. It is highly improbable that variàtions having such a high degree of correlation at widely separated sites should be due to any effect in the atmosphere, and it seems certain that most of them are extraterrestrial, and presumably solar, in origin. The evidence does not exclude residual variations due to the atmosphere.

A few measurements have been carried out on other frequencies between 3000 and 75 Mcyc./sec. The intensity observed on 3000 and 1200 Mcyc./sec. was of the order of 5×10^{-15} W m.$^{-2}$ (Mcyc./sec.)$^{-1}$, which is consistent with Southworth's results and is a little higher than the usual quiet-day 200 Mcyc./sec. intensity. An increase during the great sunspot of February was observed on 3000 Mcyc./sec., but was trivial compared with those on 200 and 75 Mcyc./sec. Observations on 75 Mcyc./sec. during disturbed periods have shown intensities considerably higher than the corresponding 200 Mcyc./sec. ones and with similar variations.

Day-to-day variations of intensity

Observations of mean intensity based on dawn observations at Collaroy began on 3 October 1945, and were carried through with few gaps to 15 February 1946. Most of the observations were carried out by Royal Australian Air Force personnel. It is now clear that, because of the violent changes in intensity which can occur, these observations, each of about a half-hour in duration, give an inadequate estimate of the daily value. Consequently in figure 3, 3-day running averages have been used in plotting the daily intensity over this period. Figure 3 also includes provisional Wolf sunspot numbers and integrated areas based on pencil sketches.* The great sunspot of February 1946 was associated with very high intensities, and both intensities and sunspot areas went off the scale of the graph. The data for this period are replotted on one-fifth scale on the right.

* The solar data throughout this paper were supplied by Dr C. W. Allen of the Commonwealth Observatory, Mt Stromlo, Canberra.

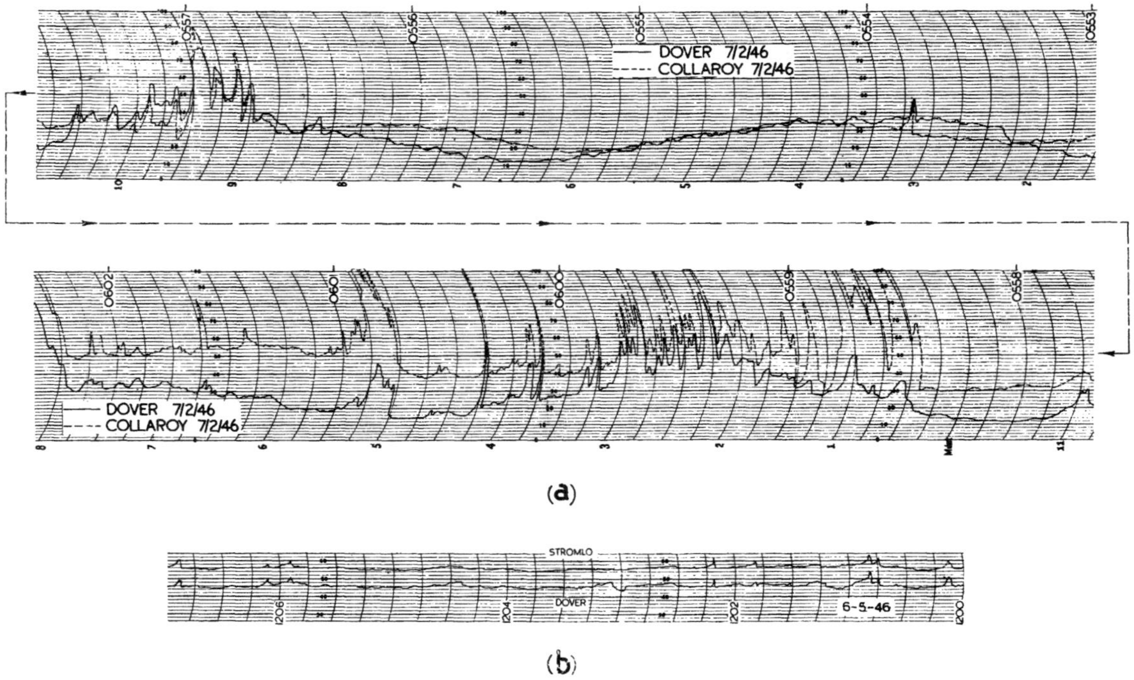

FIGURE 2. Simultaneous records of 200 Mcyc./sec. solar radiation from different places. (*a*) Dover Heights—Collaroy, separation 10 miles, sunrise interference patterns superimposed on solar variations. (*b*) Dover Heights—Stromlo, separation 160 miles, aerials pointed to sun near noon so interference pattern absent.

These data confirm a connexion between sunspots and radiation intensity. In this series of observations the correlation with areas is somewhat closer than with sunspot number but neither is exact.

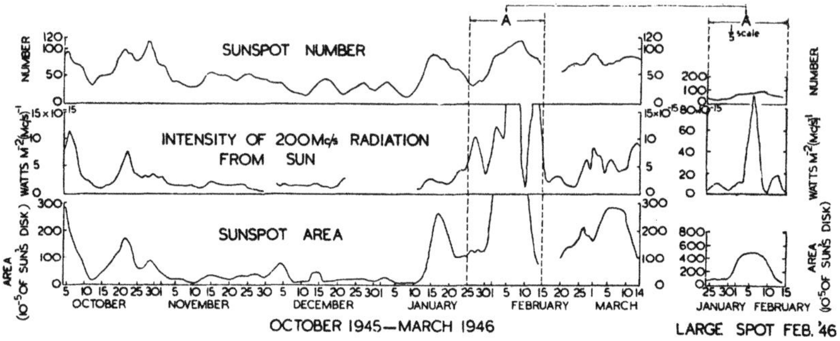

FIGURE 3. Day-to-day intensity variations (200 Mcyc./sec.), neglecting isolated bursts,
2 together with provisional Wolf sunspot numbers and integrated areas (3-day running averages).*

Figure 4 shows sketches of the sun on four typical days with corresponding values of intensity, sunspot area and sunspot number. The days chosen are:

(*a*) 5 October—two large groups of spots on meridian—intense 200 Mcyc./sec. radiation.

(*b*) 11 November—a typical day of few sunspots and little radiation.

(*c*) 26 January—the first day on which it was demonstrated that the radiation was coming from only a small part of the sun.

(*d*) 7 February—the great sunspot near the meridian—extremely intense radiation from a limited area.

Accurate location of the source of radiation using the interference pattern at dawn

An attempt was next made to elucidate the connexion between sunspots and the radiation by means of accurate directional measurements. Because an aerial of about a mile in aperture would be required to produce a beam narrow compared with the half-degree angular diameter of the sun, the direct-scanning method is not feasible. An alternative is a method involving the use of a steerable minimum. In practice, such a method may be realized rather simply by recording the intensity variations as the sun rises over the sea. Interference occurs between the direct and reflected rays, leading to a series of maxima and minima familiar in radar as 'lobes', or in optics as 'Lloyd's mirror' interference fringes. Since the angular separation
3 of the lobes on our equipment is about equal to the sun's diameter,* clearly defined maxima and minima will not be expected unless the radiating source is considerably smaller than the sun itself.

In initial observations early in October no interference pattern was observed; figure 4*a* shows that at this time there was a wide distribution of spots over the sun's

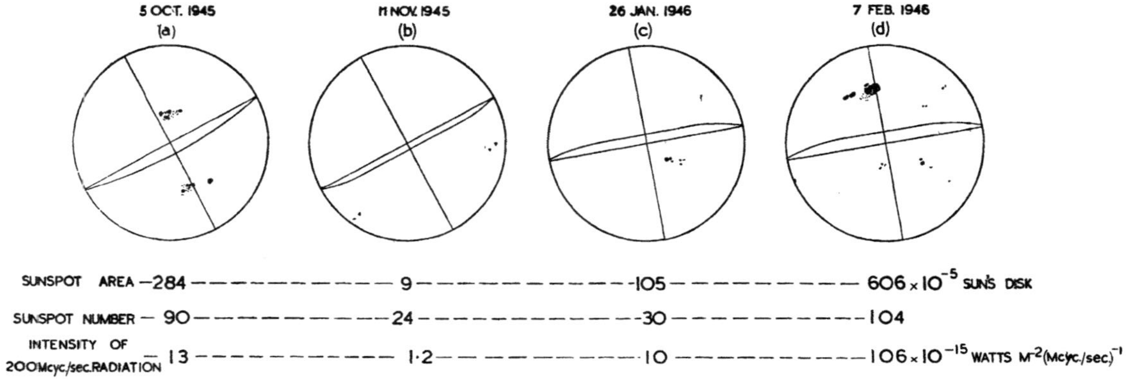

FIGURE 4. Sketches of sunspots on typical days, with values of 200 Mcyc./sec. radiation and sunspot number and area.

surface. Towards the end of January a compact sunspot group dominated the sun (figure 4*c*), and for this reason an attempt was made to detect a lobe pattern on the morning of 26 January. A regular series of maxima and minima was observed, with the expected period and with very deep minima which were less than the limit of detection (3 % of the maxima).

An ideal opportunity to extend these observations came a few days later with the appearance of the great sunspot of February 1946. Figure 5*a* shows the record obtained at dawn at Dover Heights on 7 February. The record is unmistakably an interference pattern with deep minima, complicated by the short-period variations already discussed. The section between 0553 and 0602 hr. shows the beginning of a remarkable series of bursts, an isolated one at 0554 hr., followed by an almost continuous sequence lasting beyond the end of the record. These bursts show interference maxima substantially in phase with the steadier radiation. (This is further evidence that these bursts are not due to terrestrial interference.) The received amplitude decreased towards the end of the record as the sun rose out of the beam
4 of the aerial.[*] Figure 5*b* shows the pattern obtained from Collaroy on the succeeding day. The period of the oscillation is shorter corresponding to the greater height of the Collaroy aerial, 400 compared with 278 ft. above mean sea-level. Figure 5*c* shows the record obtained on a day on which the minima were shallow, corresponding to a distributed source of radiation.

Let us examine the significance of the shape of the interference pattern, assuming tentatively a plane earth of reflexion coefficient -1. At a point of height h, the phase difference Δ between direct and reflected rays from a distant point source at a small angle of elevation θ is $4\pi h\theta/\lambda$, where λ is the wave-length. Consequently Δ is a linear measure of angle of elevation. The received power, p, is related to that due to either ray, p_0, by

$$p = p_0(2\sin\tfrac{1}{2}\Delta)^2 = 2p_0(1-\cos\Delta). \tag{1}$$

Suppose the source is a uniform strip, parallel to the horizon, of width $2W$ and located with angles of elevation as indicated in figure 6*a*. Let α_1, β_1, W_1 represent respective phase differences so that

$$\alpha_1 = \frac{4\pi h}{\lambda}\alpha, \text{ etc.} \tag{2}$$

Then, if the power contribution per unit angle of elevation is constant over the strip and zero outside, the total received power P is given by

$$P = \frac{P_0}{2W_1}\int_{\alpha_1+\beta_1-W_1}^{\alpha_1+\beta_1+W_1} 2(1-\cos\Delta)\,d\Delta,$$

where P_0 is the total power in the absence of a reflected ray. This reduces to

$$P = 2P_0 - 2P_0\frac{\sin W_1}{W_1}\cos(\alpha_1+\beta_1). \tag{3}$$

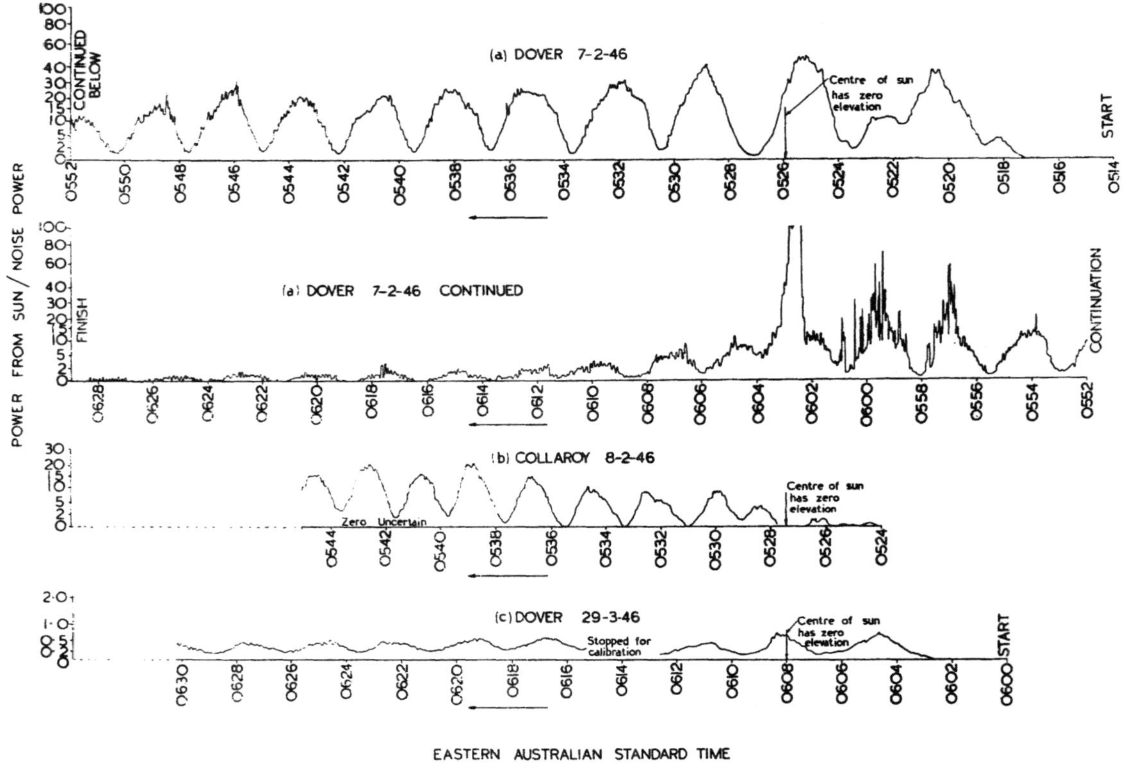

FIGURE 5. Interference patterns at dawn, aerial looking out to sea. (*a*) Great sunspot of February 1946. Intense radiation with deep minima, indicating small source. Series of large bursts about 6 a.m. (aerial height 278 ft.). (*b*) Following day at another station; aerial height 400 ft., period of interference pattern correspondingly shorter. (*c*) Record obtained on normal day. Shallow minima indicate extended source (aerial height 278 ft.). See figure 7 for appearance of sun on these days.

As the sun rises, α_1 alone varies and P fluctuates sinusoidally about twice the free-space value P_0 as shown in figure 6*b*. The record will show maxima and minima at the times for which $(\alpha_1+\beta_1)$ is equal to an odd or even multiple of π, and a ratio, R, of minimum to maximum power given by

$$R = \frac{1-(\sin W_1)/W_1}{1+(\sin W_1)/W_1}. \tag{4}$$

For small values of W_1 this beomes $\frac{1}{12}W_1^2$ and the actual width, $2W$, of the equivalent radiating strip is given by

$$2W = \frac{\lambda}{\pi h}\sqrt{(3R)}. \tag{5}$$

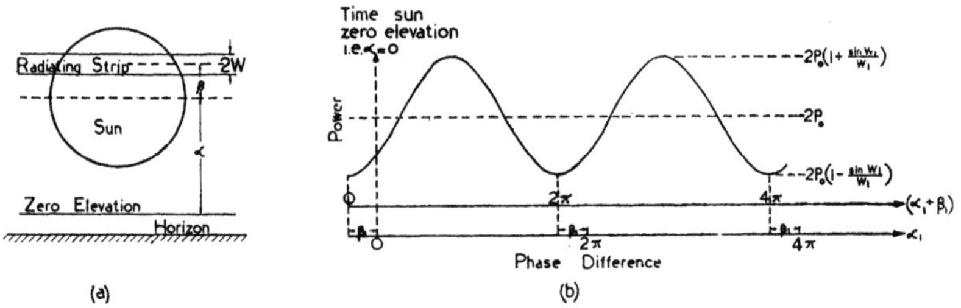

FIGURE 6. (*a*) Angles of elevation of strip source and sun. (*b*) Predicted interference pattern from strip source abscissae measured in units of phase difference, upper scale relating to centre of strip, lower to centre of sun.

If the power distribution is not rectangular but arbitrary given by $f(\beta_1)$, where β_1 measures the difference in elevation relative to the centre of the sun in terms of phase difference as before, then, substituting $\Delta-\alpha_1$ for β_1, the total power P is given by

$$\begin{aligned} P &= 2\int f(\Delta-\alpha_1)(1-\cos\Delta)\,d\Delta \\ &= 2\int f(\Delta-\alpha_1)\,d\Delta - 2\int f(\Delta-\alpha_1)\cos\Delta\,d\Delta, \end{aligned} \tag{6}$$

where the integration is taken over the significant range of Δ. The first term is twice the free-space power as in (3). The second term is in the form of a Fourier cosine transform. A little consideration indicates that, as α_1 varies, this term varies sinusoidally with an amplitude given by the modulus of the component of the Fourier transform of $f(\beta_1)$ at unit angular frequency.

Since an indefinite number of distributions have identical Fourier components at one frequency, measurement of the phase and amplitude of the variation of intensity at one place at dawn cannot in general be used to determine the distribution over the sun without further information. It is possible in principle to determine the actual form of the distribution in a complex case by Fourier synthesis using information derived from a large number of components. In the interference method suggested here Δ is a function of h and λ, and different Fourier components may be obtained by varying h or λ. Variation of λ is inadvisable, as over the necessary wide range the distribution of radiation may be a function of wave-length. Variation of h would be feasible but clumsy. A different interference method may be more practicable.

We now return to the special case in which the minima are very deep, i.e. the power is concentrated in a small range of angles, so that we feel justified in quoting the position and width of an equivalent rectangular strip. These quantities have been calculated as indicated in figure 6*b*. The width is derived from the ratio R, using equation (5), while the position relative to the centre of the sun (i.e. β) is calculated from the times of occurrence of minima, measured from the time when the centre of the sun has zero elevation. The calculations are complicated by certain factors discussed in appendix 2 and by the variable nature of the signal; errors due to the latter are reduced by taking the average value over a large number of lobes.

Such calculations have been made for some mornings on which deep minima occurred and the results are shown in figure 7. In each case the estimated accuracy of the position of the strip is a few minutes of arc. The widths given are upper limits and may be too wide by a few minutes (see appendix 2). The upper row shows sketches of the sun for each morning from 6 to 9 February with the equivalent radiating strips calculated from the observations at Collaroy. The next two sketches show corresponding results from Dover Heights on two of these days, and are in satisfactory agreement. In each case the radiating strip has a width considerably less than that of the sun's disk, being of the order of the size of the sunspot group, and passes through the group. It moves across the sun with the spots as the sun rotates. Although this method gives no information about the distribution parallel to the horizon, there seems no reasonable doubt that the source was localized in a small region in the vicinity of the spots. The observations do not provide any information as to the detailed structure of the source within this region. The great series of bursts about 0600 hr. on 7 February (figure 5*a*) have their maxima at the same times as the steady radiation and hence originate in the same vicinity.

The bottom row of figure 7 shows a similar series of sketches for 29 to 31 March. There was at this time a predominant sunspot group, but it did not completely overshadow the other spots as in the February series. The equivalent position of the source is in the vicinity of this group but moved towards the 'centre of gravity' of the spots. The width is much greater than previously, corresponding to the dispersal of spots. This series is again consistent with the assumption that energy originates in the vicinity of sunspots.

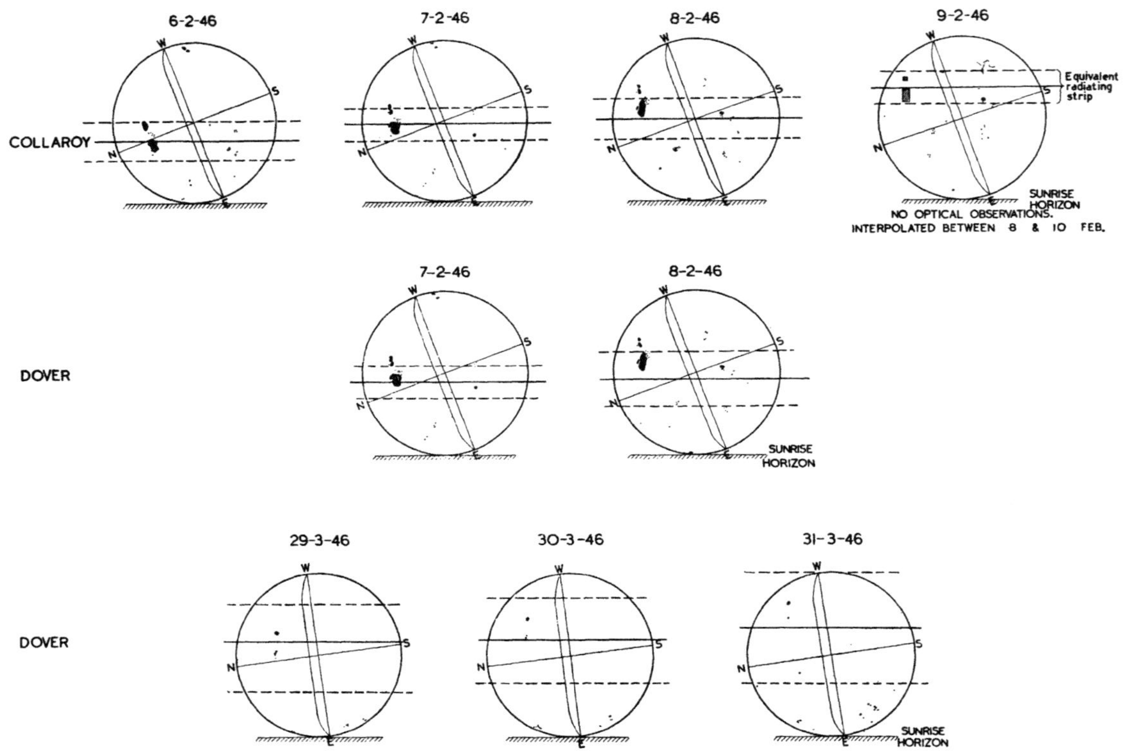

FIGURE 7. Sketches of sun with position and width of equivalent radiating strip. (Widths are upper limits only, position uncertain to a few minutes of arc.) 1st row: great sunspot, Collaroy observations. 2nd row: great sunspot, Dover Heights observations. 3rd row: sources of radiation scattered, though radiation mainly from one group, Dover Heights observations.

Interpretation of observations

The previous sections have included accounts of observations and direct inferences from them. Before discussing more speculative deductions this material is summarized as follows. The statements refer to 200 Mcyc./sec. radiation unless stated otherwise:

(1) Solar radio-frequency energy of intensity varying between observed limits of $0{\cdot}5 \times 10^{-15}$ and more than 100×10^{-15} W m.$^{-2}$ (Mcyc./sec.)$^{-1}$ is incident on the earth.

(2) The intensity varies grossly with time. Comparatively slow variations have superposed on them bursts of duration of the order of a second. These bursts may be isolated, or appear in slow or rapid succession. A large proportion of the rapid changes are extraterrestrial, and presumably solar, in origin.

(3) On days of great solar activity high levels of intensity have been observed on 75, 200 and 3000 Mcyc./sec., the relative increases being large on 75 and 200 Mcyc./sec. but small on 3000 Mcyc./sec. Observations over 6 months on 200 Mcyc./sec. showed correlation with both sunspot numbers and areas, the correlation with areas being better.

(4) The region from which energy is received varies, but at times of a single dominant sunspot or group this region is of the same order of size as, and is situated in the immediate vicinity of, the spot or group.

A consideration of intensity values yields evidence as to the mode of origin of the radiation. If it is thermal in origin, its intensity will not be greater than that given by the Planck formula for the appropriate temperature. At radio-frequencies the Planck formula reduces to the Rayleigh-Jeans form which, for constant frequency increments, is

$$S = \frac{2kT}{\lambda^3}\Omega_s \times 10^6 \text{ W m.}^{-2}\text{ (Mcyc./sec.)}^{-1},^{\star} \qquad (7)$$ 5

where S = intensity on the earth (W m.$^{-2}$ (Mcyc./sec.)$^{-1}$),

k = Boltzmann's constant ($1{\cdot}38 \times 10^{-23}$ joule/degree),

λ = wave-length (m.),

T = temperature of source (degrees Kelvin),

Ω_s = solid angle subtended by source (steradians).

If the whole disk of the sun radiates uniformly the intensities observed by Southworth (1945) on 3000 and 10,000 Mcyc./sec. correspond to temperatures of about 20,000° K,* which is three times the optical temperature. On the other hand, the present 200 Mcyc./sec. results would require temperatures of from 0·5 to 100 millions of degrees for the general level with still higher values for bursts. In addition, it has been shown that the radiation does not originate uniformly over the surface but in small regions, so that even higher temperatures are required. For example, on 7 February the intensity was 100×10^{-15} W m.$^{-2}$ (Mcyc./sec.)$^{-1}$ and the area of

* The original paper gave a temperature of about 6000° K, owing to an error subsequently corrected.

origin apparently less than $(6{\cdot}5'/32')^2$ or $\frac{1}{25}$ of the sun's disk (see figure 7). This would mean a black-body temperature of about 3000 million degrees, which is impossibly high compared with any known temperatures on the sun. The known temperatures range from 6000° at the visible surface to about a million degrees in the corona and a few tens of millions at the centre. Consequently, though thermal radiation will be present, it is overshadowed at 200 Mcyc./sec. by radiation due to some other mechanism, probably gross electrical disturbances as suggested in our previous communication (Pawsey *et al.* 1946). The occurrence of short-duration bursts favours this hypothesis. It must be borne in mind that the detailed mechanism in the partially conducting ionized atmosphere of the sun will be different from the sudden breakdown occurring in lightning flashes. An analogous terrestrial phenomenon on a much smaller scale is the generation of noise in radio valves containing gas. The appearance of bursts superposed on a more slowly varying background suggests that there may be two separate mechanisms, though it is possible that the background may consist of a large number of overlapping bursts.

The connexion between this radiation and sunspots is established by two independent lines of evidence, the correlation of intensity with sunspot area and the coincidence of direction of origin with that of sunspot groups. No evidence is yet available as to a particular visible solar phenomenon, associated with sunspots, which gives rise to the radiation.

Departing now from solar phenomena, two other aspects of this work are of interest. In the direction-finding procedures leading to the identification of the area of origin of the radiation, variable refraction effects due to the earth's atmosphere entered in as a complication reducing the inherent accuracy. The early lobes often show marked displacements and departures from sinusoidal form, e.g. the first lobe in figure 5*a*, which appear to be due to abnormal refractive effects. It seems possible to exploit the technique to investigate refraction and perhaps deduce the refractive-index structure of the atmosphere.

Cosmic noise was originally attributed to radiation from interstellar matter, rather than from stars, at a time when similar radiation from the sun had not been detected. The discovery of solar noise raises the question as to whether the cosmic noise is due to similar processes in stars. The basic difficulty remains that the intensity of cosmic noise is vastly greater than it should be if the stars emitted the same ratio of radio-frequency energy to light as does the sun. Nevertheless, the great variability of solar noise suggests the possibility of vastly greater output from stars differing from the sun and it seems that data at present available leave the question completely open.

Appendix 1

Derivation of intensity in terms of instrumental constants

The equipment is adequately described in terms of the 'effective area' A of the aerial and the receiver 'noise factor' N which is specified at an arbitrary ambient temperature T_a. Let the relative increase in power output from the receiver when the aerial is turned from clear sky to the sun be R, i.e. if P_{sky} and P_{sun} are respective power outputs, let $R = (P_{\text{sun}} - P_{\text{sky}})/P_{\text{sky}}$. Then the intensity S_1 of the component polarized so as to be accepted by the aerial is shown below to be given approximately by

$$S_1 \doteqdot R\frac{N}{A}kT_a, \tag{8}$$

where k is Boltzmann's constant. Plane-polarized aerials have been used and the total intensity S taken as twice S_1.

At the frequencies used, the limit to the smallest signal that can be detected is set by the noise power originating in the receiver itself. The power received from the sun is of the same order as this noise power, so that it is conveniently measured in terms of this power. Since, for small signals which do not overload the receiver, the amplification is constant, ratios are preserved if all signals are referred to the input terminal. The merit of a receiver can be measured by the magnitude of receiver noise power P_N relative to a standard signal. The standard signal chosen is the fluctuation noise power P_R available from a resistor of given temperature T_a. This equals $kT_a\Delta f$, where Δf is the band width accepted, independently of the value of the resistance. The noise factor N is defined by

$$N = \frac{P_R + P_N}{P_R}. \tag{9}$$

The effective area A of an aerial is defined by $A = P/I$, where P is the available power at the output terminals of the aerial due to an energy flux I, of appropriate polarization, incident on the aerial. It is related to the isotropic power gain g by the equation $A = \lambda^2 g/4\pi$. For uni-directional broadside arrays, large with respect to the wave-length, A can be made to approach, but not exceed, the actual area of the array.

If an aerial is pointed to the clear sky, a power P_C is received due to cosmic noise, and this increases to $P_C + P_S$ when turned to the sun (P_C being assumed constant) where

$$P_S = (S_1 \Delta f)A. \tag{10}$$

Consequently

$$R = \frac{P_S}{P_N + P_C}$$

or

$$P_S = R(P_N + P_R)\left\{\frac{P_N + P_C}{P_N + P_R}\right\} \doteqdot R(P_N + P_R), \tag{11}$$

since, on the wave-lengths and equipments used, it has been found that P_N swamps differences between P_C and P_R to within a few per cent. Combining (11) with (9) and (10) and writing $P_R = kT_a\Delta f$ leads to (8) and to the numerical equation for total intensity

$$S = 8{\cdot}2 \times 10^{-15} RN/A \text{ W m.}^{-2} (\text{Mcyc./sec.})^{-1}, \tag{12}$$

where A is in square metres and T_a is taken as 300° K.

If a square-law detector is used, the current indicates relative power level directly. Otherwise current values must be interpreted as relative power levels from the known detector law before deducing R.

Appendix 2

Factors complicating the analysis of the interference pattern

The analysis given in the body of this paper neglected sundry factors. Some of these are calculable while others introduce uncertainties. The factors are stated below, together with notes on the treatment adopted in deriving the results given in figure 7.

(1) *Earth curvature*

For a spherical earth of radius a, the phase difference Δ, at height h above the earth, between direct and reflected rays from a very distant point source, is approximately given by

$$\Delta \doteqdot \frac{4\pi x}{\lambda}\left\{\frac{h}{x}-\frac{x}{2a}\right\}^2, \tag{13}$$

where x is the distance of the point of observation from the point of reflexion. The angle of elevation of the direct ray at the point of observation is given by

$$\theta \doteqdot \left(\frac{h}{x}-\frac{3x}{2a}\right). \tag{14}$$

Predicted values of θ for interference minima were obtained by plotting corresponding values of Δ and θ calculated for arbitrary values of x and then reading off values of θ corresponding to integral values of $\Delta/2\pi$.

The error incurred in using the above approximate formulae becomes appreciable for large values of θ but is less than 1 min. of arc in θ up to 7° and 3 min. at 10°.

(2) *Inaccurate height data*

Corrections for tides were made from *Admiralty Tide Tables*. No allowance was made for a possible change in effective sea-level due to waves. The estimated uncertainty in h for each station is 1 %, giving rise to an uncertainty in β (figure 6*a*) of about 3 min. of arc for $\alpha = 3°$ (sixth lobe at Dover Heights) and 6 min. for $\alpha = 10°$ (twentieth lobe). The width of the equivalent strip is unaffected by height errors.

(3) *Refraction*

Refraction introduces the major uncertainty in the analysis. It differs from optical refraction owing to the greater refractive index of water vapour at radio frequencies so that known astronomical corrections cannot be used.

An unpublished formula due to T. Pearcey of this laboratory has been used. It is based on a linear variation of refractive index with height and, using the notation of figure 8, is

$$R = \frac{\tau_0}{H/a-\tau_0}\left[\sqrt{\left\{2\left(\frac{H}{a}-\tau_0\right)+\theta^2\right\}}-\theta\right],$$

where the refractive index μ at any height z is given by

$$\mu = 1+\tau_0(1-z/H) \quad \text{for} \quad z<H$$

and

$$\mu = 1 \quad \text{for} \quad z>H.$$

The formula is not in doubt at large values of θ since it then depends only on τ_0, which is known from ground-level measurements of pressure, temperature and humidity. The validity of the formula for low angles was tested as follows. Taking H as 5 miles (in conformity with the lapse rate of refractive index near the ground) and the value of τ_0 as that at 0600 hr. on the appropriate day (obtained from Sydney Weather Bureau data), the angles of elevation of the lobe minima were calculated from this formula in conjunction with equations (13) and (14). From the times of minima in the experimental record for the day the corresponding elevations of the centre of the sun were computed. For any one day there should be a constant difference between the two sets of angles of elevation, equal to the displacement of the radiating source from the centre of the sun. The actual differences show a certain scatter. Their mean value, neglecting the first few lobes, was taken as giving the true displacement. The deviations from this mean were then plotted as deviations from the theoretical curve for the average value of τ_0 on the days involved (see figure 8).

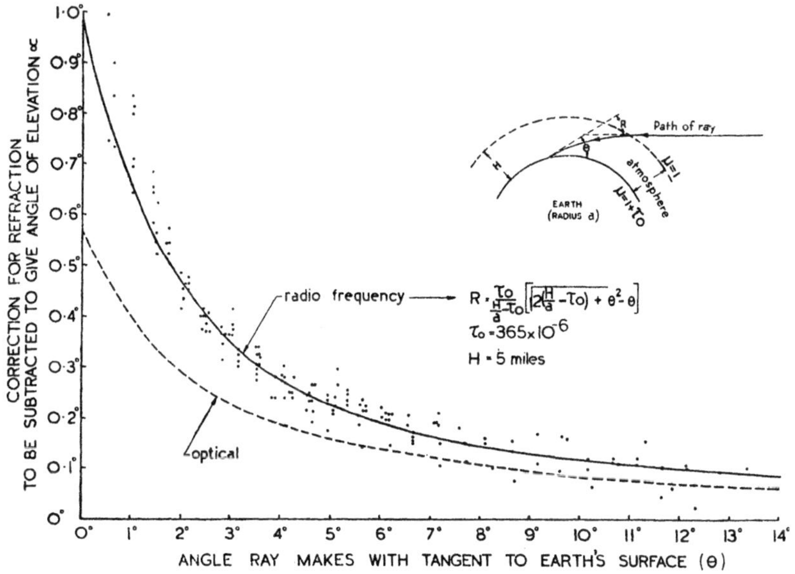

FIGURE 8. Refraction correction used at 200 Mcyc./sec. and deviation of experimental results. Also astronomical correction for refraction (dotted curve).

The agreement between the curve and the mean distribution of observations down to $\theta = 2^\circ$ is evidence for the validity of the formula in this region. Below 2° it is apparent that the amount of refraction varies considerably and that corrections greater than those given by the formula are normally required. This region has not been used in deriving the position of the equivalent radiating strip.

For comparison, figure 8 also shows the correction for refraction used in astronomy (Norton & Inglis 1943).

(4) *Imperfect reflexion from the sea due to waves*

The effect on the reflexion coefficient is believed small, and no allowance has been made for departure from unity. A reduction in the magnitude of the reflexion coefficient to 0·9 would increase the apparent effective width of the radiating strip by less than 2 min. of arc.

(5) *Receiver band width*

The finite range of frequencies accepted by the receiver leads to a corresponding range of phase differences between the direct and reflected rays. This is equivalent to an increase in angular width of the source, producing a reduction in the ratio of maximum to minimum which increases with elevation of the sun. The apparent increase in width has been allowed for by subtracting $\theta(\Delta f/f)$ from the estimated values of $2W$. The ratio of band width to mid-frequency, $\Delta f/f$, is about 1 % for receivers used.

This work has been carried out as part of the research programme of the Division of Radiophysics, Council for Scientific and Industrial Research, Commonwealth of Australia. The authors wish to acknowledge the assistance of the radar sections of the Australian Army and Air Force in providing equipment, and particularly of the personnel of Collaroy radar station in making observations; also of the staff of the Commonwealth Observatory, Mt Stromlo, Canberra, who co-operated in observations, provided sunspot data, and made many valuable suggestions.

References

Appleton, E. V. 1945 *Nature*, **156**, 534.
Hey, J. S. 1946 *Nature*, **157**, 47.
Jansky, K. G. 1933 *Proc. Inst. Radio Engrs, N.Y.*, **21**, 1387.
Norton, A. P. & Inglis, J. G. 1943 *A star atlas*. London: Gall and Inglis.
Pawsey, J. L., Payne-Scott, R. & McCready, L. L. 1946 *Nature*, **157**, 158.
Reber, G. 1944 *Astrophys. J.* **100**, 279.
Southworth, G. C. 1945 *J. Franklin Inst.* **239**, 285.

EDITOR'S NOTES AND AMENDMENTS

1. The *second detector* was the detector following the mixer ("first detector").
2. See Paper 21 for a subsequent use of these same radio monitoring data.
3. The lobe spacings in fact were ~30 arcmin at Dover Heights and ~21 arcmin at Collaroy.
4. It must be remembered that these radar antennas were constrained to move only along the horizon. It appears that the primary beam of the antenna was of vertical width ~5°–7°.
5. "λ^3" should be "λ^2".

SECTION J

THE HOT SOLAR CORONA

The great intensity of the solar bursts at first masked the presence of any contribution at meter wavelengths from the quiet, undisturbed Sun. The following pair of papers were the first to demonstrate in detail, both theoretically and observationally, the existence of thermal radio emission from a million-degree corona. It should be noted that the radio results were not actually the first astronomical observations to establish the existence of the hot corona, but they played a strong role in its eventual acceptance by the astronomical community. Three main lines of evidence had earlier provided strong indications: (1) Edlén's identifications of the "coronium" lines with transitions in highly ionized metals, (2) the extremely broad widths of these lines, and (3) the scale height of the electron distribution obtained from eclipse observations.

Not included in this collection are independent theoretical studies published in the Soviet Union several months earlier by Ginzburg (1946) and Shklovsky (1946). Ginzburg's treatment is not unlike that of Martyn, but is not as complete since his goal, motivated by the observations of Reber and Southworth, was to examine the possibility of radar reflection off the solar atmosphere. Nevertheless, Ginzburg also demonstrated that observations at meter wavelengths should yield brightness temperatures of order 10^6 K. Shklovsky arrived at similar conclusions and in particular emphasized that the solar radio radiation originated in two distinct components, "thermal" and "sporadic". In the same wide-ranging article he furthermore was the first to suggest plasma oscillations as a mechanism involved in the radio bursts.

In Paper 20, D. F. Martyn (1906–1970) explains his ideas in a clear and succinct manner and makes many definite predictions amenable to experimental verification. Almost all of his results concerning the apparent temperature of the Sun, as a function of both frequency and position on the solar disk, have now in fact been verified. Martyn only went astray in predicting differing observed temperatures for the two senses of circular polarization. This was a result of his using a value of 50 gauss (about 20 times too large) for the general magnetic field of the Sun. Martyn's theory is worked out in detail in a 1948 publication.

Paper 21, by Pawsey, immediately followed Paper 20 in the original issue of *Nature* and, in fact, there is some dispute over which author should get the credit for recognizing the importance of solar thermal radio emission (see Hey, 1973). In any case, using exactly the same radio data as shown in Figure 3 of Paper 19, but now more revealingly broken down into daily averages rather than three-day averages, Pawsey clearly demonstrates the existence of a base level, at a wavelength of 1.5 m, below which the Sun's intensity never falls. This intensity in turn corresponds to an apparent temperature of about one million degrees, as predicted.

TEMPERATURE RADIATION FROM THE QUIET SUN IN THE RADIO
1 SPECTRUM*

By Dr. D. F. MARTYN
Commonwealth Observatory, Canberra

THE radio-frequency emissivity of the sun considered as a black body is proportional to $T\lambda^{-2}$ per unit frequency increment, where T is the temperature of the radiating region, and λ is the wave-length of the radiation. The sensitivity of radio equipment is now such that it is possible to detect this radiation on the shorter wave-lengths in the radio spectrum. In particular, Reber[1] and Southworth[2] have measured it on short radio wave-lengths, using highly directive aerial systems. Appleton[3] has pointed out that it should be impossible to detect this temperature radiation at the longer radio wave-lengths, owing to the rapid falling off of solar emissivity, combined with the impracticability of using highly directive aerials on these wave-lengths. He and others[4] suggest that the radiations which *are* observed on the longer wave-lengths, and which appear to be correlated with sunspots, cannot be thermal in origin, since such an explanation would require solar temperatures of the order one million degrees and upwards. There is little doubt that these views, so far as they refer to temperatures *upwards* of 10^6 degrees, must be correct, especially in the light of recent evidence[5,6] showing that at such times the radiation comes from restricted areas in the immediate vicinity of sunspots. It is the purpose of this note to point out, however, that at such wave-lengths we should expect thermal radiation corresponding to values of T *downwards* from 10^6 degrees to the familiar surface temperature of order 10^4 degrees.

Reprinted, with permission, from 'Nature' 158, 632–634 (1946).

Studies by Baumbach[7] of the light scattering by electrons in the solar corona give reliable estimates of the electron densities in this region. Since the corona must be completely ionized, the electron collision frequencies may be calculated by the method of Chapman and Cowling[8]. It is then easy, by application of Milne's concept of optical depth, to show that solar radiations the wave-length of which is longer than about 1 metre must emanate from the corona. Now, Edlén's recent work[9], together with other spectroscopic evidence, shows that the coronal matter is normally at a temperature approaching 10^6 degrees. We should therefore expect to find black-body radiation of about 1 metre wave-length having a normal (quiet sun) intensity corresponding nearly to $T = 10^6$.

We may proceed to investigate the effective values of T at other wave-lengths by using Milne's criterion that the observed radiation comes from an average optical depth $\tau = 1$, where $\tau = -\int \varkappa_\lambda dr$, $\varkappa_\lambda$ being the absorption coefficient and r the distance from the sun's centre. When we apply this criterion for wave-lengths shorter than 1 metre it is found that, as λ decreases, the radiation begins to emanate from the chromosphere, which is at a much lower temperature than 10^6 degrees, so that the solar emissivity falls off rapidly in this region of the spectrum.

On the long-wave side of 1 metre the corona becomes a reflector and τ is less than unity. This happens because of the exponential decrease of electron density and collision frequency in the corona as r increases. Assuming local thermodynamic equilibrium, we may apply Kirchhoff's law, so that the effective temperature now becomes $\varkappa_\lambda T$. As λ increases, $\varkappa_\lambda$ decreases and the effective temperature falls off. There is thus a maximum in effective T in the vicinity of $\lambda = 1$ m. The results of detailed calculations of this kind are shown in Fig. 1 (full line). It will be noticed that the curve descends steeply on the short wave-length side of the maximum, and comparatively gradually on the long wave-length side. In the latter region the effective size of the solar disk will be considerably increased, since it is the corona which is responsible for the observed

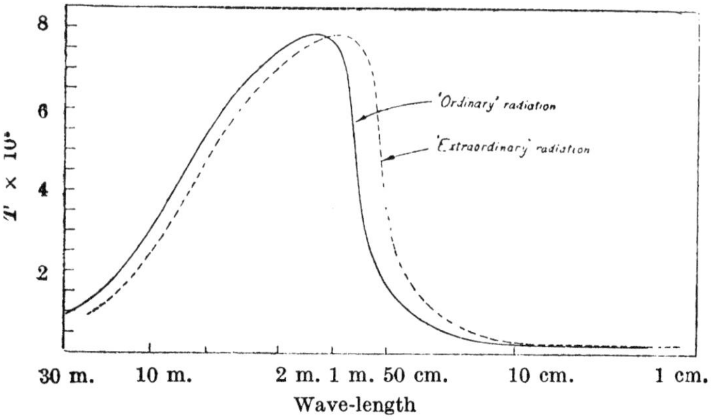

Fig. 1. EFFECTIVE TEMPERATURE OF SUN CONSIDERED AS A BLACK-BODY RADIATOR IN THE RADIO SPECTRUM

radiation. In practice, however, it will not usually be necessary to take account of this, because of pronounced 'limb-darkening', which we now consider.

So far, we have confined our attention to the effective values of T averaged over the whole disk. It is of considerable interest to examine the variation of T (the brightness) within the disk. Fig. 2 (curves *a–d*) shows this variation for a range of the longer wave-lengths. It will be noticed that there is a considerable falling off in brightness as the limb of the sun is approached. This effect is superficially similar to the 'limb-darkening' familiar to solar physicists in visible light. The causes, however, are fundamentally different. Limb-darkening in the visible spectrum is due to the fact that a rising temperature is encountered as we penetrate the photosphere. The light we see from the centre of the disk comes from lower and hotter regions than that which comes to us at the more nearly glancing angles on the limb. In the radio case the effect of temperature gradient *in the corona* is negligible compared with the falling off in κ_λ at the shallower angles of emergence as the limb is approached. For wave-lengths shorter than about 1 metre (Fig. 2, curves *d–g*) the solar atmosphere becomes optically thick over most of the limb, and a treatment more strictly analogous to that used by solar physicists may be applied. Since the temperature *decreases* rapidly as we pass from the corona to the chromosphere, it follows, however, that limb *brightening* should be observed in the radio case. This phenomenon should be very marked at wave-

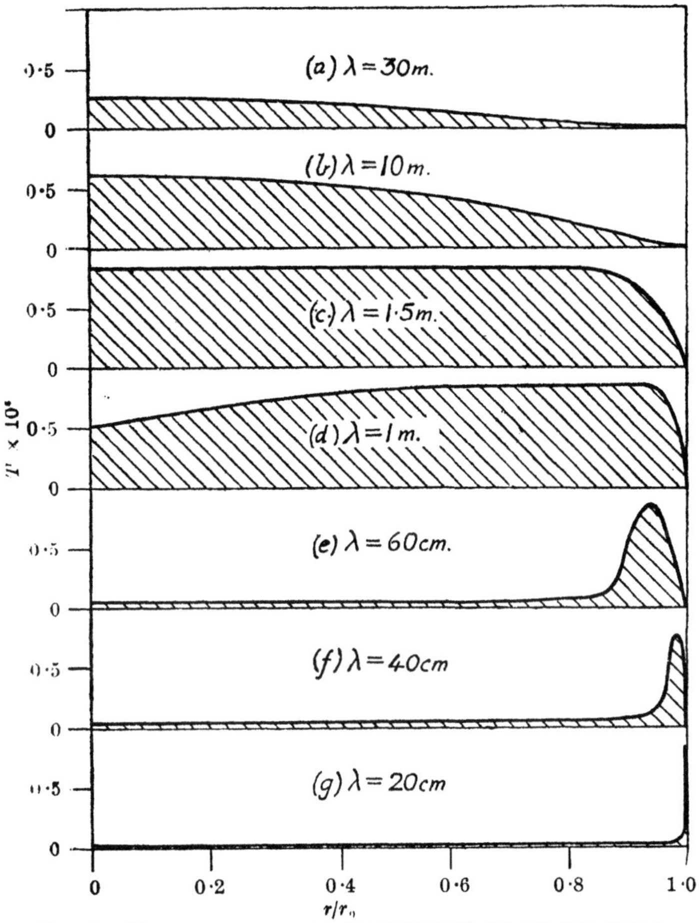

FIG. 2. VARIATION OF RADIO 'BRIGHTNESS' ACROSS THE SOLAR DISK AT VARIOUS WAVE-LENGTHS ($r = r_0$ AT LIMB)* 2

lengths of 60 cm. downwards. It should be capable of direct experimental test either by the use of highly directive aerials on centimetre wave-lengths, or at eclipses.

The discussion so far has neglected the effect of the sun's general magnetic field. Account can be taken of this factor by making a solar application of Appleton's magneto-ionic theory of the ionosphere. The total radiation now divides into two parts, the 'ordinary' and 'extraordinary', each with characteristic elliptical polarization of opposite senses of rotation. The discussion above may be taken with

sufficient accuracy as applying to the 'ordinary' radiation. The distribution of 'extraordinary' radiation over the spectrum is shown by the dashed line in Fig. 1. It is seen that the intensities of these two characteristic radiations are markedly different over
3 a wide range of wave-lengths.* At first sight it might
seem that this conclusion could be simply tested by the use of aerial systems alternately disposed to receive right- or left-handed radiations. Unfortunately, however, while the 'ordinary' radiation from the northern solar hemisphere is right-handed, the same radiation from the southern hemisphere is left-handed. For testing the above conclusion we must rely on statistical observations of the quiet sun over a period of time during which the solar axis is tilted markedly towards or away from the earth. Alternatively, observations could be made at eclipses or in regions providing a suitable horizon at sunrise or sunset.

As yet, few published observations exist which can be compared quantitatively with our conclusions. However, Southworth's original observations, which appeared to give $T = 6{,}000^\circ$ in the region $\lambda = 10$ cm., have since been corrected by him[10] to give $T = 20{,}000^\circ$, agreeing with the curve in Fig. 1.

The complete exploration of quiet sun solar radiation over the radio spectrum is probably beyond the resources of a single organisation, owing to the comparative inflexibility of the necessary equipment with respect to wave-length. It is hoped that the above conclusions, which appear to rest on well-established solar data and physical principles, may be of service to those planning regular observations in this field.

A full description of this work, which is part of the research programme of the Council for Scientific
4 and Industrial Research, will be published elsewhere.*
I am indebted to Drs. R. Woolley and C. W. Allen for much advice on solar data.

[1] Reber, *Astrophys. J.*, **100**, 279 (1944).
[2] Southworth, *J. Franklin Inst.*, **239**, 285 (1944).
[3] Appleton, *Nature*, **156**, 534 (1945).
[4] Pawsey, Payne-Scott and McCready, *Nature*, **157**, 158 (1946).
5 [5] Pawsey, Payne-Scott and McCready, in the press.*
[6] Martyn, *Nature*, **153**, 308 (1946).
[7] Baumbach, *Ast. Nach.*, **263**, 121 (1937).
[8] Chapman and Cowling, "Mathematical Theory of Non-Uniform Gases" (Camb. Univ. Press, 1939), 177.
[9] Edlen, *Ark. f. mat. Ast. Fys.*, **28 B**, No. 1 (1942).
6 [10] Southworth, *J. Franklin Inst.*, **241** (March 1946).*

OBSERVATION OF MILLION DEGREE THERMAL RADIATION FROM THE SUN AT A WAVE-LENGTH OF 1·5 METRES

By Dr. J. L. PAWSEY
Radiophysics Laboratory, Chippendale, N.S.W.

IN the preceding communication, Martyn has shown that at wave-lengths of a few metres, thermal radiation corresponding to a temperature of about a million degrees should be radiated from the sun. The detection of this radiation is complicated by the presence of a further source, which is highly variable and is associated in some way with sunspots. This source can, on occasion, yield radiation up to 100 times the expected thermal intensity. If we confine ourselves to intensity measurements, it would in fact be detectable only if the intensity due to the variable source, not infrequently fell below that of the thermal one.

Daily measurements of intensity on a wave-length of 1·5 metres over a period of about six months indicate that this condition is satisfied in this part of the spectrum. The distribution of observed intensities is shown in the accompanying histograms. This distribution is markedly skew, having a sharp cut-off on the low side at the intensity range 0·5–1·0 $\times$ 10^{-15} watt m.$^{-2}$ (Mc./s.)$^{-1}$, corresponding to effective temperatures between 0·6 and 1·2 $\times$ 10^{6} degrees Kelvin. This distribution is consistent with the co-existence of a steady source, of intensity equal to the cut-off value, and a symmetrically distributed highly variable source, which, on this wave-length, exceeds the steady value for about 60 per cent of the observations.

The agreement between the observed cut-off intensity, which corresponds to a temperature of about one million degrees Kelvin, and the effective temperature derived by Martyn, 0·8 $\times$ 10^{6} degrees Kelvin, leaves little doubt that million degree thermal radiation is being observed at this wave-length.

I am indebted to Dr. D. F. Martyn for pointing out to me the probable existence of high-level thermal

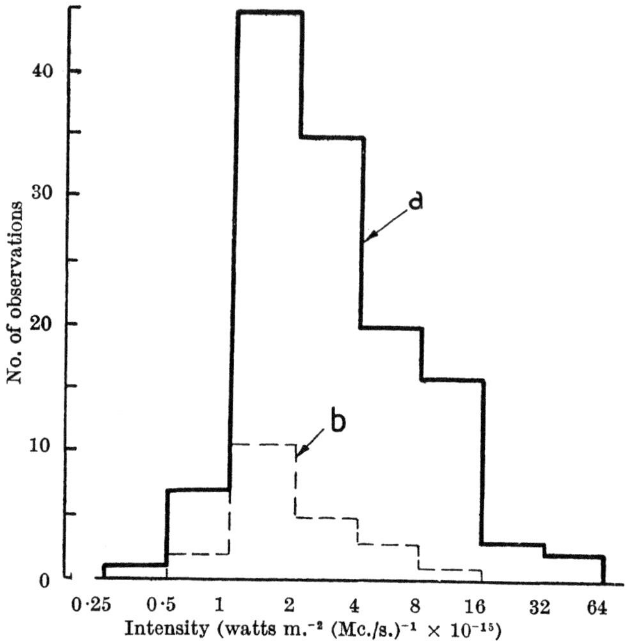

HISTOGRAMS SHOWING DISTRIBUTION OF DAILY VALUES OF SOLAR RADIATION INTENSITY AT 1·5 METRES WAVE-LENGTH (INCREASES OF A FEW SECONDS ARE NEGLECTED). (*a*) DAILY: OCT. 5, 1945–DEC. 12, 1945; JAN. 1, 1946–MARCH 15, 1946 (R.A.A.F. OBSERVERS). (*b*) SUNDRY DAYS, MARCH TO MAY 1946 (LABORATORY OBSERVERS)

radiation, and to members of the Royal Australian Air Force and of the Radiophysics Laboratory who took the observations.

This work is part of the research programme of the Radiophysics Laboratory, Commonwealth Council for Scientific and Industrial Research.

EDITOR'S NOTES AND AMENDMENTS

1. *Temperature radiation* is "thermal radiation".
2. Martyn (1948) later indicates that the definition of r_0 is rather nebulous and varies with frequency; it is *not* the optical radius. Also, the ordinate of Figure 2 is in units of 10^6 K.
3. In Figure 1 Martyn has used a value of 50 gauss (at least 20 times too large) for the general magnetic field of the sun; modern observations do not show the predicted effect.
4. Martyn (1948).
5. *Proc. Roy. Soc.* **A190**, 357 (1947).
6. Following p. 166.

SECTION K

LUNAR RADIATION

As with other wartime radio observers, the basic research described in Paper 22 was only a secondary line for the authors. The primary aim of R. H. Dicke and R. Beringer was to meet the needs of the wartime M.I.T. Radiation Laboratory; in particular, they were working on microwave receivers and atmospheric effects at these very short wavelengths. Having developed an extraordinarily sensitive comparison-type receiver (see Paper 11), however, it was only natural that Dicke and his colleagues should try some astronomical observations in their spare moments. But what dividends those few observations paid! Not only is their experimental method in Paper 22 accurate enough that their measured values agree well with those of today, but they also establish the following "radio firsts": first solar eclipse observation, first *absolute* intensity calibration, first measurement of a size for the Sun, and first measurement of the lunar brightness.[1] The lunar work was soon followed up at several microwavelengths and at all phases, most notably in the work of J. H. Piddington and H. C. Minnett (1949), who were able reliably to establish many of the electrical and thermal properties of the lunar surface.

NOTE

[1] In view of Dicke's later close connection with cosmology and the 2.7 K cosmic background radiation, it is also of interest to note a result given in a companion paper where Dicke *et al.* (1946) report an upper limit of 20 K for any "radiation from cosmic matter at the radiometer wavelengths" (which were 1.00, 1.25, and 1.50 cm).

MICROWAVE RADIATION FROM THE SUN AND MOON*

G. C. Southworth[1] has reported the measurement of microwave radiation from the sun with a conventional superheterodyne receiver and a large parabolic antenna. This note is to report similar measurements made with a microwave radiometer, which was designed primarily for the measurement of the atmospheric absorption of water vapor by observing the thermal radiation emitted by the atmosphere.[2] The instrument has also been used to observe the microwave radiation from the sun and moon. These astronomical measurements are quite incomplete because of the pressing demands of war research. It was decided, however, to report them now, since it seems unlikely that the authors will be able to continue this work.

The radiometer[3] measured the thermal radiation in a small band (0.06 per cent) of wave lengths at a wave length of 1.25 cm. Its sensitivity was such that it could detect a change of $\frac{1}{2}°$ C in the temperature of a black body. Stated more exactly, the root-mean-square fluctuation in the indicating meter was about $\frac{1}{2}°$ C when expressed in temperature units.

For the astronomical measurements the radiometer was provided with a parabolic-reflector antenna (18-inch diameter). This antenna had a measured half-power beam width of 2° and a measured gain of 6.0×10^3 times that of an isotropic radiator. The radiometer was calibrated by replacing the antenna by an artificial black body (mounted in a wave-guide), which could be heated to known temperatures. Three series of astronomical measurements were made, all at Cambridge, Massachusetts. These will be described below.

a) The microwave radiation from the sun was observed during the partial eclipse of July 9, 1945. The results are shown as points in Figure 1. The radiation intensities are expressed in arbitrary units but have been corrected for atmospheric absorption, using the results of thermal-radiation measurements made at the same time, which gave an absorption of 0.46 db/atmosphere (i.e., 0.46 db total absorption for a vertical path through the atmosphere). The curve of Figure 1 is the calculated intensity during the eclipse, assuming a uniform sun disk of optical size and the optical percentage of eclipse. It is arbitrarily adjusted to fit the measured points under the above assumptions. The agreement between this curve and the measured points indicates that the size of the sun's disk at $\lambda = 1.25$ cm is not greatly different from that at optical wave lengths. The ordinate of Figure 1 may be put into absolute units by a knowledge of the sensitivity of the radiometer and the antenna gain. The results of such a calculation are indicated in Figure 1 by the length of the arrow labeled 1.0×10^4 ° K. It is seen that the best fit with the observations is obtained with an effective black-body temperature for the sun of 1.0×10^4 ° K.

b) On July 12, 1945, the microwave radiation from the sun was measured in a similar manner. The atmospheric absorption at that time was 0.38 db/atmosphere. The effective black-body temperature of the sun was found to be 1.1×10^4 ° K.

c) On October 19, 1945, at 10:25 P.M., E.S.T., the effective black-body temperature of the nearly full moon was measured to be 292° K. It was assumed that the moon's disk

* This paper is based on work done for the Office of Scientific Research and Development under Contract OEMsr-262 with the Massachusetts Institute of Technology.

[1] *J. Frank Inst.*, **239**, 285, 1945.

1 [2] To appear shortly in *Phys. Rev.*★

2 [3] Details of the device will appear shortly in the *Rev. Sci. Inst.*★

Reprinted, with permission, from 'Astrophysical Journal' **103**, 375–376 (1946).

was a uniform radiator subtending an angle of 32′ along a diameter. The measured atmospheric absorption at that time was 0.24 db/atmosphere.

The only known sources of large error in these measurements are the calibration of the radiometer and the measurement of the antenna gain. The radiometer calibration is a direct comparison between the radiation received by the antenna and that emitted by a black body at a known temperature. The antenna gain was measured several times

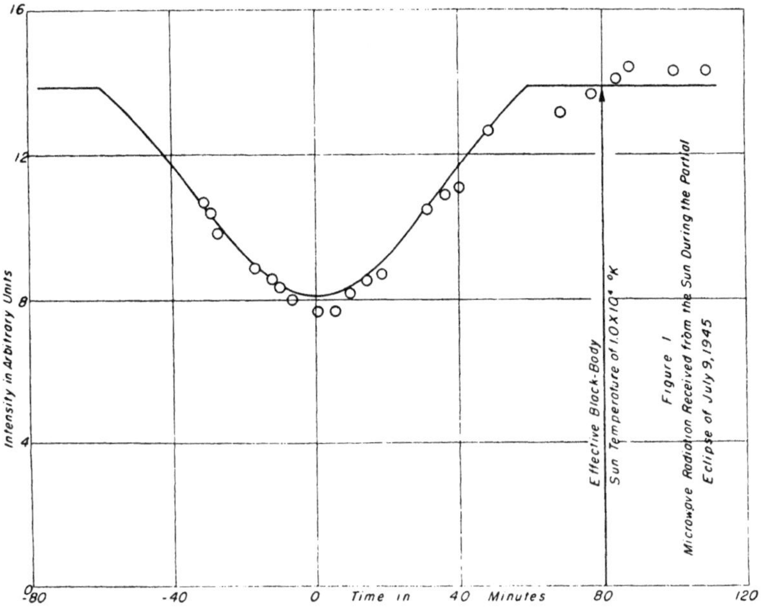

FIG. 1

by two different methods[4] and is believed to be reliable. The probable error in the effective black-body temperatures of the sun and moon is believed to be about ± 10 per cent.

ROBERT H. DICKE
ROBERT BERINGER

PALMER PHYSICAL LABORATORY
PRINCETON UNIVERSITY
SLOAN PHYSICS LABORATORY
YALE UNIVERSITY
March 30, 1946

[4] These measurements were kindly made for us by members of the antenna group of this laboratory.

EDITOR'S NOTES AND AMENDMENTS

1. Dicke *et al.* (1946).
2. Paper 11.

PART IV
DISCRETE SOURCES

The discrete radio sources have been central to the development of radio astronomy from the initial discovery of the phenomenon in 1946 to the present day. Table III gives a handy summary, based on modern values, of the data relevant to those sources which were especially important in the early years of radio astronomy. Even before the first "official" discovery of a discrete source, however, there were strong hints of intense concentrations of radiation at certain positions in the 160 MHz all-sky survey of Reber (Paper 6). In fact, one can now discern the presence in Reber's map of at least five of the sources listed in Table III. The first recognition of a discrete source came about, however, through the accidental discovery of rapid intensity variations in Cyg A by Hey, Parsons, and Phillips (Paper 23). Although these variations turned out to be not related to the source itself, being rather of an ionospheric origin, the discovery opened the gates to an entirely new line of investigation in the fledgling science of radio astronomy.

As new sources were discovered and ever improving positions determined, several identifications with optical objects became possible. The four strongest sources of small angular size were found to correspond to an amazing collection of visible objects: the aftermath of an exploded star (Tau A), a bright elliptical galaxy with a peculiar jet (Vir A), an apparent pair of colliding galaxies at a distance of 300 million light years (Cyg A), and a system of rapidly moving filaments in our own galaxy, unlike anything previously observed (Cas A). At this stage radio astronomy began to convince even the most wary optical astronomers of its worth. As the subject developed further, we find a convenient benchmark in Pawsey's "A Catalog of Reliably Known Discrete Sources of Cosmic Radio Waves", published in 1955 under the auspices of both the IAU and URSI. This catalog consisted of sources divided into three lists, ranked according to the reliability of the radio positions and fluxes as well as the proposed optical identifications. Even after eight years of observations, the first and most reliable list contained only eight sources, while the other lists added only 31 sources to the grand total! Nevertheless, the number of identified sources and those with measured

TABLE III
Radio sources important in the early history of radio astronomy

Year recognized	Radio source	Identification†	α_{1950}	δ_{1950}	~Maximum angular size	Flux (Jy*) 178 MHz	Flux (Jy*) 1400 MHz
1946	Cyg A	G	$19^h\ 57^m\ 44^s$	+40° 36′	2′	8700	1600
1947	Tau A (Crab Nebula)	SNR	05 31 30	+21 59	5′	1500	930
1947	Vir A (M87)	G	12 28 18	+12 40	5′	1000	210
1947	Cen A (NGC 5128)	G	13 22 24	−42 45	7′, 2° [a]	1000[a]	300[a]
1948	Cas A	SNR	23 21 11	+58 33	5′	11000[b]	2100[b]
1949	Pup A	SNR	08 20 30	−42 50	1°	350	120
1950	Andromeda Nebula (M31)	G	00 40 00	+41 00	2°	200	70
1950	Cyg X	H II	20 30	+41	10° [c]	~10000	~10000
1950	Sgr A (galactic center)	–	17 42 31	−28 59	3′ [d]	100	200
1953	Ori A (Orion Nebula)	H II	05 32 50	−05 25	10′	100	400
1954	Vela X	SNR	08 34	−45	4°	3000	1600

† G = galaxy; SNR = supernova remnant; H II = region of ionized hydrogen.

* 1 Jy = Jansky = 10^{-26} W m^{-2} Hz^{-1}.

a Consists of inner and outer double sources and an extended region of ~10° size; fluxes given refer only to the outer pair.

b Epoch 1950.0; flux decreases ~1% per year.

c Boundaries rather arbitrary.

d Refers only to Sgr A proper; situated in the center of a large region of intense emission.

angular sizes methodically grew, and it eventually became clear that the term "radio star" for these strong sources was a misnomer and should be abandoned.

The significance for cosmology of the identification of Cyg A with a distant faint galaxy was its implication that the bulk of the radio sources might well be at distances which would allow tests of various world models. Through various arguments, and with substantial new evidence, Ryle changed his opinion and eventually demonstrated (1955, 1958) that, as a whole, radio sources were indeed at such great distances that cosmological tests with a larger sample would be fruitful. The controversy which ensued is one of the more fascinating and significant in the history of radio astronomy, but there is not space to pursue it here (see Edge and Mulkay (1976) for a detailed account).

In the quest for an ever larger number of radio sources, gargantuan surveys ("2C" and "MSH") of the entire sky were undertaken in the mid 1950s by the Cambridge and Sydney groups (Shakeshaft *et al.*, 1955; Ryle and Scheuer, 1955; Mills and Slee, 1957). The two surveys were immediately found to be almost totally in disagreement. For almost a decade this led to much confusion and many bitter quarrels over the validity of the differing radio techniques and over the implications for cosmology of the results. Yet what appears transparent now was often a subtle effect then. The true contribution of these investigations and of the ensuing "log N - log S" controversy is not in a list of radio sources, but in the new cosmological vistas which were opened. Still today, observations of radio sources are severely testing our fundamental ideas of how both physics and the Universe work.

SECTION L

DISCOVERY AND FIRST OPTICAL IDENTIFICATIONS

The discovery of the first discrete radio source, like so many other major discoveries, was an accident. While conducting a survey of galactic radiation (Hey, Phillips and Parsons, 1946, 1948), Hey and his co-workers noticed that the radiation from one sector of the sky seemed to vary on a time scale of seconds in an erratic fashion. Further investigation showed this to be a localized *source* of radiation at a fixed position on the sky. In Paper 23, the investigators in fact quite accurately measure the position of what we now know as Cyg A (cf. Table III) and establish that the "source of disturbance" is $\leqslant 2^\circ$ in diameter. Besides the discovery itself, this paper is notable for its deduction that the fast time variations imply that their origin lies in "a small number of discrete sources" (presumably of a size much less than 2°). We know today that the authors' analogy with solar bursts is incorrect, but the inference of small angular size is sound. These fluctuations of Cyg A actually arise in ionospheric irregularities rather than in any mechanism intrinsic to the source itself, although it took several years firmly to establish this result (Stanley and Slee, 1950; Smith, 1950; Little and Lovell, 1950).

It was over a year before further work was published on the nature of this enigmatic discrete source. Paper 24, by J. G. Bolton and G. J. Stanley, raises a question which was to be critical in the ensuing study of radio sources: "If this Cygnus source is so intense and yet of such a small size at radio wavelengths, why do we see no striking optical analogue on photographs?" Using a sea-cliff interferometer, Bolton and Stanley determine the size to be less than 8 arcmin in extent. In a longer companion paper (1948b) describing these results in detail, they published the first overlay of a radio source position on a photograph, and concluded that "the radio noise received from this region is out of all proportion to the optical radiation". Their quest for an optical identification was doomed to failure for two reasons: (1) their measured position was in error by over one degree in declination (cf. Table III; the error was due to a faulty refraction correction), and (2) even if the position had been accurate, Cyg A turns out to be associated with an inconspicuous 17^m galaxy which most probably would have been missed. It would, in fact, later require the combined power of a radio position accurate to ~ 1 arcmin and the 200 inch telescope finally to identify this source in 1951 (Papers

28 ana 29). Part of this delay was caused by the confusing fact that Cyg A is by chance situated near the galactic equator, as well as near a large complex of galactic HII regions called Cyg X (see Table III).

Within a short time, Bolton (1948) compiled a catalogue of six intense radio sources visible from Australia. Unfortunately, almost all of the positions were in error by 5° to 10°, once again primarily due to errors in the correction for refraction. The paper's main legacy is in its introduction of the nomenclature Cyg A, Tau A, Her A, Cen A, etc. The source Virgo A, however, was shifted by a sufficient amount to place it in a neighboring constellation where it became temporarily known as Coma Berenices A!

The intense competitive atmosphere between the Australian and English groups is indicated in Paper 25 where Ryle and F. G. Smith use the word *teams* when referring to the various scientific groups; nevertheless, correspondence between the Hemispheres about latest results and technical details was frequent at this time. The Cambridge workers use a Michelson interferometer to establish that the radiation from Cyg A is not significantly polarized and to study further the nature of the intensity fluctuations; because of uncertainty over whether or not its position is constant, no position determination for Cyg A is given. More importantly, they discover the strongest radio source in the sky, Cas A. The power of Cas A becomes evident when one realizes that the source which they describe as a "less intense but more variable source in Ursa Major" was, in fact, a spurious result due to the passage of circumpolar Cas A through the back-lobes of the interferometer at lower culmination. Finally, the fact that the intensity fluctuations of Cas A were measured to be much smaller than those of Cyg A was an unfortunate red herring which strengthened the erroneous belief that the variations were intrinsic to the sources.

Paper 26, by Bolton, Stanley and O. B. Slee, continues the work on discrete sources at the CSIRO Radiophysics Laboratory. The erroneous positions given in Bolton (1948) have been repaired and the $\sim 10'$ error bars for the new positions are now sufficient to suggest three possible optical identifications. The improved positions arose from a painstaking series of observations from cliffs in both Australia and New Zealand; the portability of the converted military radars was indeed being used to full advantage! And the identifications, all of which were later confirmed, immediately proclaimed that the study of discrete sources would be an exciting and profitable pursuit. Tau A is identified with the Crab Nebula, only just previously established in a convincing manner as the remnant of a supernova recorded by Chinese observers in 1054 AD (Duyvendak, 1942; Mayall and Oort, 1942). The characteristics known at that time of the "central star" showed it also to be very unusual, but not until 20 years later was it discovered that this star is, in fact, a pulsar, a neutron star which blinks on and off 30 times every second. Vir A is identified with an elliptical galaxy and Cen A with a bright peculiar galaxy most notable for its band of absorption. Although the authors are here reluctant to accept these two nebulae as being truly extragalactic, it soon became clear that they were indeed at enormous distances. The implied radio luminosities were just as stunning as they were unavoidable.

Fluctuations in Cosmic Radiation at Radio-Frequencies

IN a previous publication[1] we described the results of an investigation into the spatial distribution of cosmic electromagnetic noise radiation at 5 metres wave-length. We have recently been engaged in an attempt to make a more detailed determination by using a more sensitive receiver of narrower beam-width. An interesting new feature which has emerged from these latter experiments is the occurrence of short-period irregular fluctuations which have been found to be associated with the direction of Cygnus. This region, which is a secondary peak in the cosmic noise distribution, appears to be unique in being characterized by short-period variations of marked amplitude in the intensity of power flux.

A watch on this region has been kept intermittently during the last four months. The receiving apparatus, situated in Richmond Park,
1 has an aerial beam rotatable in bearing* but fixed in elevation at an angle of 12°. The region of the fluctuations ascended and descended through the aerial beam on bearings 30° and 330° respectively. The corresponding times were 0100 hr. and 1900 hr. G.M.T. in February, when the watch was commenced, while in June they were 1800 hr. and 1200 hr. G.M.T. Care was taken to avoid including recordings taken in daylight periods when the powerful solar noise emission associated with the great sunspot in February was also present. Since the observations covered a wide range of bearings and solar times, we were able to rule out the possibilities of terrestrial or solar causes, and the interpretation of the results was consistent only with an origin in the direction of Cygnus.

It is not easy to determine the bearing of a source of irregular disturbance with a high order of accuracy unless an exceptionally narrow beam is used. The aerial of the equipment has a beam width of approximately ± 6° to half power in bearing and elevation, and the average of a large number of observations indicated a source of disturbance subtending an angle not exceeding 2°. There may be other areas
2 of occasional fluctuation in the immediate vicinity (within 8°).*

The average amplitude of the fluctuation is 15 per cent of the mean power received. If the disturbed area be assumed to extend over a circle of angular diameter 2°, then this solid angle is 1/36 of that for the equivalent acceptance cone of the aerial beam. The variations in power per unit solid angle therefore correspond to more than five times the mean power per unit solid angle for the whole beam. The centre of the region is approximately R.A. 2000 hr., Decl. + 43°. The type of fluctuation, which itself varies from day to day, is illustrated in the accompanying figure. The noise from Sagittarius would, by comparison, appear as a straight line on a diagram of this scale.

It appears probable that such marked variations could only originate from a small number of discrete sources. This suggests at once the analogy with the radio-frequency sunspot radiation[2,3,4]. The solar radio noise from sunspots is also characterized by strong fluctuations. A recording of these solar radiations, taken on July 2, is also shown

Reprinted, with permission, from 'Nature' 158, 234 (1946).

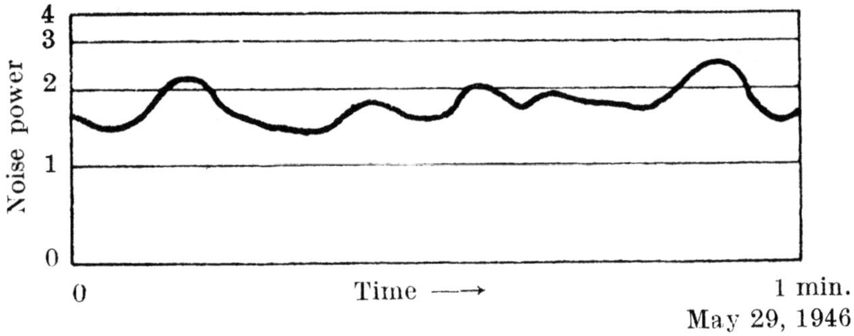

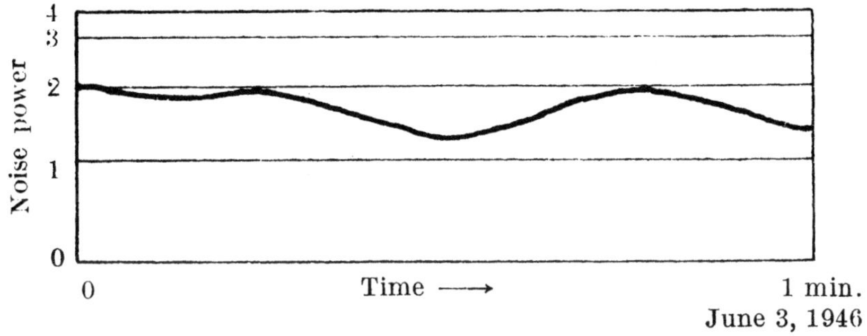

Fluctuations of radio noise from the direction of Cygnus

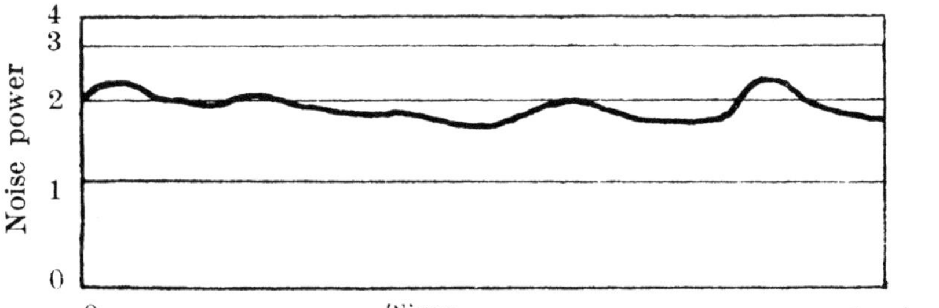

Fluctuations of solar radio noise July 2, 1946

in the figure. On the other hand, Greenstein, Henyey and Keenan[5] have recently pointed out the difficulties in attempting to account for the magnitude of cosmic radiations in terms of the solar phenomena; further, they direct attention to the close agreement between experimental observations of cosmic noise intensity and their calculations of the expected interstellar radiation arising from free transitions of electrons in the field of protons. A theory in terms of widely distributed interstellar matter does not, however, appear readily to account for the localized disturbances just described. These fluctuations therefore appear of special importance in that they may prove particularly relevant to the explanation of the origin of cosmic radiations at radio frequencies.

We are indebted to the Director General of Scientific Research and Development (Defence), Ministry of Supply, for permission to publish this communication.

J. S. HEY
S. J. PARSONS
J. W. PHILLIPS

Ministry of Supply.
July 4.

[1] Hey, Parsons and Phillips, *Nature*, **157**, 296 (1946).
[2] Appleton, *Nature*, **156**, 534 (1945).
[3] Hey, *Nature*, **157**, 47 (1946).
[4] Pawsey, Payne-Scott and McCready, *Nature*, **157**, 158 (1946).
[5] Greenstein, Henyey and Keenan, *Nature*, **157**, 805 (1946).

EDITOR'S NOTES AND AMENDMENTS

1. *Bearing* means "azimuth".
2. This confusion was probably caused by the nearby Cyg X complex.

Variable Source of Radio Frequency Radiation in the Constellation of Cygnus

COSMIC or galactic noise was discovered by Jansky[1] in 1931; but its exact origin has remained uncertain. It is generally supposed to originate from collisions in interstellar matter[2]; but there are divergencies between existing theory and experimental results, particularly at lower radio frequencies[3]. Hey, Parsons and Phillips[4] discovered variations in the intensity of galactic noise from the direction of the constellation of Cygnus, with a period of about one minute—suggesting that this particular radiation has its origin in a discrete source.

During the past three months, we have made a study of this region, mainly on 100 Mc./s., but also occasionally on 60, 85 and 200 Mc./s. The technique employed was to observe the region rising over the sea with aerials situated on a high cliff, as described by Pawsey, Payne-Scott and McCready[5]. Due to interference between the direct ray and the ray reflected from the sea, a lobe pattern is obtained which gives rise to a succession of maxima and minima. An estimate of the size of the source can be made from the relative heights of maxima and minima, and an accurate position found from the times of occurrence of minima.

Small aerial arrays were used—one or two Yagis—and considerable care was taken with receiver stabilization enabling receiver noise to be balanced out and the input signal amplified for presentation on a recording milliammeter. At 100 Mc./s. receiver fluctuations represented an input signal of less than 3×10^{-24} watts $m.^{-2}(c./s.)^{-1}$.

The general results of our investigations are:

(1) The variations from Cygnus originate in an area of less than 8′ angular width. This is an upper estimate and the source may well be effectively a point.

(2) The location of the source is R.A. 19 hr. 58 min. 47 sec. ± 10 sec. Dec. +41° 41′ ± 7′.

Reprinted, with permission, from 'Nature' 161, 312–313 (1948).

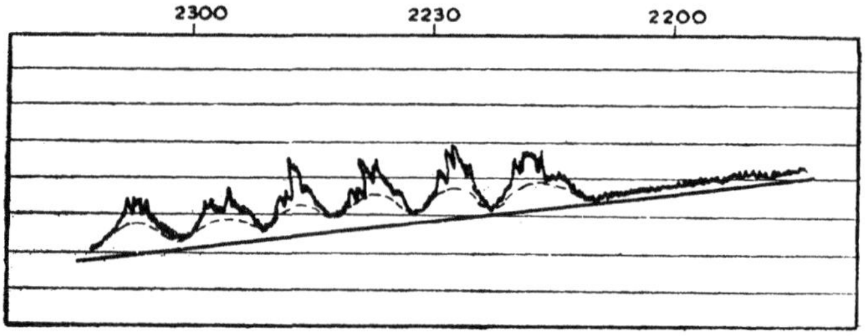

1 Fig. 1. Record of source in Cygnus rising*

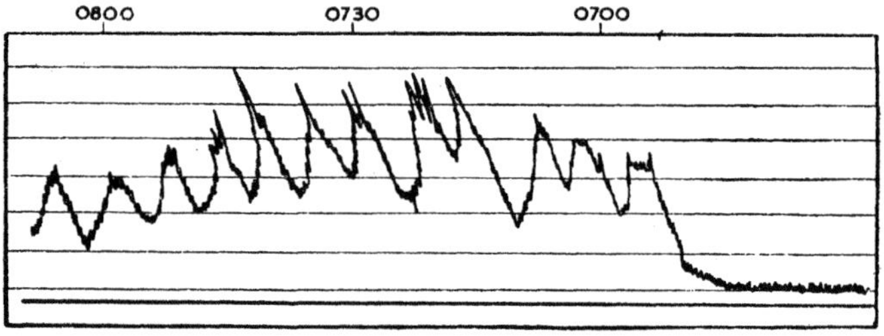

Fig. 2. Record of 'quiet' sun rising

(3) The radiation consists of two components, one constant and the other showing considerable variation over short and long periods. In Fig. 1, a typical record of the source rising, the two components are separated by a dotted line. A record of the 'quiet' sun rising is shown for comparison in Fig. 2. The frequency in both cases was 100 Mc./s. ; intensity scales are the same on both records.

(4) Observation on different frequencies shows that the intensity of the constant component has a shallow peak in the region of 100 Mc./s., whereas that of the variable component increases rapidly with decreasing frequency. No variable component has so far been detected on 200 Mc./s.

(5) The periodicity of the variations decreases with decreasing frequency. No close correlation has been found between short-period variations on frequencies as close as 85 and 100 Mc./s., but the general activity of the variable component shows good correlation on different frequencies.

The determined position does not coincide with any outstanding stellar object, but is close to two seventh magnitude stars. It is notable that Hey[6] and Moxon[7] found a secondary maximum of galactic noise in Cygnus. The present evidence indicates that the noise from this region comes from a discrete source superimposed on a smooth continuum.

Taking the size of the sources as less than 8′, the constant component represents an effective temperature of more than 4×10^6 degrees Kelvin at 100 Mc./s., which makes a thermal origin of the noise doubtful. A mechanism similar to that proposed by Martyn[8] to account for enhanced solar radiation on metre wave-lengths (association of moving ionized matter and magnetic fields) may account for the phenomenon, though further evidence is required to throw more light on the problem.

Our thanks are due to Dr. J. L. Pawsey for his help and encouragement during this work, which is part of the programme of the Radiophysics Laboratory of the Council for Scientific and Industrial Research. Full details are being published elsewhere.* 2

J. G. Bolton
G. J. Stanley

Division of Radiophysics,
Council for Scientific and Industrial Research,
Commonwealth of Australia.
Dec. 4.

[1] *Proc. Inst. Radio Eng.*, **20**, 1920 (1932).
[2] *Astrophys. J.*, **91**, 625 (1940).
[3] *Astrophys. J.*, **105**, 235 (1947).
[4] *Nature*, **158**, 234 (1946).
[5] *Nature*, **157**, 158 (1946).
[6] *Nature*, **157**, 297 (1946).
[7] *Nature*, **158**, 758 (1946).
[8] *Nature*. **159**, 26 (1947).

EDITOR'S NOTES AND AMENDMENTS

1. The slope in the baseline of this trace is caused by the drift of the galactic center region out of the direction of the primary beam. Note also the differing fringe visibilities of Cyg A in Figure 1 and the sun in Figure 2 (the fringe spacing is ~40 arcmin).
2. Bolton and Stanley (1948b).

A NEW INTENSE SOURCE OF RADIO-FREQUENCY RADIATION IN THE CONSTELLATION OF CASSIOPEIA

By M. RYLE and F. G. SMITH
Cavendish Laboratory, Cambridge

THE discovery by Hey, Parsons and Phillips[1,2] of a variable source of radio-frequency radiation in the constellation of Cygnus suggested that at least part of the galactic radiation was due to very intense discrete sources. Afterwards a detailed study of this region was made by Bolton and Stanley[3], using an interference system of greater resolving power ; their results showed that the angular diameter of the source was smaller than the resolving power of their apparatus (8 minutes of arc), and the more precise determination of its position enabled them to show that the source does not coincide with any outstanding star. In a more recent note Bolton[4] describes the extension of this work and the discovery of other discrete sources.

The similarity between the fluctuations observed from the source in Cygnus and from the disturbed sun has suggested to both teams that the mechanism for the emission of intense radiation from discrete galactic sources is likely to be the same as that of sunspot emission. The results obtained by many workers have shown that sunspot radiation is circularly polarized, and it is therefore of importance to measure the polarization of the radiation emitted by discrete galactic sources.

Recently we have carried out a series of experiments on a frequency of 80 Mc./s. to determine the polarization of the radiation from the source in Cygnus. During the course of this work we have found a new and more intense source situated in the constellation of Cassiopeia, as well as several minor sources having a smaller intensity than Cygnus.

Reprinted, with permission, from 'Nature' 162, 462–463 (1948).

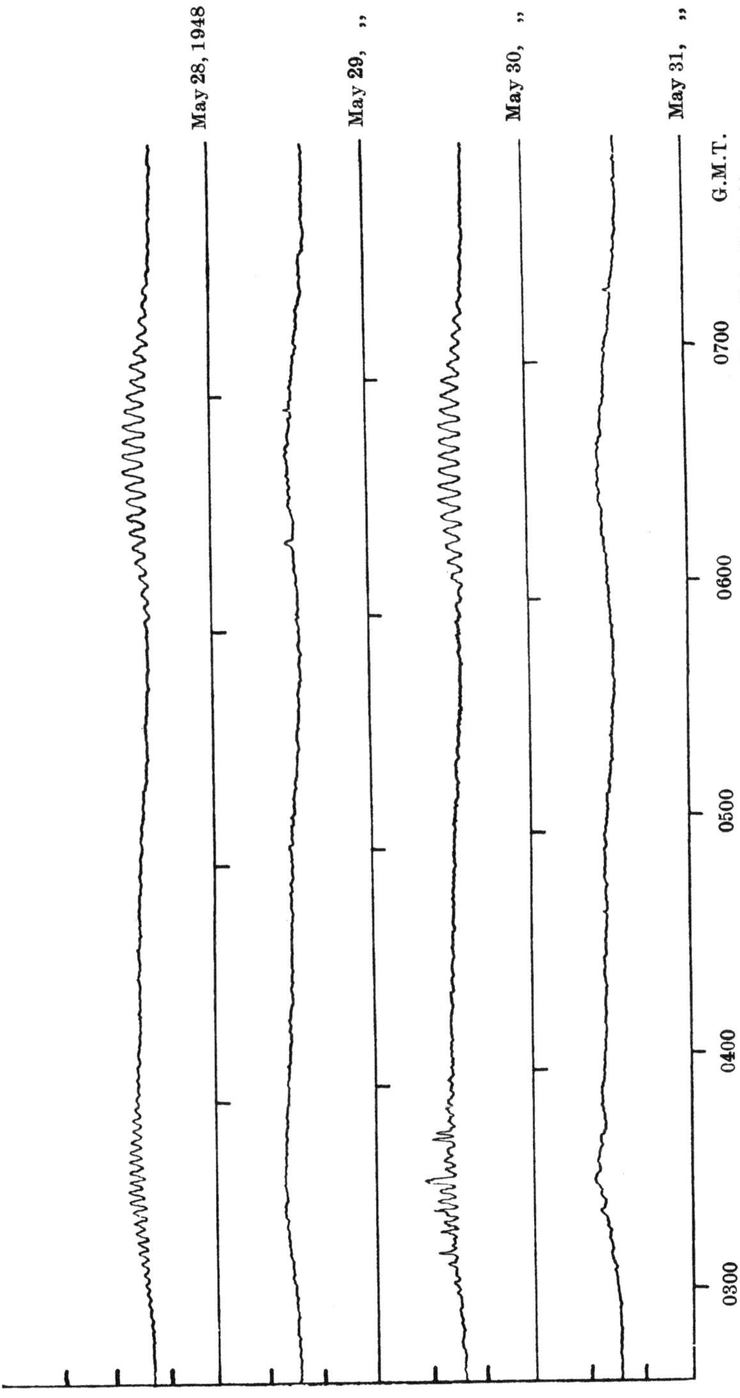

Records of galactic radiation using the spaced aerial system (units of intensity are 5×10^{-23} watts/m.^{2} (c./s.)). May 28, aerials parallel; (2) May 29, aerials mutually perpendicular; (3) May 30, aerials parallel; (4) May 31, aerials mutually perpendicular

In our experiments we have used the radio version of Michelson's stellar interferometer which we have previously described[5,6]. When making observations with a large resolving power, this type of interferometer has advantages over the method used by Bolton and Stanley, since no corrections are necessary for refraction in the troposphere or ionosphere. In order to identify the central maximum on the record we have limited the total width of the interference pattern by using two aerials of considerable directivity. By noting the times of the adjacent minima it is then possible to determine the Right Ascension of the source.

It can be shown that the periodicity of the interference pattern produced on the record by such a system depends on the declination (δ) of the source, and is given by $t_0 \sec \delta$, where t_0 represents the period of the central interference pattern for a source in the equatorial plane. It may therefore be seen that by measuring the period of the interference pattern the declination of the source may be found from the same record, without the need for an aerial system having a large resolving power in a north-south plane.

If the two halves of the aerial system are now mounted with mutually perpendicular planes of polarization, the system may be used to determine the polarization of the radiation from a galactic source. For a wave which is circularly polarized, the amplitude of the interference pattern produced will be the same as with the aerials parallel, but the phase of the pattern will be displaced by a quarter period. If the radiation is only partially polarized, the amplitude of the interference pattern will be reduced, and by comparing the amplitudes obtained with the two aerials parallel and perpendicular to one another, it is possible to determine the degree of polarization. (The full analysis of a complex wave is possible by this method, as has already been mentioned[6].)

We have used an aerial spacing of about 0·5 km. at a frequency of 80 Mc./s. An accuracy in Right Ascension of $\pm$ 10 sec. should be possible with this system for sources as intense as that in Cygnus, although our accuracy has been limited by the accuracy of time calibration to $\pm$ 20 sec. The determination of declination (δ) becomes inaccurate for small values of δ, but for δ greater than 30° an accuracy of $\pm$ 20′ becomes possible.

Observations have been made between declinations of 0° and + 75°. In addition to the source in Cygnus, a more intense source was located in Cassiopeia, the radiation from which has been notable for showing considerably smaller fluctuations than that from the source in Cygnus.

Some typical records are shown in the accompanying figure (curves 1 and 3), in which the interference patterns produced by the sources in Cygnus (0335) and Cassiopeia (0640) can be seen.

The source in Cassiopeia has been found from these records to have a Right Ascension of 23 hr. 17 min. 50 sec. and a declination of 58° 0′ (1948 co-ordinates). The total intensity (twice the value obtained with aerials which accept only one component of polarization) has been found to be 23×10^{-23} watts/m.^{2}/(c./s.), with 'bursts' of radiation having an intensity of up to 6×10^{-23} watts/m.^{2}/(c./s.). The intensity of the source in Cygnus has been found to be 14×10^{-23} watts/m.^{2}/(c./s.) with 'bursts' of intensity up to 20×10^{-23} watts/m.^{2}/(c./s.). The aerial system has not yet been fully calibrated, but although the absolute values of intensity may be in error by as much as $\pm$ 30 per cent, the relative values are reliable.

A less intense but more variable source in Ursa Major (R.A. 12 hr. 20 min., declination 58°) has been found to give a maximum intensity of 5×10^{-23}
watts/m.^{2}/(c./s.).* 1

The angular diameters of all the sources which have so far been observed are less than the resolving power of the apparatus (about 6 minutes of arc for the two intense sources). The observation of 'bursts' of duration less than 20 sec. coming from the source in Cygnus has shown that the linear dimensions of the source of the variable component cannot greatly exceed the distance travelled by electromagnetic waves in 20 sec. If a source of this diameter (about 10^7 km.) is situated at the distance of the nearest fixed star, the observed intensity would correspond to radiation from a black-body source at a temperature greater than 10^{14} °K.

Observations made with the two halves of the aerial system mutually perpendicular have given records in which the amplitude of the interference pattern is less than 5 per cent of that with the aerials parallel. Some typical records are shown in curves 2 and 4. The results show that, within the limits of experimental error, the steady component of the radiation from both sources is randomly polarized. The determination of the polarization of the variable component of the radiation (see curve 4) is difficult, since the 'bursts' last for a time comparable with the period of the interference pattern, but a study of such records suggests that this component is also randomly polarized.

The detection of intense radiation from discrete galactic bodies has suggested that the radiation is due to processes similar to those occurring in sunspots. The observation that the radiation is not circularly polarized, however, suggests that the mechanism is somewhat different, and may not involve an appreciable magnetic field in the emitting region. The mechanism may possibly be more nearly analogous to the radiation from the undisturbed sun.

[1] Hey, Phillips and Parsons, *Nature*, **158**, 234 (1946).
[2] Hey, Parsons and Phillips, *Proc. Roy. Soc.*, A, **192**, 425 (1948).
[3] Bolton and Stanley, *Nature*, **161**, 312 (1948).
[4] Bolton, *Nature*, **162**, 141 (1948).
[5] Ryle and Vonberg, *Nature*, **158**, 339 (1946).
[6] Ryle and Vonberg, *Proc. Roy. Soc.*, A, **193**, 98 (1948).

EDITOR'S NOTES AND AMENDMENTS

1. As noted in the introduction, this "source" was actually the result of circumpolar Cas A passing daily through the *back* lobes of the interferometer. The declination, as determined from fringe periodicity, thus was reported to be the same as that of Cas A; the right ascension was not found to be exactly 12 hours different from Cas A, however, since the interferometer axis was somewhat different from east-west.

Positions of Three Discrete Sources of Galactic Radio-Frequency Radiation

In a recent communication[1] an account was given of the discovery of a number of discrete sources of galactic radio-frequency radiation. Accurate measurements of the position of three of these sources have since been made from sites on the east and west coasts of New Zealand and on the east coast of Australia. The technique employed was to observe the sources at rising or setting, with an aerial on a high cliff overlooking the sea. These observations, when corrected for atmospheric refraction, allow the path of a source above the horizon to be plotted, and the time of its rising and setting—and hence its celestial co-ordinates—to be determined.

It is found that all three sources correspond within limits of experimental error to positions of certain nebulous objects. The positions of the sources determined and details of these objects are given in the accompanying table.

The position of the source Taurus A agrees, within the limits of experimental error, with that of the Crab nebula. This is believed to be the expanding shell of the supernova of A.D. 1054. The present dimensions of the nebula are $4' \times 6'$, and photographs taken with different wave-length ranges show considerable fine structure, including $H\alpha$ filaments. The emission from the nebulosity is some 600 times that of the central star identified by Baade[2]. Minkowski[3] considers that most of the radiation from the central star is in the far ultra-violet and estimates the surface temperature as 500,000° K. He suggests a temperature of 50,000° K. for the nebulosity and explains a low ratio of $H\alpha$ to [N II] emission as being due to a deficiency of hydrogen. It should be pointed out, however, that a much higher temperature would have a similar effect on this ratio and at the same time increase the electron density. Intensity measurements we have made at 100 Mc./s., assuming a source size of 5′, give an apparent temperature of 2,000,000° K. for Taurus A. The present estimates of density and temperature in the Crab nebula would fall well

Reprinted, with permission, from 'Nature' **164**, 101–102 (1949).

Source	Position (Epoch 1948)		Possible associated visible object		
	Right ascension	Declination	Object	Spectrum	Remarks
Taurus A	$5^h\ 31^m\ 00^s \pm 30^s$	$+22°\ 01' \pm 7'$	N.G.C. 1952 (Messier 1)	Continuous. Weak emission lines of H, He, forbidden lines of N, O and Si	The Crab nebula, expanding shell of an old super-nova
Virgo A	$12^h\ 28^m\ 06^s \pm 37^s$	$+12°\ 41' \pm 10'$	N.G.C. 4486 (Messier 87)	Continuous	Spherical nebula—unresolved
Centaurus A	$13^h\ 22^m\ 20^s \pm 60^s$	$-42°\ 37' \pm 8'$	N.G.C. 5128	Continuous. Weak emission lines, H_β, H_γ, H_δ, and $\lambda\,4686$	Unresolved nebula crossed by a marked obscuring band

short of explaining this result by strictly thermal processes. Non-thermal components resulting from the expansion of the nebula do not, however, seem unlikely.

Both N.G.C. 5128 and N.G.C. 4486 are generally classed as extra-galactic nebulæ, though Paraskevoulos[4] considers that the position of the obscuring band splitting the disk of N.G.C. 5128 into halves is unusual. Such obscuring bands are normally seen in nebulæ viewed on edge. Neither of these objects has been resolved into stars, so there is little definite evidence to decide whether they are true extra-galactic nebulæ or diffuse nebulosities within our own galaxy. If the identification of these objects with the discrete sources of radio-frequency energy can be accepted, it would tend to favour the latter alternative, for the possibility of an unusual object in our own galaxy seems greater than a large accumulation of such objects at a great distance.

Full details of the technique and observations involved in the determination of the positions of these three discrete sources will be published elsewhere[5].

This work is part of the research programme of the Division of Radiophysics, Commonwealth Scientific and Industrial Research Organisation, Australia.

J. G. BOLTON
G. J. STANLEY
O. B. SLEE

Radiophysics Laboratory,
Commonwealth Scientific and Industrial
Research Organisation,
Australia. May 2.

[1] Bolton, J. G., *Nature*, **162**, 141 (1948).
[2] Baade, W., *Astrophys. J.*, **96**, 188 (1942).
[3] Minkowski, R., *Astrophys. J.*, **96**, 199 (1942).
[4] Paraskevoulos, J. S., *Harvard Bull.* 890, 1 (1935).
[5] Bolton, J. G., and Stanley, G. J., *Aust. J. Sci. Res.*, A (in the press).* 1

EDITOR'S NOTES AND AMENDMENTS

1. 2, 139 (1949); 3, 234 (1950).

SECTION M

SYNCHROTRON RADIATION

Paper 27 by H. Alfvén and N. Herlofson is important as the first instance in which synchrotron radiation, the radiation emitted by relativistic electrons spiraling in a magnetic field, is suggested as the possible mechanism for extraterrestrial radio emission. The theory is applied to a model for discrete radio "stars" which we now know to be incorrect,[1] but which nevertheless was important because it suggested that the emitting regions could well be much larger than the usually assumed stellar dimensions. Alfvén's model for the region around a radio star very naturally grew out of his similar picture for the nature of our own solar system where he considered cosmic rays to originate in the Sun and then to be made isotropic by an interplanetary magnetic field.

Later work, in particular on the Crab Nebula (Tau A), was to lead to the general acceptance of the mechnism for a wide variety of radio sources. Shklovsky (1953c) first suggested that the *optical* radiation from the Crab might also be synchrotron and Vashakidze (1954) and Dombrovsky (1954) verified that the light was indeed polarized. On the other hand, detection of the polarization of the *radio* emission was an extremely difficult technical problem, not solved until the 3 cm observations of Mayer, McCullough, and Sloanaker in 1957.

The synchrotron mechanism has withstood observational tests over the ensuing years and become one of the most efficient tools for theorists. Its use has by now in one way or another infiltrated models for an amazing panoply of astronomical objects, ranging from the Sun to quasars.

NOTE

[1] See, for example, Ginzburg in Paper 10 for some early modifications. Hutchinson (1952) also worked at an early stage on the possibility of synchrotron emission from discrete radio sources.

Cosmic Radiation and Radio Stars* 1

H. Alfvén and N. Herlofson
Royal Institute of Technology, Stockholm, Sweden
April 17, 1950

THE normal radio wave emission from the sun amounts to 10^{-17} of the heat radiation, and increases during bursts[1] to as much as 10^{-13}. If a radio star, e.g., the source in Cygnus, is situated at a distance of 100 light years, its radio emission is of the order of 10^{-4} of the heat radiation of our sun. It is very unlikely that the atmosphere of any star could be so different from the sun's atmosphere as to allow a radio emission which is 10^9 to 10^{13} times greater, and it seems therefore to be excluded that the source could be as small as a star. The recent discovery[2] that the intensity variations of radio stars is a "twinkling" makes it possible to assume larger dimensions.

Ryle has suggested that there should be a connection between radio stars and cosmic radiation.[3]* 2

According to a recent development of Teller and Richtmyer's theory of cosmic radiation, the sun should be surrounded by a "trapping field" of the order 10^{-6} to 10^{-5} gauss, which confines the cosmic rays to a region with dimensions of about 10^{17} cm (0.1 light year).[4]*It is likely that almost every star has a cosmic radiation of its own, trapped in a region of similar size. We suggest that the radio star emission is produced by cosmic-ray electrons in the trapping field of a star.

Electrons with an energy $W \gg m_0c^2$ moving in a magnetic field H radiate at a rate

$$-dW/dt = (2e^2/3c)\omega_0^2\alpha^2, \tag{1}$$

where[5] $\omega_0 = eH/m_0c$ is the gyro-frequency corresponding to the rest mass m_0, and $\alpha = W/m_0c^2$. Most of the energy is emitted with a frequency of the order

$$\nu = \omega_0\alpha^2/2\pi = 2.8 \times 10^6 H\alpha^2 \text{ sec.}^{-1}. \tag{2}$$

As soon as the energy is much higher than the rest energy the emitted frequency becomes much higher than the gyro-frequency, a phenomenon which is observed in large synchrotrons, where the electron beam emits visual light.[6]

According to (2) an emission of radio waves of 100 Mc/sec. requires

$$H\alpha^2 = 36 \text{ gauss}. \tag{3}$$

Reprinted, with permission, from 'Physical Review' 78, 616 (1950).

The acceleration process of cosmic radiation should accelerate electrons as well as positive particles. In the solar environment the electron component is eliminated by Compton collisions with solar light quanta as discussed by Feenberg and Primakoff.[7] In the neighborhood of a star which does not emit much light, the electrons would be accelerated until their energy is so high that they radiate. The wave-length falls in the meter band if, for example, $\alpha=300$ ($W=1.5\times10^8$ ev) and $H=3\times10^{-4}$ gauss. This field is about 100 times the estimated strength of the sun's trapping field. As the strength is determined by the "interstellar wind," a radio star should be situated in an interstellar cloud moving rather rapidly relative to the star.

In order to account for the total energy emitted by a radio star we must suppose either that the radio emission is a transitory phenomenon, lasting a time which is short compared to the lifetime of cosmic rays in the trapping field (10^8 years), or that the cosmic-ray acceleration close to the star is supplemented by a Fermi process[8] further out in the trapping field.

According to the views presented here, a radio star must not emit very much light, and should be situated in an interstellar cloud. This would explain why it is so difficult to find astronomical objects associable with the radio stars.

[1] T. S. Hey,* M. N. R. A. S. **109**, 179 (1949). 3
[2] Smith, Little, and Lovell, Nature **165**, 424 (1950).
[3] M. Ryle, Proc. Phys. Soc. London **A62**, 491 (1949).
[4] Alfvén, Richtmyer, and Teller, Phys. Rev. **75**, 892 (1949); R. D. Richtmyer and E. Teller, Phys. Rev. **75**, 1729 (1949); H. Alfvén, Phys. Rev. **75**, 1732 (1949) and **77**, 375 (1950).
[5] D. Iwanenko and I. Pomeranchuk, Phys. Rev. **65**, 343 (1944); J. Schwinger, Phys. Rev. **75**, 1912 (1949).
[6] Elder, Langmuir, and Pollock, Phys. Rev. **74**, 52 (1948).
[7] E. Feenberg and H. Primakoff, Phys. Rev. **73**, 449 (1948).
[8] E. Fermi, Phys. Rev. **75**, 1169 (1949).

EDITOR'S NOTES AND AMENDMENTS

1. *Cosmic radiation* means "cosmic rays".
2. Ryle hypothesized that the origin of cosmic rays might be explained by the existence of radio stars having electric fields of order 1 volt per cm.
3. Should be "J. S. Hey".

OPTICAL IDENTIFICATION OF CAS A AND CYG A

The first substantial catalogs of radio sources were published in 1950 by Stanley and Slee (22 sources in the southern sky) and Ryle, Smith, and Elsmore (50 northern sources in the "1C" survey). The importance of these surveys was chiefly in establishing the isotropic distribution of the "radio stars"; unfortunately, they are too lengthy for inclusion in this selection. Other notable contributions of Stanley and Slee, besides the catalog itself, were the first good measurements of spectra for radio sources (Cyg A, Tau A, Vir A, and Cen A) and a study of the intensity fluctuations of Cyg A.

The greatest disappointment and frustration for these cataloguers undoubtedly lay in the almost complete lack of reliable optical identifications resulting from the accumulated list of sources. Ryle *et al.*, with typical positional errors of 1° to 3°, correctly inferred indications of radio emission from the Andromeda Nebula (M31) and the Whirlpool Nebula (M51), but incorrectly identified sources with the additional spiral galaxies M33 and M101. After nevertheless demonstrating that such normal nearby galaxies could not possibly account for all of the catalogued sources, they concluded that the isotropic distribution of sources must represent either a class of object quite close to the sun or at extremely large distances. Furthermore, they strongly supported the former alternative since it avoided the "theoretical difficulties encountered in attempting to explain the very great intensities of the more intense radio stars". Their final conclusion was then "that the radio star represents a hitherto unobserved type of stellar body, distributed widely throughout the galaxy, and one which is equally numerous in other spiral nebulae".

By reducing their errors down to $\sim 0\overset{\circ}{.}5$ with a 218 ft fixed paraboloid, Hanbury Brown and Hazard (1950, 1951) confirmed the existence of radiation from the Andromeda Nebula and showed it to arise from a region of similar angular extent to the optical image of the galaxy. In accordance with the Cambridge observers, they concluded that such normal galaxies could not account for the brightness of the extended background radiation, which at that time was generally taken as arising from the integrated effect of large numbers of unresolved radio stars.

Other experiments designed to unravel the mystery of the radio stars led only to negative results. For instance, there were projects to measure parallaxes and proper motions for the strongest sources (Smith, 1951b), and to detect long-period intensity variations in approximately one hundred sources (Ryle and Elsmore, 1951).

It was becoming clear that further optical identifications of the discrete sources were critical to any further progress in the field. Paper 28 presents the results of F. G. Smith's all-out direct attack on the primary weakness of the new science of radio astronomy, the very limited accuracy of determinations of absolute position (also see Smith (1952b)). The goal of course was to measure extremely accurate positions so as to reduce significantly the size of the "error box" inside which optical astronomers were obliged to search. For the four strongest sources from the 1C catalog, Smith reduces the areas of the error boxes by factors ranging from 10 to 60.[1] If anyone still held doubts at that time, the new positions positively confirm the earlier identifications by Bolton *et al.* (Paper 26) of Tau A and Vir A. More importantly, the error boxes for Cyg A and Cas A were now deemed small enough that precious time on the Mt Palomar 200 inch telescope could be brought to bear on the problem. Within a couple months after Smith's communication of his new positions, Walter Baade (1893–1960) and Rudolph Minkowski (1895–1976) in 1951 were finally unambiguously able to identify, several years after their discovery, the two strongest radio sources in the sky.[2]

Paper 29 recounts in great detail this photographic and spectroscopic evidence concerning the identifications of Cas A, Pup A, and Cyg A (see Table III). Although the basic results of this work were generally known and widely quoted well before the appearance of this article, Baade and Minkowski chose to bide their time and proceeded very cautiously in analyzing their observations over a period of almost two years before publishing. Paper 29 was, in fact, only a part of a major effort attempting to understand the nature of the sample of ~15 identified radio sources available at that time (Baade and Minkowski, 1954b; Minkowski and Aller, 1954; Minkowski and Greenstein, 1954).

Cas A is found to consist of a network of filaments, many exhibiting broad emission lines indicating extremely high velocities of order $\leqslant 2000$ km s^{-1}. The apparent lack of pattern in the velocities and positions of the filaments, however, leads the authors to believe that there exists no systematic expansion and, therefore, to reject the idea of a supernova remnant. This would only be corrected after several more years' worth of plates allowed the radial expansion of the nebula to be deduced from the proper motions of the filaments. It must also be remembered that the available sample of optical supernova remnants for comparison was extremely limited (primarily the Crab Nebula and Kepler's Remnant). The second source studied, Pup A, is also found to belong to the same "new type of galactic nebulosity" as Cas A.[3]

But the source with the most far-reaching implications is Cyg A, for it turns out to be a 17th magnitude galaxy whose radio power exceeds that at optical wavelengths. The class of objects known as *radio galaxies* is born and the notion of the radio star must needs fade away. It is clear to Baade and Minkowski that the observed isotropic distribution of radio sources represents a probe into regions of the Universe beyond the reach of even the largest optical telescopes. And while the idea of colliding galaxies as the cause of the radio emission is no longer in favor, it remains true that, thirty years later, we still lack an adequate theory for the origin of the observed energy. In fact, the situation is yet worse today, since for the past 20 years we have also not been able to satisfactorily account for the even more exceptional luminosity of the closely-related class of objects known as quasars.

NOTES

[1] Shortly after Smith's work, Mills (1952c) obtained corroborating results of a similar accuracy for the positions of the strongest sources visible from Australia.

[2] As will be seen in the notes to Paper 29, D. W. Dewhirst (1951) was also successful in provisionally identifying both Cas A and Cyg A, but the limitations of his telescope (a 36 inch reflector) and location (Cambridge, England) prevented detailed work. The identification for Cyg A was also suspected by Mills and Thomas (1951), but their radio position was not of sufficient accuracy for a definite determination.

[3] Despite these conclusions, Hanbury Brown (1954) and Shklovsky (1954) argued at an early stage that a significant fraction of the extended sources in the galactic plane were indeed associated with the remnants of supernovae.

An Accurate Determination of the Positions of Four Radio Stars

THE use of an interferometer aerial system on a wave-length of 3·7 m. to measure the positions of a number of radio stars has already been described[1]. The present communication contains the results of recent more accurate determinations of the positions of four intense radio stars with this interferometer, together with results for the same stars obtained by using another interferometer the individual aerials of which employ parabolic reflectors. This latter interferometer has been used at wave-lengths of 3·7 m. and 1·4 m.

With the new interferometer, it is possible to determine the 'collimation error' either by the interchange of the individual interferometer aerial elements, or by observations of both upper and lower culmination of the circumpolar star 23·01 [1]. In addition, the electrical centres of the aerials are defined more precisely than those of the broadside arrays originally used at 3·7 m.

Declination has been determined from the periodicity of the record, as in the previous observations, and also by two new independent methods. In the first method the times of transit of the star across two 'collimation planes' displaced on either side of the axial plane of the interferometer are determined : the interval between these times is related to the declination. These displaced collimation planes are produced by increasing the electrical path to one or other of the aerials. In the second method, the time-interval between the transits at the upper and lower culminations of a circumpolar star provides a measure of the declination, since the axis of the interferometer is inclined at a considerable angle to the east–west line. The collimation error must here be determined by interchanging the interferometer aerials. A more detailed account of these and other methods of
1 position-finding will be published elsewhere.*

Table 1 gives the positions of the four stars, with their probable errors, as determined from the time of transit and the periodicity of the record obtained

Reprinted, with permission, from 'Nature' 168, 555 (1951).

Table 1

Star	Constellation	Wave-length and method	Right Ascension	Declination
			h. m. s.	
05·01	Taurus	3·7 m., broadside	05 31 35 ±5	22° 05′ ±40′
		3·7 m., paraboloid	05 31 35 ±5	22° 11′ ±20′
		1·4 m., paraboloid	05 31 34·5±2	22° 09′ ±5′
		1·4 m., displaced collimation plane	— — —	22° 01′ ±3′
12·01	Virgo	3·7 m., broadside	12 28 17 ±5	12° 50′ ±1°
		3·7 m., paraboloid	12 28 18 ±5	12° 46′ ±20′
		1·4 m., paraboloid	12 28 18·8±2	12° 33′ ±15′
		1·4 m., displaced collimation plane	— — —	12° 31′ ±15′
19·01	Cygnus	3·7 m., broadside	19 57 45 ±5	40° 37′ ±5′
		3·7 m., paraboloid	19 57 45 ±5	40° 32′ ±5′
		1·4 m., paraboloid	19 57 45·3±1	40° 34·5 ±1′
		1·4 m., displaced collimation plane	— — —	40° 35·3′±0·7′
23·01	Cassiopeia	3·7 m., broadside	23 21 13 ±5	58° 30′ ±3′
		3·7 m., paraboloid	23 21 12 ±3	58° 32′ ±2′
		1·4 m., paraboloid	23 21 12·0±1	58° 32·1′±0·7′
		1·4 m., displaced collimation plane	— — —	58° 32·9′±0·5′
		1·4 m., double transit	— — —	58° 30′ ±3′

with the three interferometer systems. It also shows results obtained by the two new methods, which have been used at 1·4 m. Allowance has been made for atmospheric refraction. The co-ordinates refer to the epoch of 1950.

It has been pointed out[1] that, in addition to the instrumental errors, the position may be in error due to the radiation from other nearby weaker stars. It can be shown that, in the present observations, this factor is unlikely to cause an error greater than ¼ sec. in the R.A. of the stars 19·01 and 23·01, or greater than about 2 sec. in the R.A. of the less-intense stars 05·01 and 12·01.

The final positions are given in Table 2.

Table 2

Star	Constellation	Right Ascension	Declination
		h. m. s.	
05·01	Taurus	05 31 34·5±3	22° 04′ ±5′
12·01	Virgo	12 28 18 ±3	12° 37′ ±10′
19·01	Cygnus	19 57 45·3±1	40° 35·0′±1′
23·01	Cassiopeia	23 21 12·0±1	58° 32·1′±0·7′

This work was carried out at the Cavendish Laboratory as part of a programme of radio research supported by the Department of Scientific and Industrial Research. Thanks are due to Mr. M. Ryle for his most valuable advice and help, and to Mr. J. E. Jackson of the Department of Geography for assistance in surveying the positions of the aerials of the new interferometer.

F. G. SMITH

Cavendish Laboratory,
Cambridge.
Aug. 6.

[1] Ryle, M., Smith, F. G., and Elsmore, B., *Mon. Not. Roy. Astro. Soc.*, **110**, 508 (1950).

EDITOR'S NOTES AND AMENDMENTS

1. Smith (1952b).

IDENTIFICATION OF THE RADIO SOURCES IN CASSIOPEIA, CYGNUS A, AND PUPPIS A

W. Baade and R. Minkowski
Mount Wilson and Palomar Observatories
Carnegie Institution of Washington
California Institute of Technology
Received June 19, 1953

ABSTRACT

The radio sources in Cassiopeia and Puppis A are identified with a new type of galactic emission nebulosity. The outstanding features of these nebulosities are very large internal random velocities. The radio source Cygnus A is an extragalactic object, two galaxies in actual collision.

Only very few individual sources of cosmic radio emission have been identified with conspicuous astronomical objects.[1] Although the sources in Cassiopeia[2] and Cygnus A[3] are among the brightest and earliest-known radio sources of the sky, all attempts to identify them with astronomical objects in the visible range have failed so far. In the
fall of 1951,★ F. G. Smith, of the Cavendish Laboratory, communicated to us in advance 1
of publication new positions for both sources which were very much more accurate than any previous data. Using an interferometric method, Smith had reduced the uncertainties in the positions to $\pm 1^s$ in right ascension and $\pm 40''$ in declination. As it turned out, the new positions were accurate enough for an unambiguous identification of both radio sources on plates taken in September, 1951, at the 200-inch telescope. The Cassiopeia source coincides with a galactic-emission nebulosity of a new type, whereas Cygnus A is an extragalactic affair, two galaxies in collision. Quite similar to Cassiopeia is one of the fainter radio sources, Puppis A.

THE CASSIOPEIA SOURCE, RYLE 23.01[4]

On blue-sensitive plates, only an archlike nebulosity, $2'.8$ long, appears. Its detailed structure hardly leaves any doubt that we are dealing with an emission nebulosity (see Fig. 1). It is close to, but does not coincide with, the radio position which lies some 2′ farther to the south.[5] Much more revealing are the red-sensitive plates (Fig. 2) which cover the $H\alpha$ region ($\lambda\lambda$ 6400–6700). They show, besides the arc and mainly to the south of it, broken bits of nebulosity of a most remarkable kind. Some are elongated streaks up to 25″ long, others have almost stellar appearance. In intensity they range from very bright objects to mere smudges just above the limit of the plates. Not a single one of these broken bits of nebulosity registers on the blue-sensitive plates. Although there are strong indications that interstellar reddening affects the field, there can be no doubt that the broken bits of nebulosity are intrinsically very red. A good illustration is the brightest of these, marked *2* on Figure 3. Not a trace of it appears on the blue-sensitive plate,

[1] For a general discussion see the following paper by the authors (p. 215).★ 2

[2] M. Ryle and G. Smith,★ *Nature*, **162**, 462, 1948. 3

[3] J. G. Bolton and G. J. Stanley, *Nature*, **161**, 312, 1948.

[4] Since no generally accepted system exists, the designation of the sources in the list of Ryle, Smith, and Elsmore (*M.N.*, **110**, 508, 1950) and in the list of B. Y. Mills (*Australian J. Sci. Res., A*, **5**, 266, 1952) will be given.

[5] About an attempt by D. W. Dewhirst to identify the Cassiopeia source see *Observatory*, **71**, 212, 1951.★ 4
The object noted as peculiar by Dewhirst is actually a blend of a star and some brighter spots of the northern arc of nebulosity.

Reprinted, with permission, from 'Astrophysical Journal' 119, 206–214 (1954).

FIG. 1.—Nebulosity in Cassiopeia. λλ 3600–5000; 200-inch; 1 mm = 4″8

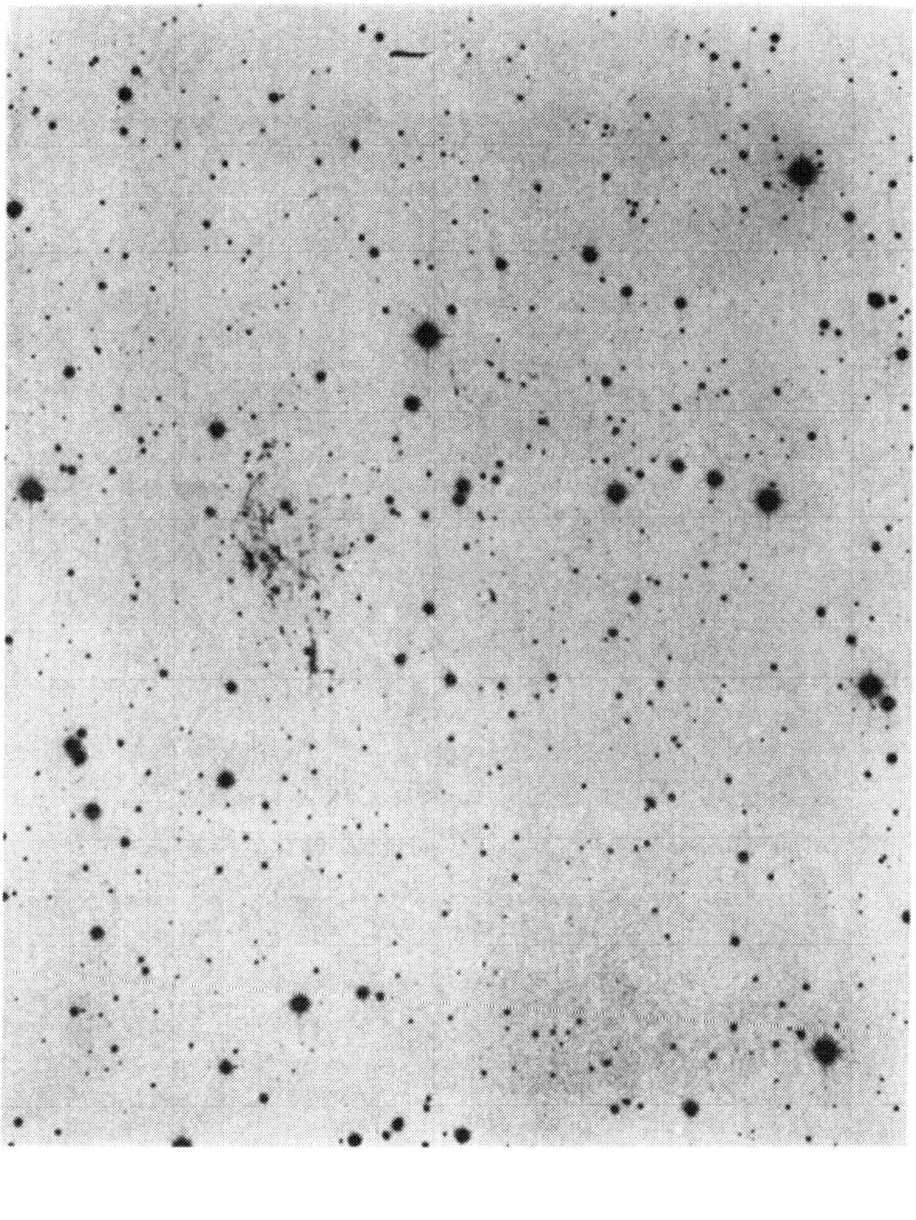

FIG. 2.—Nebulosity in Cassiopeia. λλ 6400–6700; 200-inch; 1 mm = $4''.8$

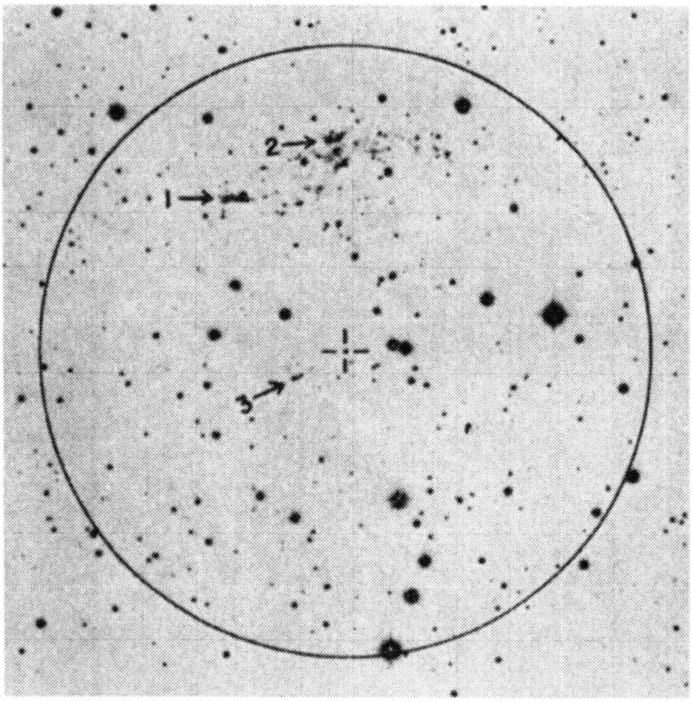

FIG. 3.—Nebulosity in Cassiopeia. λλ 6400–6700; 200-inch; 1 mm = 4″.9

whereas the nebulosity of the arc in which it is imbedded appears almost equally bright on the blue and the red plates. There can be no doubt, therefore, that in these broken bits of nebulosity the emission in the blue is abnormally faint compared to that in the *Hα* region. So far as we know, it is for the first time that we encounter the type of nebulosity just described. The total area covered by these nebulosities is the circle of $6\overset{'}{.}3$ diameter in Figure 3. Its center, marked by a cross, has the position 1950.0, $23^h21^m11\overset{s}{.}38$, $+58°31'52\overset{''}{.}9$, compared with F. G. Smith's radio position[6] of 1950.0, $23^h21^m12\overset{s}{.}0 \pm 1^s$, $+58°32\overset{'}{.}1 \pm 0\overset{'}{.}7$. Within the uncertainty of the present radio position, visible nebulosity and radio source coincide perfectly. All positions which have been obtained for the source are in Table 1.

TABLE 1

POSITION OF THE SOURCE IN CASSIOPEIA, RYLE 23.01

α (1950)	δ (1950)	Method	Observer
$23^h21^m12^s \pm 10^s$	$+58°32' \pm 4'$	Radio interferometer	Ryle, Smith, and Elsmore*
$23^h21^m12\overset{s}{.}0 \pm 1^s$	$+58°32\overset{'}{.}1 \pm 0\overset{'}{.}7$	Radio interferometer	Smith†
$23^h21^m36^s \pm 30^s$	$+58°38' \pm 10'$	Radio paraboloid	R. Hanbury Brown‡
$23^h21^m11\overset{s}{.}38$	$+58°31'52\overset{''}{.}9$	Photographic	Baade

* *M.N.*, **110**, 508, 1950.
† *Nature*, **168**, 555, 1951.
‡ *M.N.*, **113**, 123, 1953.

Although the center of the visible nebulosity closely coincides with a star of photographic magnitude 18.9, there is no reason to identify this star with the exciting source of the nebulosity, since in no way does it stand out by its color. The same is true of the other stars near the center, so that there is no visible source of excitation for the nebulosity.

Obviously, the identification of the radio source with the nebulosity just described would be strengthened very much if it could be shown that the radio source has the same diameter as the visible nebulosity. At our request, F. G. Smith[7] measured the diameter of the Cassiopeia source interferometrically. It turned out that the radio source has indeed a measurable diameter. From his measures Smith derived $d = 5\overset{'}{.}6 \pm 0\overset{'}{.}2$ under the assumption of a uniformly radiating disk, or $d = 5\overset{'}{.}0 \pm 0\overset{'}{.}2$ if the shell is optically thin. Observations by Hanbury Brown, Jennison, and Das Gupta[8] indicate a slightly elliptical area of $3' \times 4'$, with the major axis approximately in position angle 25°. The values for the diameter of the radio source are in such good agreement with the diameter of the visible nebulosity that there can be no further doubt about the identification. Both in position and in diameter the two objects agree within the errors of measurements.

Exact agreement of the radio position with the center of the nebulosity and of the radio diameter with the diameter of the nebulosity can be expected only if all parts of the nebulosity contribute equally to the radio emission. The present evidence suggests strongly that this is true and that no direct relation exists between optical and radio emission. But the accuracy of the radio measures does not yet seem to be adequate to draw final conclusions. Of all sources identified until now, the Cassiopeia source seems to offer the best opportunity for studying this question.

[6] *Nature*, **168**, 555, 1951.

[7] *Ibid.*, **170**, 1064,★ 1952. 5

[8] *Nature*, **170**, 1060,★ 1952. 5

Added October, 1953: Interferometric measures by Jennison and Das Gupta, reported at the Symposium on Radio Astronomy at the Jodrell Bank Experimental Station in July, 1953, show that the source emits as a uniformly radiating disk with some limb darkening.★ 6

That in the Cassiopeia source we have for the first time come across a galactic nebulosity of extraordinary properties was strikingly shown by the spectroscopic investigation. Spectra of three filaments, marked on the chart in Figure 3 by arrows in the position angles of the slit, have been obtained with the nebular grating spectrograph at the prime focus of the 200-inch telescope.[9] A grating with 7500 lines/inch blazed for the third-order red was used with the 1.4-inch camera; for emission objects this grating can be used to observe the third-order red together with the overlapping fourth- and fifth-order green and violet, respectively, without serious interference by blending of lines from different parts of the spectrum. The dispersion is 220 A/mm in the third-order red, 165 A/mm in the fourth-order green, and 132 A/mm in the fifth-order blue.

The spectrum of filament *1* in Figure 4 is unique in regard to composition and to internal motions. The only lines recorded are λ 4959 and λ 5007 of [*O* III] and λ 6300 and λ 6364 of [*O* I]. The absence of *Hα* indicates a large intensity ratio [*O* III]/*Hα*, but not necessarily a larger one than that found in those planetary nebulae which show the highest value of this ratio. The [*O* I] lines, however, are abnormally strong relative to *Hα* as well as to [*O* III]. The unusual weakness of the hydrogen lines relative to the forbidden lines may be explained by the assumption that the excitation is by collisions rather than by radiation.[10] It should be noted that the situation is different from that in the Crab Nebula, where the relatively low intensity of the *H* lines compared to the permitted lines of both *He* I and *He* II suggests low hydrogen abundance.[11, 12]

Even more peculiar than the composition of the spectrum are the internal motions in the filament. Each line appears to have one strong component with three fainter components of greater wave lengths. Closer inspection suggests that these components are merely intensity maxima in a broad band. The velocities within this band range from −1000 to at least +2200 km/sec; faint extensions to about +3000 km/sec are suspected for λ 5007. Details of the intensity distribution are not the same in the [*O* I] lines as in the [*O* III] lines. This suggest differences of excitation in different parts of the filament.

A possible interpretation of the line structure is that the apparent filament is really a thin sheet of matter which is spreading out rapidly in the radial direction. The maxima in the lines would then correspond to regions of high intensity within this sheet. If this interpretation is correct, high tangential velocities comparable to the radial velocities should be expected, and, unless the object is at a distance larger than 500 psc, changes in the appearance of the filament should become noticeable after a few years. But other models for the motions in the filament are kinematically possible.

The spectrogram of filament *2* is in Figure 5. It shows *Hα* flanked by the [*N* II] lines λ 6548 and λ 6584. The relative intensities are 6548:*Hα*:6584 = 1.5:1:4.5. The [*O* I] line λ 6300, blended with the same line from the sky, is present. Spectra of this composition and with similar relative intensities are shown by some planetary nebulae. The radial velocity is −45 km/sec at the east end and +36 km/sec at the west end of the filament. The width of the lines corresponds to a velocity spread of about 400 km/sec.

Filament *3* shows a similar spectrum as filament *2*. Owing to the faintness of this filament, only the condensation of the southeast end is recorded in all three lines, and the rest shows only faintly in λ 6584 [*N* II]. The radial velocity is −170 km/sec at the southeast end. The values for the center and the northwest end, based on measures of one line and correspondingly uncertain, are −125 and −170 km/sec, respectively. The spectrogram is too faint to permit a reliable estimate of the width of the lines.

Filaments *1* and *2* are near the edge of the nebulosity; filament *3* is near the center. There are no indications that the nebulosity as a whole is expanding. Any possible veloci-

[9] J. S. Bowen, *Ap. J.*, **116**, 7, 1952.

[10] See the following paper by R. Minkowski and L. H. Aller, *Ap. J.*, **119**, 232, 1954.

[11] R. Minkowski, *Ap. J.*, **86**, 199, 1942; *Mt. W. Contr.*, No. 666.

[12] R. Minkowski and J. L. Greenstein, *Ap. J.*, **119**, 238, 1954.

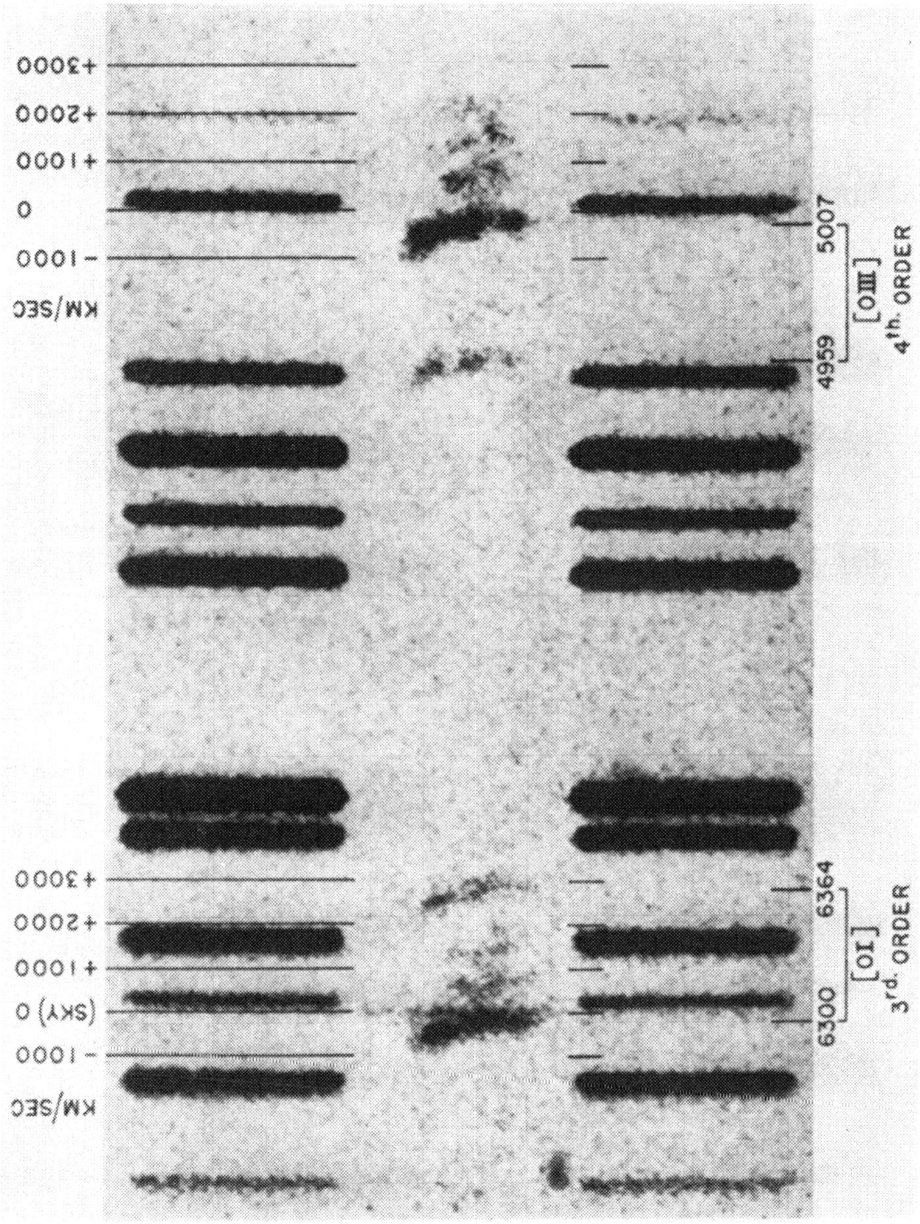

FIG. 4.—Spectrum of filament 1 of the nebulosity in Cassiopeia

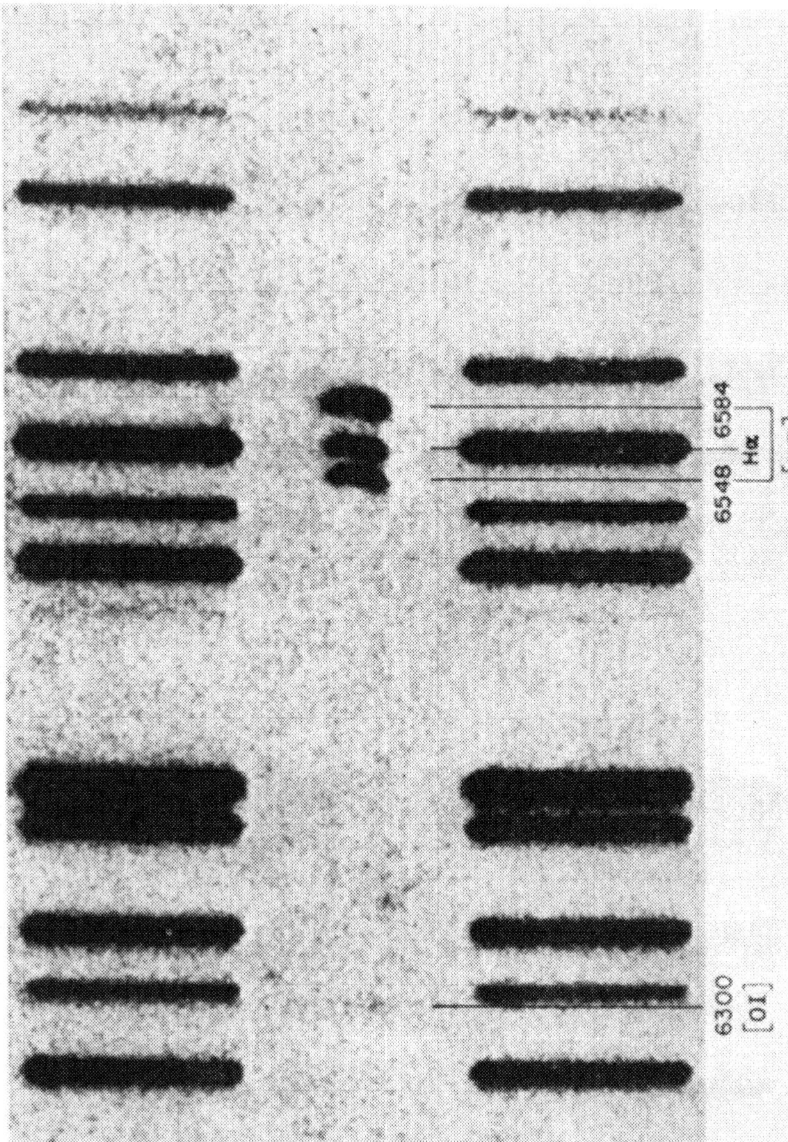

FIG. 5.—Spectrum of filament 2 of the nebulosity in Cassiopeia

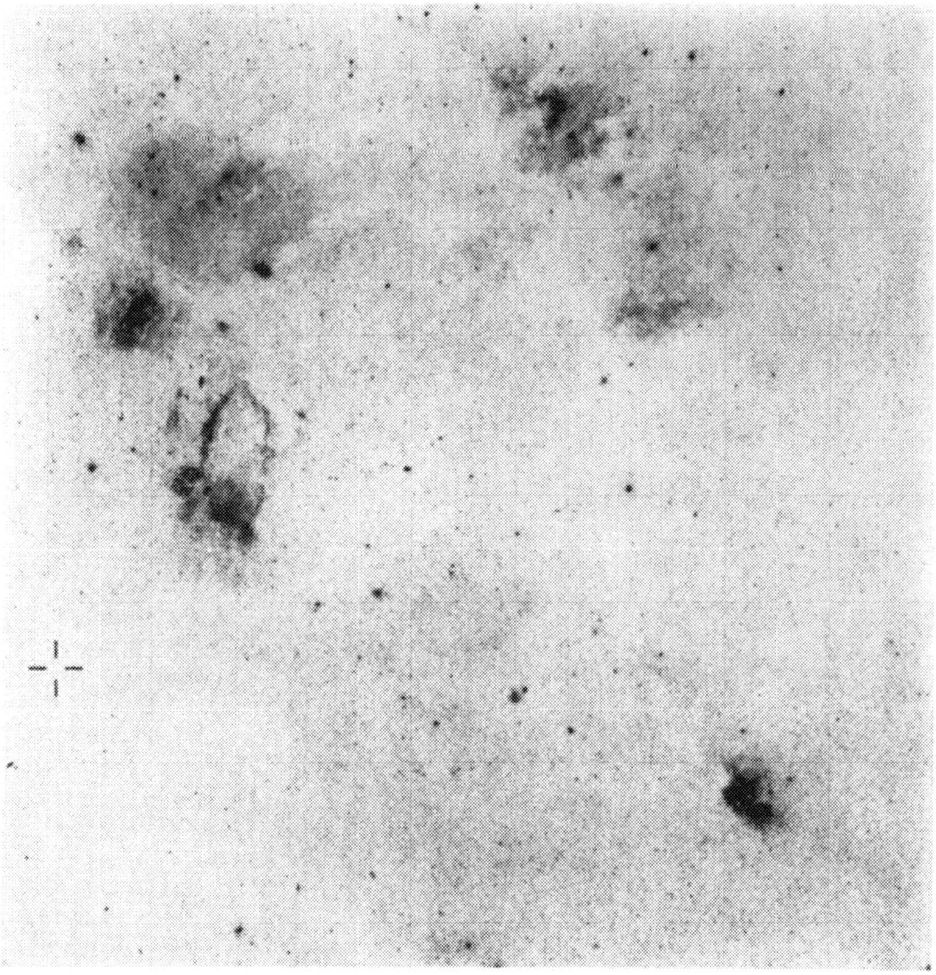

FIG. 6.—Field in Cassiopeia. Center (1950) 23^h40^m, $+61°24'$; λλ 6300–6700; 48-inch; 1 mm = 3'.1

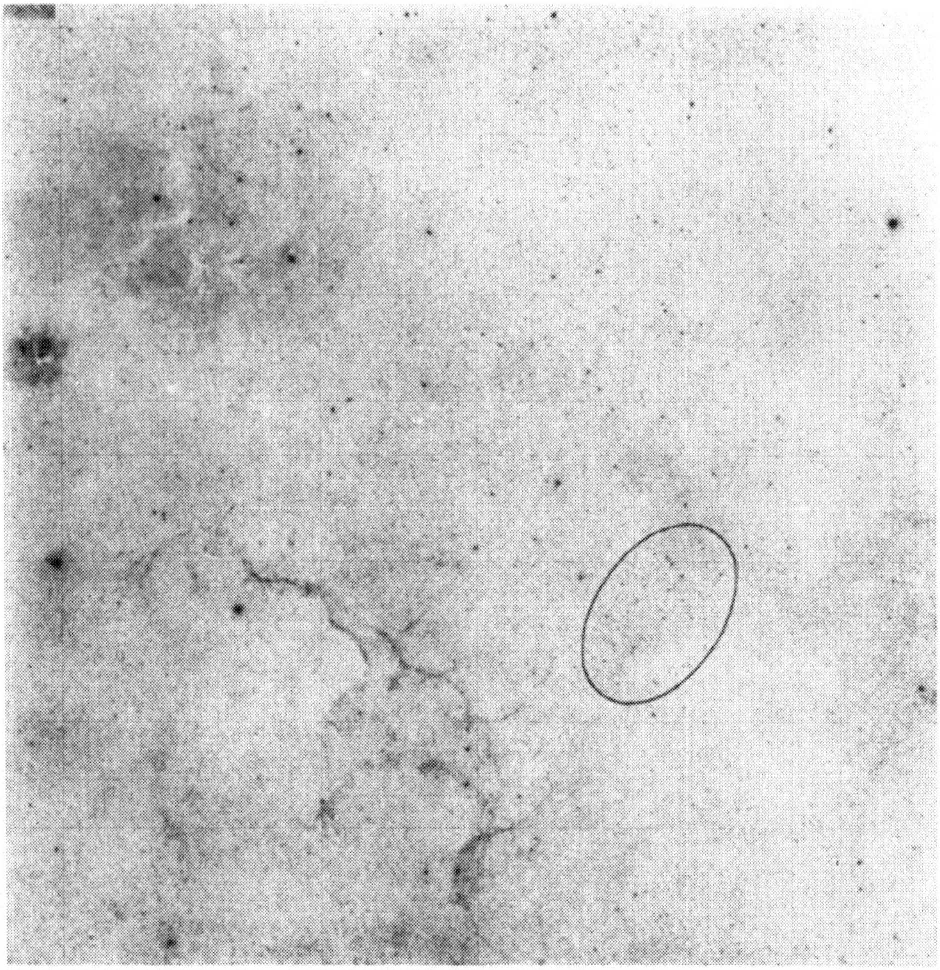

FIG. 7.—Field in Puppis. Center (1950) 8^h27^m, $-42°0'$; λλ 6300–6700; 48-inch; 1 mm = 3'.1

FIG. 8.—Filaments in Puppis. λλ 6200–6700; 48-inch; 1 mm = 29″

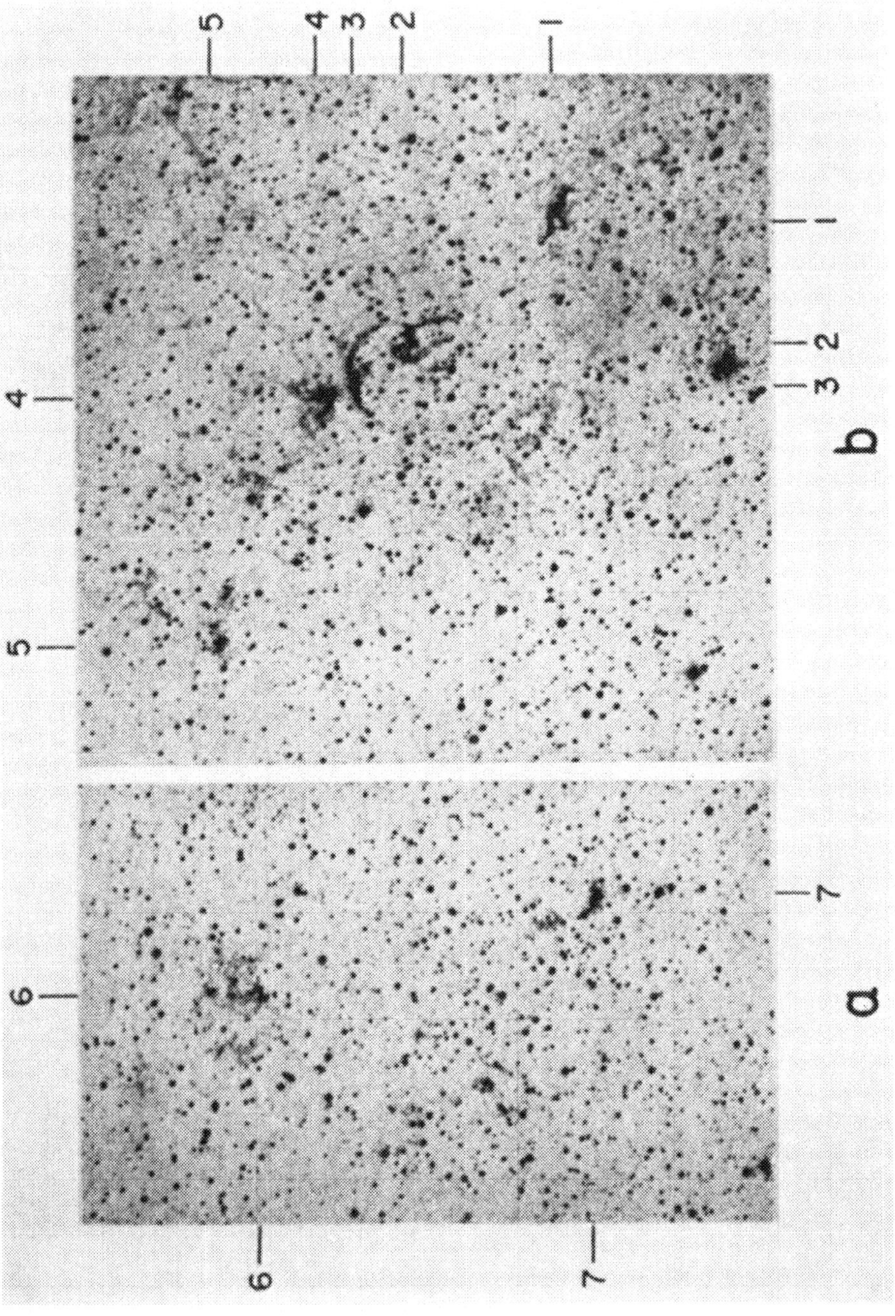

FIG. 9.—Filaments in Puppis. λλ 6200–6700; 48-inch; 1 mm = 15″

ty of expansion is obviously small compared to the random motions. This is quite different from the conditions found in shells of novae and supernovae and clearly shows that the nebulosity is not a shell of this type.[13]

The outstanding characteristic of the filaments is the large internal velocity dispersion. It seems plausible to assume that the unusually large internal motions in this nebulosity are directly connected with the strong radio emission. The present observations leave open the question of whether the filaments are being formed or dissolved. Unless the distance of the source is much larger than 500 psc, it should be possible after a few years to decide by observation between the alternatives.

The field containing the source is shown in Figure 6, reproducing a plate taken with the 48-inch Schmidt telescope. The position of the source is marked by the cross in the lower left corner; owing to the reduced scale, the faint nebulosity is not visible in Figure 6. The field is rich in large diffuse nebulosities and obscuring clouds, but none of these features coincides with the radio source, which is strikingly inconspicuous.

Added October, 1953: Intercomparison of plates taken in the fall of 1953 with the earlier plates of 1951 show marked motions and intensity changes in the filaments making up the northern arc of the nebulosity. In contrast, the red broken bits of nebulosity, such as filaments *2* and *3* in Figure 3, show neither perceptible motions nor intensity changes. Additional spectra have been obtained of seven more condensations. The diffuse filaments of the arc all have large radial motions of the order of 2000 km/sec, while the sharp red bits of nebulosity have small radial velocities around 50 km/sec.

PUPPIS A (MILLS 08−4)

When, in November, 1951, J. G. Bolton sent a list of certain sources with appreciable angular diameters, a search for these sources was made by Minkowski on 48-inch Schmidt plates. An extended visual object was found only in one of the positions, that of Puppis A.

A Schmidt plate containing the source is shown in Figure 7, reduced 2.8 times. The field is rich in large diffuse nebulosities. Not connected with these conspicuous nebulosities is a loose mass of filaments which greatly resemble the filaments of the Cassiopeia source. These filaments are within an approximately elliptical area of about $50' \times 80'$, with the major axis in position angle 145°; this area is marked in Figure 7.*The filaments 7
are not visible in Figure 7, owing to the reduced scale. The area containing the filaments is shown in Figure 8 on a larger scale. Parts of this area are still more enlarged in Figure 9, where those filaments are marked of which spectra have been observed. The centers of Figure 9, *a* and *b*, are marked in Figure 8 by a cross and a circle, respectively.

The two radio positions of the source in Table 2 agree in declination but not in right ascension. According to private communications from J. G. Bolton and B. Y. Mills, the position by Stanley and Slee may be more nearly correct. The position of the filamentary nebulosity, whose center cannot be determined with high accuracy, agrees with that by Stanley and Slee in right ascension, but the center of the nebulosity is 48′ south of the radio position. The northern border of the nebulosity at −42°13′ is, however, close to the radio position. Mills[14] found 33′ for the equivalent angular size of the source in right ascension. This is somewhat smaller than the extent of the nebulosity, but of the correct order. Radio source and nebulosity seem to overlap. In view of the faintness of this radio source and the low accuracy of its position, the agreement in regard to both position and size may be considered satisfactory. The possibility that the radio emission is stronger near the northern end of the nebulosity should not be disregarded. Exact agreement between optical and radio postion is therefore not to be expected.

[13] Since there is every reason to believe that the Cassiopeia source has nothing to do with supernovae, the attempt by I. S. Shklovsky (*Astr. J. U.S.S.R.*, **30**, 26, 1953) to identify the source with a new star of A.D. 369 is beside the point.

[14] *Australian J. Sci. Res.*, *A*, **5**, 266, 1952.

Spectra of seven filaments marked in Figure 9 have been obtained with a nebular spectrograph at the Newton focus of the 100-inch telescope. The spectrograph is similar to that used at the 200-inch telescope.[9] The same grating was used with a $F/0.67$ thick-mirror Schmidt camera; the dispersion at $H\alpha$ is 238 A/mm. Owing to the large zenith distance and the consequent high intensity of spectral features from city lights, only the strongest lines, $H\alpha$ and the [N II] lines λ 6548 and λ 6584, were observed; the presence of other lines is possible. The velocities and relative intensities are in Table 3. In several

TABLE 2

POSITION OF THE SOURCE IN PUPPIS, MILLS 08−4

α (1950)	δ (1950)	Method	Observer
$8^h18^m \pm 13^m$	$-42° \pm 40'$	Radio interferometry	Stanley and Slee*
$8^h35^m \pm 4^m$	$-42° \pm 20'$	Radio interferometry	B. Y. Mills†
8^h20^m3	$-42°48'$	Photographic	Minkowski

* *Australian J. Sci. Res.*, *A*, **3**, 234, 1950.
† *Australian J. Sci. Res.*, *A*, **5**, 266, 1952.

TABLE 3

VELOCITIES OF FILAMENTS IN PUPPIS

FILAMENT	KM/SEC $H\alpha$	KM/SEC [N II]	6548/$H\alpha$	6584/$H\alpha$	REMARKS
1	+ 8	− 47	1.5	4.5	
2	− 12	− 18	3	9	Lines diffuse
3	+113	+124	0.5–1.5	1.5–4.5	Lines wavy
4	+ 57	+ 76	2	6	*
5	− 19	− 8	1.5	4.5	Lines diffuse
6*a*	+ 84	+ 70	1.5	5	Lines diffuse†
6*b*	....	+124	....	....	
7	+ 22	− 9	0.8	2.5	

* The tabulated velocity is for the strong condensation. Velocity at the east end of the filament is +42 km/sec; at the west end, +87 km/sec.

† Filament 6 contains a sharp filament approximately in position angle 20° superposed on fainter, more irregular, filaments. The velocity for 6*a* is for the sharp filament, that for 6*b* for the background. $H\alpha$ is present in 6*b* but is too faint to be measured.

filaments the lines are noticeably diffuse; the velocity dispersion indicated is of the order of 150–200 km/sec.

The conditions shown by the spectra resemble greatly those in the Cassiopeia source, but on a reduced scale in regard to the velocities. The velocity dispersion is large, and any possible velocity of expansion is smaller than the random velocities. If the radio emission is connected with large velocity dispersion in filaments, a smaller velocity spread in Puppis A than in the Cassiopeia source is to be expected, since the radio emission of Puppis A is substantially smaller than that of the Cassiopeia source. The general similarity of the conditions in the Puppis nebulosity to those in the Cassiopeia nebulosity in regard to appearance, spectrum, and motions strengthens the identification as the radio source.

CYGNUS A (RYLE 19.01; MILLS 19 + 4)

Except for the first positions by Bolton and Stanley,[15] all positions of the source given in Table 4 are within an area of 18′ in right ascension and 10′ in declination. This area

[15] *Nature*, **161**, 312, 1948.

is marked in Figure 10, which shows the field surrounding the source. No bright star or any other conspicuous object is near the position of the source. Faint outlying parts of the nebulosities near γ Cygni cover the field. But interstellar absorption is low, as shown by the fact that numerous faint galaxies appear in the field.

Previous observers[16] had already noted that the region around Cygnus A must be quite transparent in spite of its low galactic latitude ($b = -4°$)★ because extragalactic nebulae 8
shine through in large numbers. The 200-inch exposures clearly show the reason why, aside from the general transparency of the field, extragalactic nebulae are so frequent. The radio source lies amid a rich cluster of galaxies, the brightest members of which are of about the seventeenth photographic magnitude. In conformity with the general rule, the dominating nebular types in the cluster are E- and So-systems. The radio source

TABLE 4

POSITION OF CYGNUS A (RYLE 19:01; MILLS 19+4)

α (1950)	δ (1950)	Method	Observer
$19^h58^m47^s \pm 10^s$	$+41°41' \pm 7'$	Radio interferometry	Bolton and Stanley*
$19^h58^m14^s \pm 60^s$	$+40°36' \pm 10'$	Radio interferometry	Stanley and Slee†
$19^h57^m46^s \pm 5^s$	$+40°30' \pm 7'$	Radio interferometry	Ryle, Smith, and Elsmore‡
$19^h57^m37^s \pm 6^s$	$+40°34' \pm 3'$	Radio interferometry	Mills and Thomas§
$19^h57^m44^s \pm 2^s.5$	$+40°35' \pm 1'.5$	Radio interferometry	Mills‖
$19^h57^m22^s \pm 25^s$	$+40°22' \pm 16'$	Radio paraboloid	Hanbury Brown and Hazard#
$19^h57^m45^s.3 \pm 1^s$	$+40°35'.0 \pm 1'$	Radio interferometry	Smith**
$19^h57^m44^s.49$	$+40°35'46''.3$	Photographic	Baade

* *Nature*, **161**, 312, 1948. This position has been superseded by the value given by Stanley and Slee.
† *Australian J. Sci. Res.*, *A*, **3**, 234, 1950.
‡ *M.N.*, **110**, 508, 1950.
§ *Australian J. Sci. Res.*, *A*, **4**, 158, 1951.
‖ *Australian J. Sci. Res.*, *A*, **5**, 456, 1952.
\# *M.N.*, **111**, 576, 1951.
** *Nature*, **168**, 555, 1951.

coincides in position with one of the brightest members of the cluster:[17] Cyg A (radio position by F. G. Smith[7]), 1950.0, $19^h57^m45^s.3 \pm 1^s$, $+40°35'.0 \pm 1'$; position of nebula, 1950.00, $19^h57^m44^s.49$, $+40°35'46''.3$.

In the center of this nebula (Fig. 11, *a*) are two bright condensations separated by about 2″ in position angle 115°. The bright central region of about 3″ × 5″ is surrounded by much fainter outer parts of elliptical outline, about 18″ × 30″, with the major axis in position angle 150°.

At first sight, this nebula is a very curious object which seems to defy classification. The clue to a proper interpretation lies in the fact that it has two nuclei which are tidally distorted and that hence we are dealing with the superimposed images of two galaxies. Both are late-type systems, judging by the low density gradients of the two disks. Spatially they are oriented face to face, they are slightly decentered, and we look upon them

[16] B. Y. Mills and A. B. Thomas, *Australian J. Sci. Res.*, *A*, **4**, 158–171, 1951.

[17] This coincidence was already noted in 1951 by Mills and Thomas (see n. 16), who had obtained for the radio position of Cyg A: 1950.0, $19^h57^m37^s \pm 6^s$, $+40°34' \pm 3'$; but it seemed unlikely at that time that a distant galaxy could be the radio emitter. Moreover, the coincidence established by them was not convincing, since, besides the nebula in question, three of the brighter members of the cluster fall into the area defined by the uncertainty of the position. Minkowski therefore wrote Mills that he did not think it was permissible to identify the source with one of the faint extragalactic nebulae in the area and emphasized that what was wanted was a more accurate radio position. The accuracy of Smith's position was needed to make the identification among the cluster members unambiguous. After Smith's position became known, the coincidence of the nebula with the radio source was also noted by D. W. Dewhirst (*Observatory*, **71**, 212, 1951). B. Y. Mills later gave the improved position: 1950.0, $19^h57^m44^s \pm 2\frac{1}{2}^s$, $+40°35' \pm 1\frac{1}{2}'$.

at an angle not far from 45°. Actually, the two systems must be in close contact because of the strong signs of tidal distortions which the nuclei show. This suggests that we are dealing with the exceedingly rare case of two galaxies which are in actual collision. The main features of such a collision have been discussed by Spitzer and Baade.[18] On the cosmical time scale, collisions of galaxies are a rather frequent phenomenon in the rich clusters of galaxies. As far as the stars of the colliding systems are concerned, such a collision is an absolutely harmless affair. The average distance between two stars is so large that the two galaxies penetrate each other without any stellar collisions. The situation is very different for the gas and dust imbedded in the two systems. Because of the much shorter free paths of the gas and dust particles, the collision of the two galaxies means a real collision of the imbedded gas and dust, which are heated up to very high temperatures, since the collisional velocities range from hundreds to thousands of kilometers per second.

It is obvious that this behavior of the gas and dust provides a beautiful test of our hypothesis that the two galaxies which we identify with the radio source Cygnus A are actually colliding; for we should expect a very unusual spectrum for our nebula. Besides the continuous spectrum provided by the stars of the two systems, the emission spectrum of the colliding gases should be visible. In fact, this emission spectrum should be quite strong, since we are dealing with a face-on collision of the two galaxies, which makes it certain that maximum amounts of the gas are in collision simultaneously. Moreover, on account of the high velocities involved in the collision of two galaxies, the resulting emission spectrum should be one of unusually high excitation and the lines themselves should show broadening. Spectra taken at the 100-inch and 200-inch telescopes largely confirmed these predictions.

Spectra of the bright nuclear region have been obtained with the nebular spectrographs at the prime focus of the 200-inch telescope and at the Newtonian focus of the 100-inch telescope. The spectrum in Figure 12 was obtained with the grating mentioned earlier, which permits the simultaneous observation of the third-, fourth-, and fifth-order red, green, and violet, respectively. For the photographic region, a grating with 15,000 lines/inch blazed for violet has also been used with a dispersion of 435 A/mm.

The spectrograms show that the nebula is indeed a very peculiar, if not unique, object. Strong forbidden-line emission is recorded, while the continuous spectrum barely shows. Photoelectric measures by Dr. W. A. Baum indicate that more than 50 per cent of the total light of the condensations is in the emission lines. The emission lines and the nebular red shift derived from them are in Table 5. From the red shift, the distance of the nebula
9 is 3.3×10^7 psc, with the value 540 km/sec per 10^6 psc for the red-shift constant.*

The large red shift of 16,830 km/sec places all lines in regions of the spectrum where the plate sensitivity differs from its normal values for the lines; reliable estimates of relative line intensities, therefore, cannot be given at this time. It is obvious that $H\alpha$ is relatively faint compared to the adjacent [*N* II] lines. The absence of $H\beta$ from the observed lines indicates relatively low intensity compared to the neighboring [*O* III] lines. The lines are diffuse, but the dispersion is too low to determine reliable values of the width. Individual lines seem to differ in width, but at least part of this is caused by intensity differences. The velocity spread is of the order of 400 km/sec.[19] The [*O* II] line λ 3727 extends through the full length of the slit in Figure 12. The corresponding diameter of more than 30″ in position angle 90° is slightly larger than the visible size of the nebula.

[18] *Ap. J.*, **113**, 413, 18, 1951.

[19] The spectrum of the nebula resembles to a certain degree the spectra of those relatively rare nebulae whose nuclei show strong emission lines (C. J. Seyfert, *Ap. J.*, **97**, 28, 1943; *Mt. W. Contr.*, No. 671), but there are marked differences. In NGC 1068 and NGC 4151, the outstanding representatives of the emission nuclei, only 13 and 22 per cent of the light of the nucleus, respectively, is in the emission lines. The outstanding difference between Cygnus A and the nebulae with emission nuclei is that in Cygnus A the area of very strong emission has a diameter of 800 psc, whereas in the other nebulae the emission is restricted to diameters of the order of 25 psc.

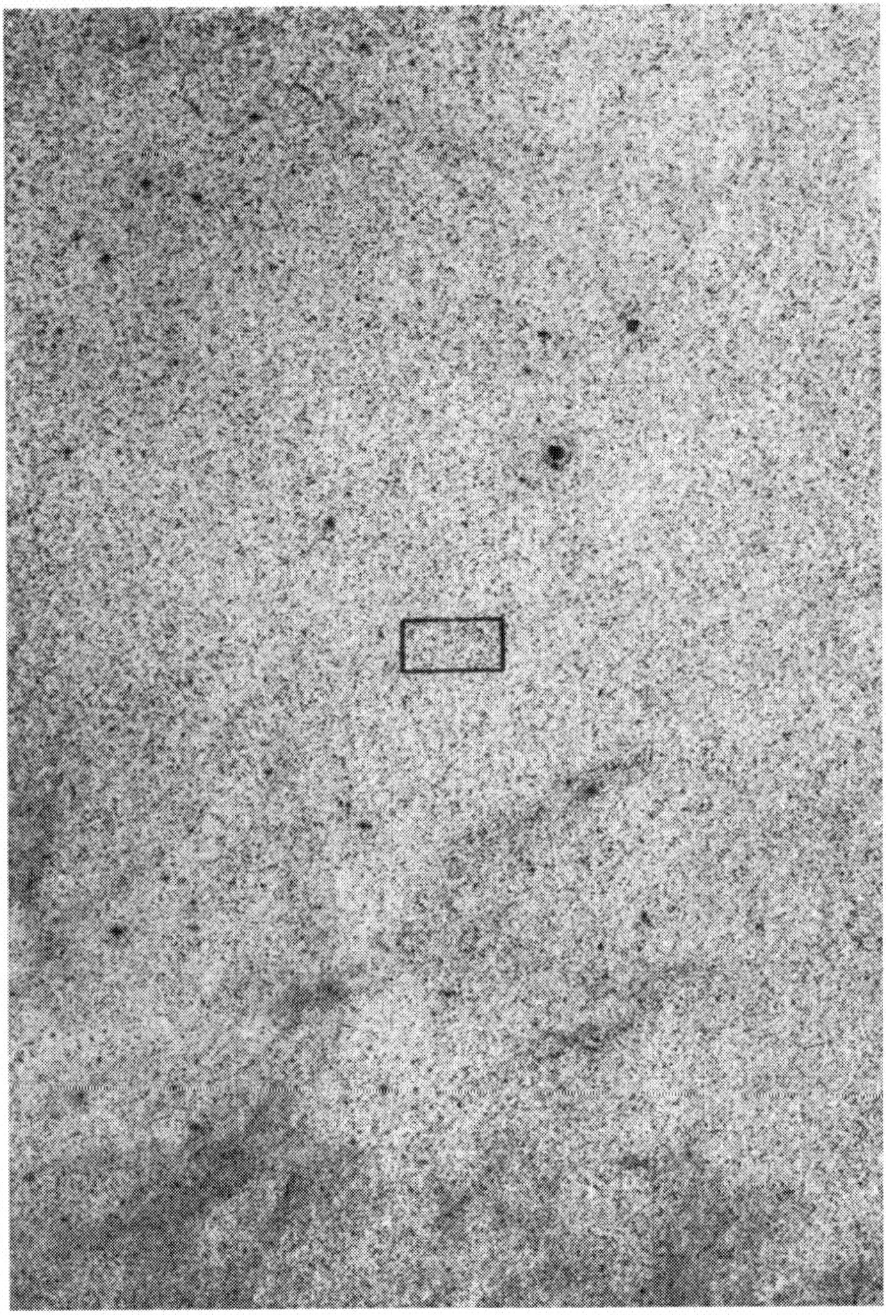

FIG. 10.—Field in Cygnus. Center (1950) $19^{h}57^{m}.6$, $+40°35'$; 48-inch; 1 mm = 67″

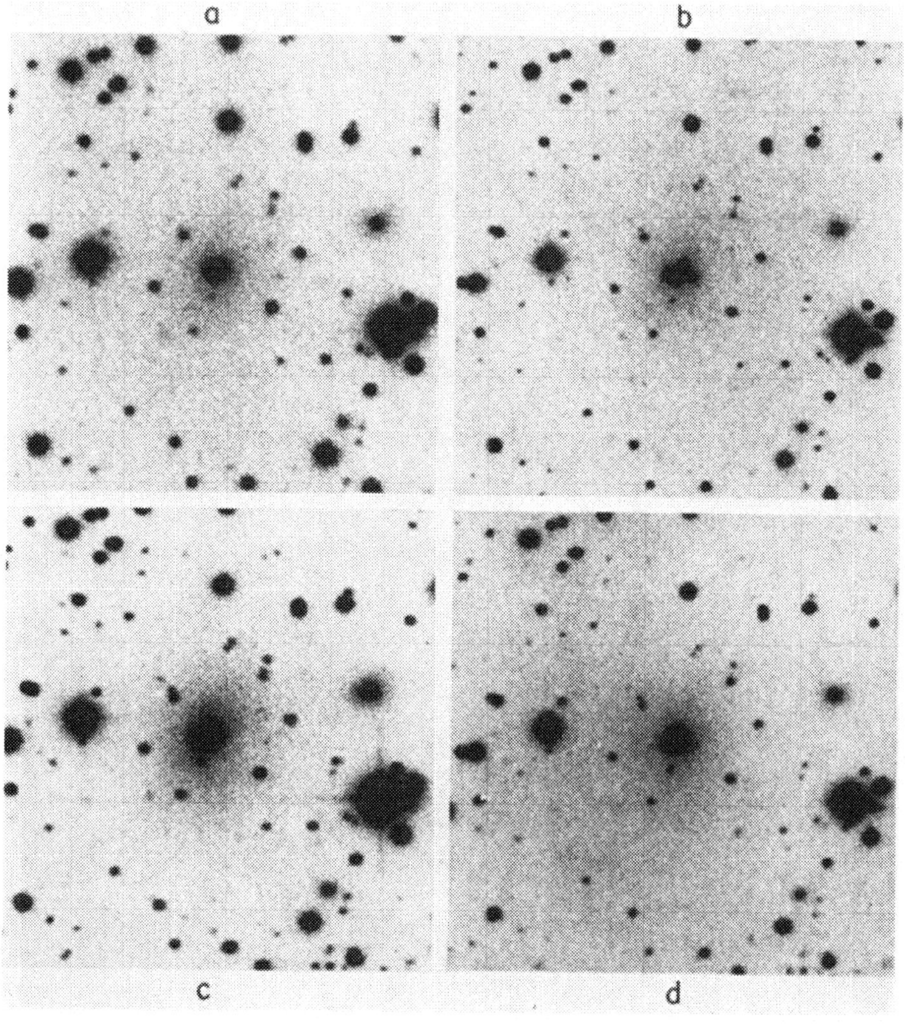

FIG. 11.—Extragalactic nebula in Cygnus. 200-inch; 1 mm = 1''.2. *a*, λλ 3600–5000 [*Ne* V], [*O* II], [*Ne* III]; *b*, λλ 5000–5400 [*O* III]; *c*, λλ 5000–6200 [*O* III]; *d*, λλ 6700–7500 *Hα*, [*N* II].

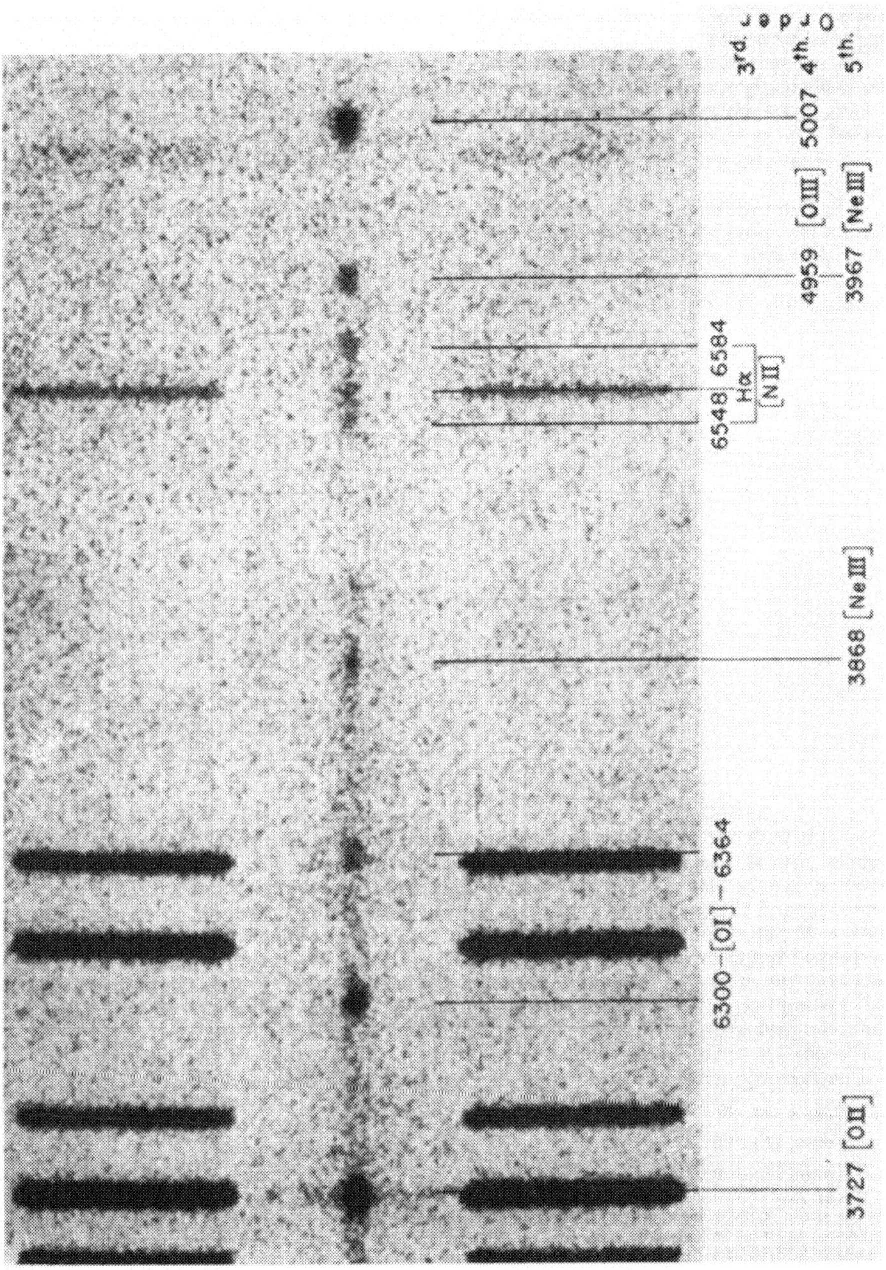

Fig. 12.—Spectrum of the extragalactic nebula in Cygnus

Inclination of the line indicates rotation of the nebula; the south-following side is receding relative to the center.

On account of the great intensity of some of the emission lines, it is possible, by selecting sufficiently narrow filter ranges, to photograph the emission regions. As Figure 11 shows, there are marked differences in the pictures, depending on the emission lines selected. Owing to the nebular red shift, the emission lines do not appear in their normal wave-length region; the main lines contained in each photograph are listed in the caption of Figure 11.

The determinations of the size of the radio source confirm that the source is not stellar, as had been assumed originally, but has a small diameter measurable with refined interferometric methods. Measures by Mills[20] give an effective size of 1′.1. Smith[21] finds a diameter of about 3′.5. Both values should apply to the east-west direction. Measures by Hanbury Brown, Jennison, and Das Gupta[22] suggest an elliptical outline of about

TABLE 5

EMISSION LINES IN CYGNUS A

λ Observed	Identification		$cd\lambda/\lambda_0$ (Km/Sec)
3619.9......	3425.86	[*Ne* V]	17,010
3937.2......	3726.06/8.82	[*O* II]	16,930
4087.5......	3868.77	[*Ne* III]	16,930
4189.6......	3967.48	[*Ne* III]	16,820
5234.3......	4958.91	[*O* III]	16,660
5284.6......	5001.85	[*O* III]	16,955
10 6642.5★.....	6300.27	[*O* I]	16,770
6718.6......	6363.88	[*O* I]	16,720
6916.6......	6548.06	[*N* II]	(16,870)*
6928.0......	6562.66	*Hα*	(16,300)*
6949.2......	6583.43	*N* II]	16,670
Mean...	....		16,830

* λλ 6916.6 and 6928.0 are blended: the accuracy of the measured wave lengths therefore is low, and the values of the red shift for both lines have been excluded from the mean.

1′ × 2′.1, with the major axis in a position angle close to that of the major axis of the nebula. This is three to four times larger than the visible size, which, as is usual for faint nebulae, is only that of the main body of the nebula. The emission line λ 3727 [*O* II] extends beyond the main body, and it is quite possible that the fainter outer parts may have a diameter much larger than the main body. If, however, a large part of the disagreement has to be explained by an essential contribution of the outer parts to the radio emission, the radio and the optical emission cannot be related. It seems improbable that this assumption entirely suffices to explain the discordance. In view of the difficulty of the radio measures, however, the lack of agreement in regard to the precise extent seems less important than the agreement as to the order of magnitude of the size.

Photoelectric magnitude and color of Cygnus A have been measured by Dr. W. A.

[20] *Nature*, **170**, 1063, 1952.

5 [21] *Nature*, **170**, 1064,★ 1952.

5 [22] *Nature*, **170**, 1060,★ 1952.

Added October, 1953: Interferometric measures by Jennison and Das Gupta, reported at the Symposium on Radio Astronomy at the Jodrell Bank Experimental Station in July, 1953, show that the radio emission originates in two areas separated by 82″ and 43″ long in position angle 95° with a width of 30″. No object of this description can be seen on any of our plates which cover the wave-length range from the
11 ultraviolet to the infrared.★

Baum. With a diaphragm of 11″.5 diameter, the magnitudes and colors reduced to the international system are $m_{pg} = 17.90$; $m_{pv} = 16.22$; $CI = +1.68$. To determine the absolute magnitude, corrections have to be applied for interstellar absorption and for the size of the diaphragm. By comparison with nebulae of similar appearance, a conservative estimate is that the nebula is 0.75 mag. brighter than the part admitted by the diaphragm. The interstellar absorption cannot be determined from the color index of the nebula, since the color is affected by the emission lines. But, on the assumption that the elliptical nebula 25″ north-preceding is a member of the cluster to which Cygnus A belongs and therefore at the same distance, this apparently normal nebula may be used. Its magnitudes and colors are $m_{pg} = 18.95$; $m_{pv} = 17.42$; $CI = +1.53$. The color of an elliptical nebula with a red shift of 17,000 km/sec is[23] $CI = +1.08$. The color excess of $+0.45$ mag. between this value and the measured color index is due to interstellar reddening. The ratio of photographic absorption to international color excess is[24] somewhat less than 5. The photographic absorption, therefore, is 2.1 mag. Thus we have for Cygnus A the values in the accompanying tabulation.

	Mag.	
Apparent m_{pg}	17.90	
Correction for incompleteness	0.75	
Photographic absorption	2.1	
Corrected m_{pg}	15.05	
Distance modulus $m - M$	32.6	(from red shift)
M_{pg}	−17.5	

The system, therefore, is a giant system which may be expected to have a diameter of more than 10,000 psc. The major axis of 30″ of the visible main body of the nebula is only 5000 psc. The assumption is justified that the nebula has a larger diameter than this; but a diameter of 20,000 psc, which would correspond to the radio interferometer measures, would call for strong radio emission from the outermost parts of the nebula, where the average density is exceedingly low.

Since the distance of the nebula is known, the total energy emitted in the radio region may be computed. The flux received from the nebula is of the order of 1.3×10^{-22} watts m^{-2} $(c/sec)^{-1}$ in a band of an equivalent width of 500 Mc/sec. The total flux, therefore, is of the order of 6×10^{-14} watts m^{-2} or 6×10^{-11} ergs cm^{-2} sec^{-1}. With the source at the distance of 3.3×10^{7} psc, the total energy emitted in the radio region is 8×10^{42} ergs/sec. This is larger than the total optical emission of 5.6×10^{42} ergs/sec and larger than the energy contained in the line emission. The source of energy for the radio emission may be the relative kinetic energy of the colliding nebulae, which is of the order of 10^{59} ergs for a relative velocity of 500 km/sec.

We want to express our sincere thanks to all members of the radio-groups in Sydney, Cambridge, and Manchester for their generous co-operation.

[23] J. Stebbins and A. E. Whitford, *Ap. J.*, **108**, 413, 1952.

[24] J. L. Greenstein, in *Astrophysics: A Topical Symposium*, ed. J. A. Hynek (New York: McGraw-Hill Book Co., 1951), chap. xiii, p. 538.

EDITOR'S NOTES AND AMENDMENTS

1. This letter from Smith to Baade was in fact dated 22 August 1951.
2. Baade and Minkowski (1954b).
3. Should be "F. G. Smith".
4. This attempt involved infrared plates taken on the Cambridge 36-inch reflector; see the introduction.
5. Should be "1065" in footnotes 7 and 21 and 1061 in footnotes 8 and 22.
6. See *Observatory* **73**, 185 (1953).
7. The large complex of filaments to the southeast of the ellipse are part of the Vela X supernova remnant.
8. The correct latitude for Cyg A is $b^{\mathrm{I}} = +5°$.
9. The modern value for the "red-shift constant", or Hubble constant, is 5 to 10 times smaller, placing Cyg A at 5 to 10 times the quoted distance and increasing its luminosity by a factor of 25 to 100.
10. Observed wavelength should be 6652.5.
11. See Paper 33.

SECTION O

SOURCE STRUCTURE

During this revolutionary period in the investigation of the discrete sources, the most compelling argument against the idea of radio *stars* was the recognition of the finite angular extent of the most intense sources. The extent of the radio-emitting region also was useful in that it provided a second criterion for optical identifications, e.g., as used for the Andromeda Nebula by Hanbury Brown and Hazard (1950, 1951) and for Cas A by Baade and Minkowski (Paper 29). Mills (1952a, c) was actually the first to make such measurements on the strong sources, finding angular widths at 100 MHz of ~20 arcmin for both Cen A and For A and ~33 arcmin for Pup A. The center of attention was nevertheless focussed on the two strongest sources, Cas A and Cyg A, and in Papers 30 through 33 we find reported the first measurements of sizes for these sources. The first three of these papers appeared together in the 20 December 1952 issue of *Nature* and provide a showcase for the differing techniques employed, as well as the intense rivalry at that time among the three major groups in radio astronomy. In fact these results had all been previously simultaneously reported at the same URSI meeting in Sydney in 1952.

The Jodrell Bank results reported by R. Hanbury Brown, R. C. Jennison, and M. K. Das Gupta are the first published observations with the intensity interferometer (see the introduction to Part II). They are important both for that reason and for the fact that Cyg A is found to be highly asymmetric. In the second experiment, Mills, in Paper 31, achieves a long baseline (10 km or ~ 3300 λ) through the use of a radio link which transmits both the local oscillator signal and the intermediate-frequency signal from the portable antenna. With this equipment, he is able to measure quite accurately the east-west extents of Cyg A, Tau A, Vir A, and Cen A. Finally, F. G. Smith reports his observations made with a pair of Michelson phase-switched interferometers which measure the *ratio* of fringe amplitudes at two different baselines. Although his maximum achievable baseline is only 0.6 km, this is sufficient to establish on a reasonable model the angular extents of Cyg A and Cas A.

The real surprise comes in Paper 33 when Jennison and Das Gupta continue their investigation of the structure of Cyg A and find that it is not best modeled as a single elongated source, but rather as two components, well separated and of approximately equal intensity. This deduction is only possible through use of data taken at close intervals over a wide range of spacings.[1] Although this first "map" of a discrete source seems crude by modern standards, comparison with today's aperture synthesis maps reveals it to be a faithful representation of Cyg A. This first double radio source then of course immediately raised the question of why the optical image and the radio brightness distribution should be so different. And today we are still asking that same question.

NOTE

[1] It can be seen that Mills' lack of spacings between 1 and 5 km in Paper 31 created a critical gap which, in fact, denied him the opportunity of first detecting this double structure.

APPARENT ANGULAR SIZES OF DISCRETE RADIO SOURCES

Observations at Jodrell Bank, Manchester

THE existence of discrete sources of extra-terrestrial radio-frequency radiation is now well established[1,2] and the positions of more than one hundred sources have been published[3-5]. Attempts to identify these sources with any particular class of visual object have so far failed, and the origin of the radiation remains unexplained. One of the fundamental requirements in the study of these sources is a knowledge of their apparent angular size, and although attempts to make this measurement have been made by several observers[1-3], it has proved to be beyond the resolving power of their equipment. The present communication gives a preliminary account of a successful attempt to measure the angular size of the two most intense sources the positions[6] and intensities[4] of which are given in Table 1.

Table 1. CO-ORDINATES AND INTENSITY OF THE TWO MOST INTENSE RADIO SOURCES

Source	Right Ascension epoch 1950	Declination epoch 1950	Intensity* at 81 Mc./s. (watts/sq.m./c.p.s.)
Cygnus	19h. 57m. 45·3s. ±1s.	N 40° 35·0′ ±1′	$13 \cdot 5 \times 10^{-23}$
Cassiopeia	23h. 21m. 12·0s. ±1s.	N 58° 32·1′ ±0·7′	$22 \cdot 0 \times 10^{-23}$

* The intensity given is twice that observed in one plane of polarization.

The co-ordinates are those given by Smith (ref. 6), and the intensities are those given by Ryle, Smith and Elsmore (ref. 4).

Previous measurements have been made with interferometers. Bolton and Stanley[1], using an interferometer mounted on a cliff and operating in a manner analogous to 'Lloyd's mirror', have shown that the apparent size of the source in Cygnus is less than 8 minutes of arc. More recently, Stanley and Slee[3], using a similar instrument, claim to have reduced this limit to $1\frac{1}{2}$ minutes of arc. Ryle and Smith[2], using an instrument analogous to

Reprinted, with permission, from 'Nature' **170**, 1061–1065 (1952).

Michelson's stellar interferometer and with a base-line of 500 metres, have shown that the source in Cassiopeia has an angular size of less than 6 minutes of arc.

The resolving power of an interferometer depends primarily on the ratio of the wave-length to the base-line, and the limits quoted above represented the best performance obtained with the instruments. These limits cannot be reduced significantly without a corresponding extension of the base-line.

In 1950 it was decided at Jodrell Bank to attempt to measure the angular size of the two sources shown in Table 1, or at least to reduce the upper limits given for their size. It was assumed that this angular size might lie anywhere between the limit of a few minutes of arc and the diameter of the visible stars, and an instrument of the highest possible resolving power was therefore sought. While it appeared to be possible to extend the base-lines of existing interferometers to lengths of the order of 10–50 km., it was considered that for much longer base-lines the problem of maintaining an adequate stability of phase in the transmission of signals along the base line would prove to be difficult. For this reason an interferometer of completely new design was developed.

Fig. 1 shows a simplified block diagram of the instrument. Two aerial systems A_1 and A_2, each with an aperture of 500 square metres, are connected to two independent superheterodyne receivers R_1 and R_2. These receivers are both tuned to 125 Mc./s. and have a band-width of 200 kc./s. The intermediate frequency output of receiver R_1 is rectified in a square-law detector and is fed to a low-frequency filter the pass-band of which extends from 1 to 2 kc./s. The output from receiver R_2 is treated in an identical manner and the two low-frequency outputs are then multiplied together in a 'correlator' and their 'cross-correlation coefficient' is obtained. When a long base-line is used, one low-frequency output is transmitted over a radio-link by modulating a high-frequency carrier. The time of arrival of the two low-frequency outputs at the 'correlator' is equalized by inserting in one output a delay equal to the time of transmission along the base-line. It is essential

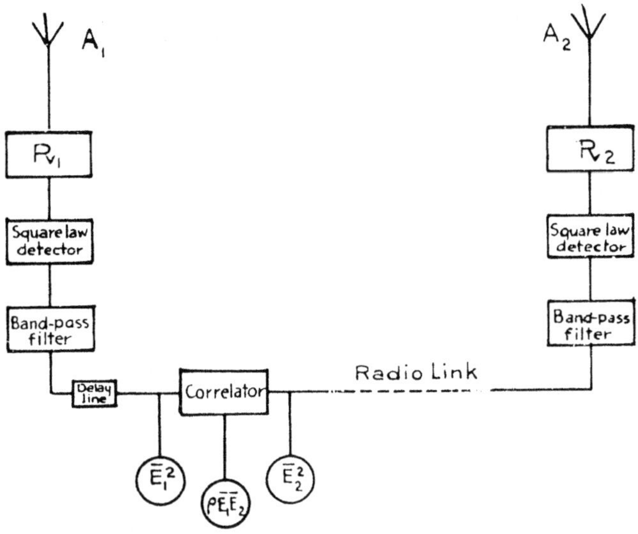

Fig. 1. Schematic diagram of the equipment

that all the components should preserve the relative phase of the two low-frequency signals. In practice, this problem is considerably simpler than the corre-ponding one of preserving the relative phase of the radio-frequency signals in a conventional interferometer, and hence it should be easier to use the new type of instrument with long base-lines.

In operation, the two aerials (A_1 and A_2) are fixed in azimuth (due north or south) and at an elevation corresponding to the declination of the source. As the source transits the aerial beam, the increase in the mean square value of the two low-frequency outputs ($\overline{\varepsilon_1^2}$ and $\overline{\varepsilon_2^2}$) is recorded together with their product ($\rho \overline{\varepsilon_1 \varepsilon_2}$). The value of the 'cross-correlation' coefficient' (ρ) is then found from an analysis of the three records.

The theory of the instrument is involved, and it will be given in detail elsewhere[7]. It can be shown that the value of the cross-correlation coefficient (ρ) is given by an expression similar to that for the visibility of the fringes in a Michelson stellar interferometer :

$$\rho \approx \frac{\sin^2(\pi\alpha b/\lambda)}{(\pi\alpha b/\lambda)^2}, \tag{1}$$

where α is the angular width of an equivalent rectangular source of constant surface intensity ; b is the effective length of the base-line ; and λ is wavelength.

Equation (1) shows that by a suitable choice of the base-line it is possible to obtain a value for the angular width of an equivalent rectangular source of constant surface intensity. The shape and angular width of the actual source, together with the distribution of brightness across its disk, can only be found from a large number of observations made with base-lines of different length and orientation.

A model of the instrument, built in 1950, was used to measure the apparent diameter of the sun at 125 Mc./s. As the results of this test were satisfactory, a full-scale instrument was built in 1951, and this has since been applied to the two sources shown in Table 1. Measurements have so far been made on both sources with four different base-lines and the results are given in Table 2. The length and orientation of the base-lines are given in the table, together

Table 2. EXPERIMENTAL RESULTS

Base-line			Cygnus		Cassiopeia	
	Length* (km.)	Bearing†	Correlation coefficient	Angular width of equivalent strip	Correlation coefficient	Angular width of equivalent strip
A	0·30	349·5°	0·99 ± 0·10	< 5′	0·96 ± 0·09	3′ 40″(< 5′ 50″)
B	2·16	113·0°	0·30 ± 0·03	2′ 10″ ± 4″	0·08 ± 0·02	2′ 55″ ± 10″
C	2·16	235·5°	0·79 ± 0·08	1′ 00″ ± 7″	< 0·01	≮ 3′ 30″
D	3·99	177·0°	0·79 ± 0·07	0′ 34″ ± 8″	0·07 ± 0·01**	

* The value given is the actual distance between the two stations. The effective length of the base-line is calculated from the elevation of the source and the orientation of the base-line.

† The bearing is measured east from north, and is the relative bearing of the base-line from the fixed station at Jodrell Bank.

** This value is not taken into account in Fig. 3, since the other results indicate that the point may lie on a secondary maximum of the curve relating correlation coefficient to base-line.

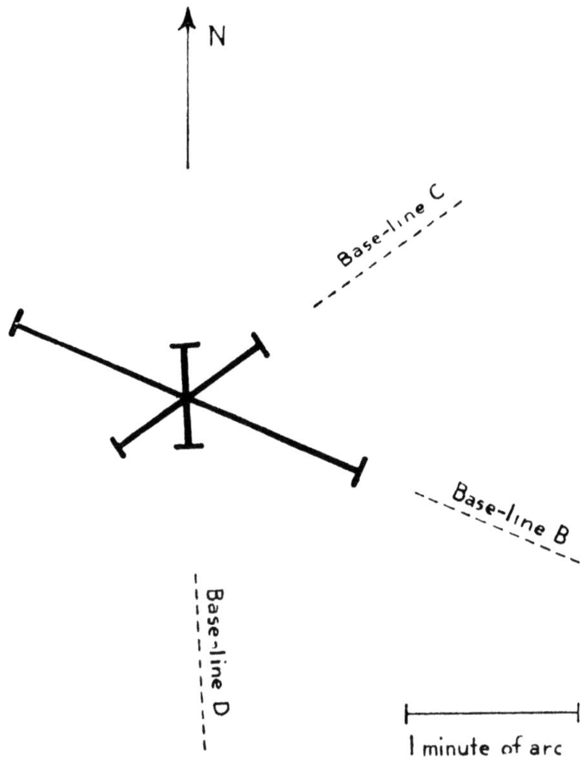

Fig. 2. Equivalent angular width of the source in Cygnus observed from different base-lines. The value shown is the width of an equivalent rectangular strip of constant surface intensity* 1

with the observed values of the cross-correlation coefficient (ρ). The angular size of the equivalent rectangular strip has been calculated for each base-line and the results are also shown in Table 2.

The interpretation of these results in terms of the precise actual size and shape of the sources must be treated with caution, since the three extended base-lines differed in length; and until more results are available it is uncertain what effect variations in the length of the base-line may have on the equivalent angular size of the source.

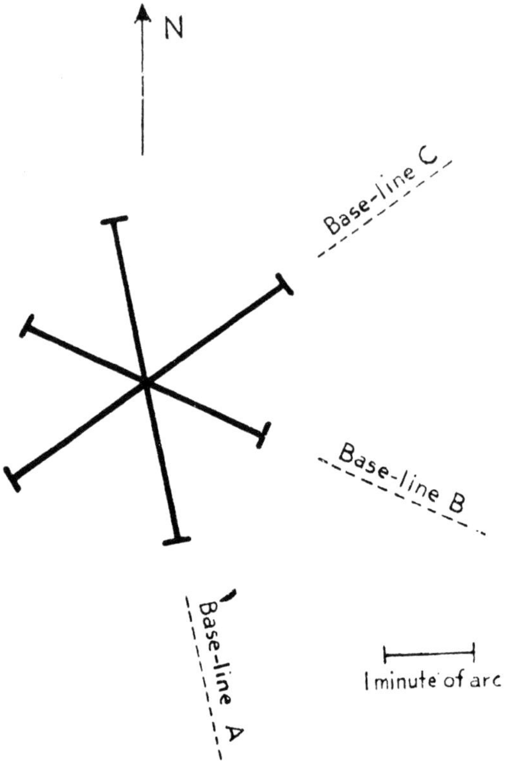

Fig. 3. Equivalent angular width of the source in Cassiopeia observed from different base-lines. The value shown is the width
1 of an equivalent rectangular strip of constant surface intensity*

Figs. 2 and 3 show a plot of the measurements given in Table 2. Fig. 2 shows clearly that the source in Cygnus exhibits a pronounced asymmetry. The value of its equivalent angular size varies from 35″ to 2′ 10″ of arc with the orientation of the base-line, and the major axis of the source appears to be inclined at an angle of between 90° and 120° to the celestial meridian. A preliminary analysis indicates that the results are incompatible with a source of simple elliptical shape and constant surface intensity, and that a more complicated model must be used. Further progress in this analysis must await more observations. The results in Fig. 3 suggest that the source in Cassiopeia is not markedly asymmetrical. The equivalent angular size of the source appears to be about 4′ of arc.

These preliminary measurements establish two major points. First, the apparent angular size of the two most intense radio sources is thousands of times greater than that of the visible stars and is of the order of a few minutes of arc. Secondly, the source in Cygnus exhibits a pronounced asymmetry in angular size, whereas the source in Cassiopeia appears to be roughly symmetrical. The measurements are not yet adequate to define satisfactorily the shape of the sources or the distribution of intensity across their disks. Further observations are now being made with the present apparatus using different base-
lines, and the results will be published later.* 2

The work was carried out at the Jodrell Bank Experimental Station of the University of Manchester. The construction of the apparatus was made possible by a grant from the Department of Scientific and Industrial Research. We wish to thank Dr. R. Q. Twiss for his assistance with the mathematical theory, and Prof. A. C. B. Lovell for making the necessary facilities available and for his interest in the investigation. One of us (R. H. B.) is indebted to Messrs. Imperial Chemical Industries for a fellowship, and one (M. K. D. G.) to the Government of India for a scholarship and to the Department of Scientific and Industrial Research for a maintenance grant.

R. Hanbury Brown
R. C. Jennison
M. K. Das Gupta

Jodrell Bank Experimental Station,
Holmes Chapel,
Cheshire.

[1] Bolton, J. G., and Stanley, G. J., *Nature*, **161**, 312 (1948).
[2] Ryle, M., and Smith, F. G., *Nature*, **162**, 462 (1948).
[3] Stanley, G. J., and Slee, O. B., *Aust. J. Sci. Res.*, A, **3**, 234 (1950).
[4] Ryle, M., Smith, F. G., and Elsmore, B., *Mon. Not. Roy. Astro. Soc.*, **110**, 508 (1950).
[5] Mills, B. Y., *Aust. J. Sci. Res.*, A, **5**, 266 (1952).
[6] Smith, F. G., *Nature*, **168**, 555 (1951).
[7] Hanbury Brown, R., and Twiss, R. Q. (in preparation).* 3

Observations at Sydney

The first observers of the discrete sources of cosmic noise were unable to determine their size, and many considered that they might be of stellar dimensions. However, recent observations have shown that a number of these apparently 'point' sources have angular extensions of the order of $\frac{1}{2}°$ [1-3]. Position measurements have also suggested that the stronger sources might be identified with nebulæ of an angular size ranging from half a minute to several minutes of arc[2,4,5].

In order to check the hypothesis that the discrete source radiation originates in nebulæ rather than in individual stars, an interferometer with a lobe spacing as small as 1 minute of arc was considered necessary. At a frequency of 100 Mc./s., which is convenient because small low-gain aerials may be used, a Michelson-type interferometer requires an aerial separation of 10 km. to achieve this resolution. Such an interferometer was constructed using one small, easily portable aerial and one large fixed aerial. The signal from the portable aerial is transmitted back to the main receiver by a radio link. A simplified diagram of the equipment is shown in Fig. 1.

The local oscillator frequency must be transmitted together with the converted signal, as it is necessary to preserve the phase of the reconstituted signal at the receiving end of the link. Also a time-delay must be inserted in the signal-path from the fixed aerial to equalize the propagation times from each aerial to the mixing point. This time-delay is obtained with a mercury-filled acoustic delay line. Automatic gain controls are used to ensure equality of the signal-levels from each aerial and pre-amplifier system before they are combined in the receiver. The equipment has shown a variation in sensitivity of about 15 per cent over the period of six months that it has been in operation.

Preliminary measurements have been made on the four most intense sources observable from Sydney with various aerial spacings in an east–west direction. With the range of aerial spacings used, the four

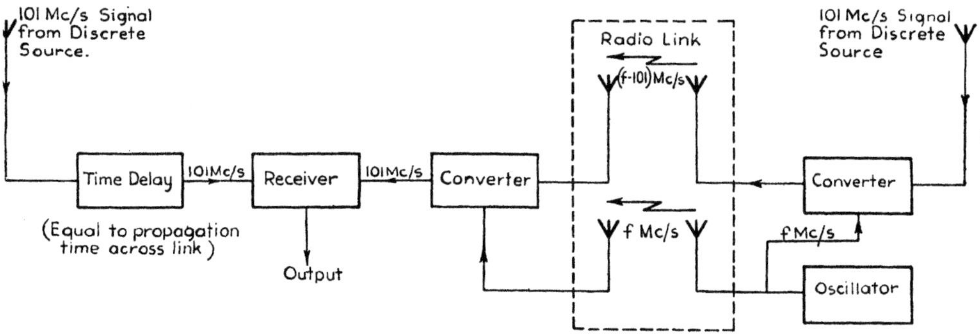

Fig. 1. Simplified diagram of the interferometer

sources show a reduction in the amplitude of their interference patterns as the aerial spacing is increased, indicating that all four sources have been resolved by the equipment. The effects of aerial separation on the amplitudes of the interference patterns are summarized in Table 1.

The probable errors in the relative amplitudes are thought to be about 10 per cent. The errors will be discussed in detail elsewhere; but it is worth mentioning one potential source of error. This is phase instability in the equipment and propagation link. It was checked using 101 Mc./s. transmissions from a small transmitter, which were received by both aerials and compared in phase at the mixing point. The resulting error was found to be negligible.

Table 1

Aerial spacing	0·29 km.	1·02 km.	5·35 km.	10·01 km.
Relative amplitude of interference pattern for Cygnus A	1	1	0·3	0·05
Taurus A	1	0·55	*	*
Virgo A	1	0·4	*	*
Centaurus A	1	0·3	*	*

* In these cases the amplitudes of the interference patterns were less than the noise fluctuations of the equipment and could not be observed. The relative amplitudes are less than 0·1–0·2.

An inspection of the table reveals that the amplitudes of the Taurus A, Virgo A and Centaurus A sources are all significantly reduced with an aerial spacing of 1 km., whereas a spacing of 5 km. is required to produce a comparable reduction in the amplitude of the Cygnus source. The first-mentioned sources are all, therefore, of similar size and approximately five times the size of Cygnus A.

It has been shown that the amplitude and phase of the interference pattern produced by a source when observed with a given aerial spacing is related to the radio brightness distribution across the source by a Fourier transform[6] (the brightness being measured as the integrated emission across a strip at right angles to the interferometer axis). Thus, if the amplitudes and phase angles are known at a sufficient number of spacings, the brightness distribution may be obtained by Fourier synthesis. In the case of an interferometer with an east–west axis as we have here, this synthesis is given by

$$B_t = \int_0^\infty A_\omega \cos(\theta_\omega + \omega t)\, d\omega,$$

where B_t is the brightness distribution in right ascension in arbitrary units, ω denotes the pulsatance of the sinusoidal interference pattern, corresponding to the aerial spacing, and A_ω and θ_ω the amplitude and phase of the pattern.

The present interferometer cannot be easily adapted to measure phase, so that the true distribution B_t is not easily obtained. If, however, the distribution is assumed to be symmetrical, θ_ω is then zero and an 'equivalent symmetrical distribution' can be obtained. This equivalent distribution is merely the cosine transform of the amplitude–pulsatance spectrum.

For the Cygnus source we have four points from which to determine this spectrum. In Fig. 2(*a*) these points are shown and a smooth curve is drawn through them. Because of the absence of observations in the region between 1 and 5 km., there is some uncertainty in the exact shape of the curve, but its general features are clear. The equivalent brightness distribution corresponding to this spectrum is shown in Fig. 2(*b*). Because of the uncertainty in the spectrum, there is a corresponding uncertainty in the shape of this curve, particularly with regard to the skirts.

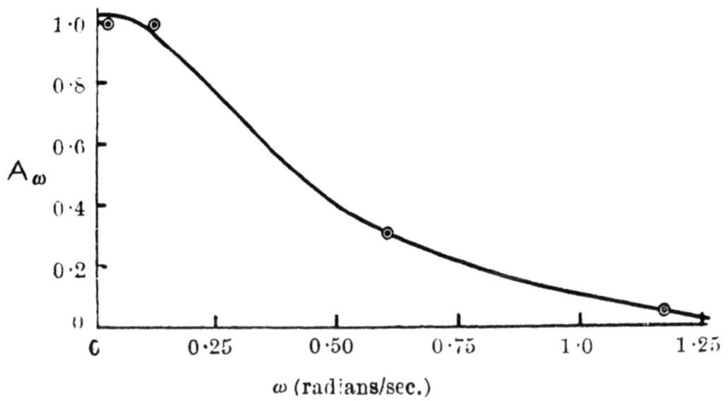

(*a*) Amplitude–pulsatance spectrum of Cygnus A

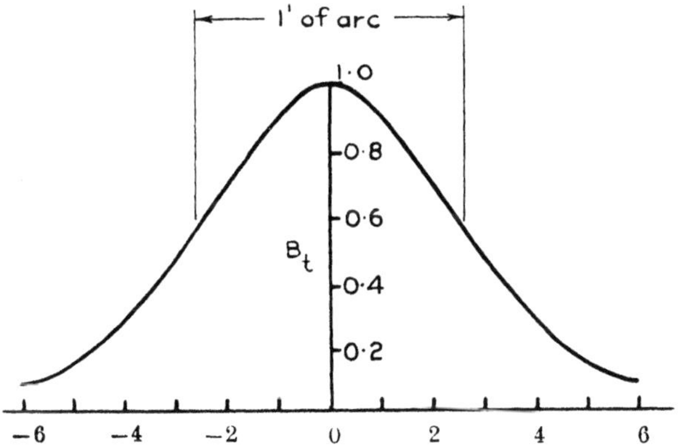

(*b*) Derived equivalent brightness distribution across Cygnus A

Fig. 2

In the case of the other sources, there are not enough points to determine the spectra ; but, if the reasonable assumption is made that their brightness distributions are similar to that of Cygnus A, it is possible to estimate their sizes. Defining an 'effective' size as the angle between half-brightness points on the equivalent distribution curve, we have the results of Table 2.

Table 2

Source	Effective size (E–W) (minutes of arc)
Cygnus A	1·1
Taurus A	4
Virgo A	5
Centaurus A	6

The size of the Centaurus source has been given previously as about $\frac{1}{3}°$ [1,2]. It has been shown, however, that it consists of an extended source of nearly 2° in size with a strong concentration near its centre[3]. The size quoted in Table 2 is that of the central concentration. Measurements with very close spacings have confirmed the interpretation in terms of two apparently quite distinct contributions.

The four sources have been found to have positions very close to nebulæ, and it is now found that their angular sizes are comparable with those of the nebulæ in question. We may therefore say that the suggested identifications of the four sources concerned have been strengthened, almost to the point of certainty. The confirmation is particularly welcome in the case of Cygnus A, as when it was first pointed out that this nebula was near the radio position of the source[5] it appeared to be a typical galaxy, too faint to be responsible for the radiation. Minkowski has now shown, however, that it has very peculiar features (personal communication). Taking these sources in conjunction with the others for which angular sizes have been obtained, there is now strong evidence that the discrete sources are largely, if not entirely, nebulæ. Both galactic and extra-galactic nebulæ have been identified.

This work is continuing and will be described in more detail elsewhere.* 4

B. Y. Mills

Division of Radiophysics,
Commonwealth Scientific and
Industrial Research Organization,
Australia.

[1] Mills, B. Y., *Aust. J. Sci. Res.*, **A5**, 266 (1952).
[2] Mills, B. Y., *Aust. J. Sci. Res.*, **A5**, 456 (1952).
[3] Bolton, J. G., "Extended Sources of Galactic Noise", U.R.S.I. Report (1952).
[4] Bolton, J. G., Stanley, G. J., and Slee, O. B., *Nature*, **164**, 101 (1949).
[5] Mills, B. Y., and Thomas, A. B., *Aust. J. Sci. Res.*, **A4**, 158 (1951).
[6] McCready, L. L., Pawsey, J. L., and Payne-Scott, Ruby, *Proc. Roy. Soc.*, A, **190**, 357 (1947).

PAPER 32

Observations at Cambridge

At the recent meeting in Sydney of the International Union of Scientific Radio, reports were given of three separate experiments designed to measure the angular diameter of radio stars by the use of interferometers of large resolving power. Two of the experiments, by Hanbury Brown and by Mills, have been extended and are the subject of the present communication. The third, carried out at Cambridge in the first half of this year, has been described elsewhere[1]; since a rather different range of apertures was used in this experiment, it is thought appropriate to describe the results briefly for comparison with the other accounts.

It was first pointed out by McCready, Pawsey and Payne-Scott[2] that interferometer aerials could be used to determine the angular distribution of radio 'brightness' across a source of radio waves. They showed that this distribution could be resolved into Fourier components by the use of different aerial spacings. The application of this principle to the determination of the distribution of radio brightness across the sun has already been described[3,4]; it was shown that a determination of the complete variation of the amplitude of a recorded signal as the aerial spacing was changed allowed the radial distribution of a circular source to be derived.

Similar observations have been made in an attempt to determine the distribution of brightness across the intense radio stars in the constellations of Cygnus and Cassiopeia. Owing to the smaller angular diameter of these sources, it was necessary to determine the amplitudes for the smaller spacings with considerable accuracy. It was not possible to determine the whole amplitude spacing curve with the maximum aerial separation available; the fine structure in the distribution of radio brightness across the source could not, therefore, be determined.

In these observations, which were made at a wave-length of 1·4 m., the phase-switching method of detection[5] was used with aerials spaced along an east–west line up to distances of 400 wave-lengths. This method of detection allows the use of pre-amplifiers at the aerials and, therefore, enables large spacings to be used without difficulties due to the attenuation in the transmission lines. In order to avoid the necessity for long-term stabilization of the gain of a movable aerial system, a new system of observation was devised in which measurements were made of the ratio of the amplitudes of the records obtained with two different aerial spacings. These two spacings were obtained by using the movable aerial alternatively with two fixed aerials, and it was then no longer necessary for its gain to remain constant. Each observation then provides a measure of the ratio of two Fourier terms; although it was not possible to derive the separate Fourier components uniquely from the results of these experiments, a fair approximation to the actual curve of Fourier components could be derived. The results obtained by this method are therefore comparable with those obtained in the two other methods; and although the present method cannot conveniently be used to make measurements at very great apertures, it has advantages for determining the Fourier terms related

Measured Ratio of Fourier Components

Aerial spacing in wave-lengths*		Cygnus	Cassiopeia
37	423	$0{\cdot}79 \pm 0{\cdot}05$	$0{\cdot}52 \pm 0{\cdot}05$
131	330	$0{\cdot}83 \pm 0{\cdot}02$	$0{\cdot}70 \pm 0{\cdot}02$
36	234	$0{\cdot}94 \pm 0{\cdot}03$	$0{\cdot}84 \pm 0{\cdot}02$
13	186	—	$0{\cdot}92 \pm 0{\cdot}03$
75	124	—	$0{\cdot}97 \pm 0{\cdot}02$

to the larger-scale structure of the brightness distribution.

The results of the Cambridge observations on the radio stars in Cygnus and Cassiopeia are shown in the accompanying table.

A possible interpretation in terms of Fourier components is shown in the diagram.

The measurements are consistent with those to be expected from a uniformly bright disk of $3{\cdot}5'$ diameter for Cygnus, and $5{\cdot}5'$ diameter for Cassiopeia. The measurements at larger resolving powers reported at the Sydney meeting of the International Union of Scientific Radio by Hanbury Brown and by Mills showed that there is also some structure of smaller angular width, and that there might be some considerable departure from circular symmetry. It is

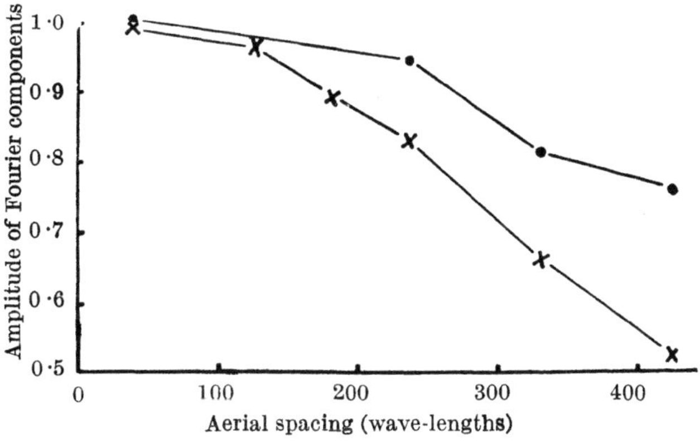

●—●—, Cygnus ; ×—×—, Cassiopeia

Measured Fourier components of the distribution of radio 'brightness' across the radio stars in Cygnus and Cassiopeia

of interest to compare these results with those that might be expected from the visible nebulæ which have been identified with those two radio stars (Baade and Minkowski, in course of publica-
6 tion).*

The Cygnus nebula is extragalactic. It has a central concentration about $\frac{3}{4}'$ of arc across, surrounded by a more diffuse region. The nebula is faint, so that it is difficult to estimate the extent of this region, but it is at least 2′ of arc in diameter.

The Cassiopeia nebula is a gaseous nebula in which the brightness is distributed unevenly over a roughly circular region about 5′ of arc in diameter. There are some particularly bright regions concentrated in an area extending 2′ of arc east–west and 1′ of arc north–south.

It appears that the radio measurements are in good agreement with the brightness distributions which might be expected from these nebulæ; this conclusion further strengthens the evidence for their identification with the two radio stars.

F. G. SMITH

Cavendish Laboratory,
Cambridge.

[1] Smith, F. G., *Proc. Phys. Soc.*, B., **65**, 971 (1952).
[2] McCready, L. L., Pawsey, J. L., and Payne-Scott, Ruby, *Proc. Roy. Soc.*, A, **190**, 357 (1947).
[3] Stanier, H. M., *Nature*, **165**, 354 (1950).
[4] Machin, K. E., *Nature*, **167**, 889 (1951).
[5] Ryle, M., *Proc. Roy. Soc.*, A, **211**, 351 (1952).

EDITOR'S NOTES AND AMENDMENTS

1. East is to the right.
2. Paper 33.
3. *Phil. Mag.* Ser. 7, **45**, 663 (1954).
4. *Austral. J. Phys.* **6**, 452 (1953).
5. The two columns in the first box of the table refer to the spacings of the two fixed aerials with respect to the movable aerial; "ratio of Fourier components" refers to the ratio of the fringe visibility at the longer baseline to that at the shorter.
6. Paper 29.

Fine Structure of the Extra-terrestrial Radio Source Cygnus I

THE investigation of the nature of the intense localized sources of extra-terrestrial radio emission which were discovered in 1948 has been hampered by the technical difficulties of achieving sufficient resolution to measure their dimensions. Partial solutions of these difficulties were reached simultaneously in 1952 by independent developments at Cambridge, Jodrell Bank and Sydney[1-3], as a result of which preliminary measurements of the angular diameters of the intense sources in Cassiopeia and Cygnus were made. These measurements were not in complete agreement, particularly in the case of the Cygnus source, where the Jodrell Bank results indicated that the source might be markedly asymmetrical. During the past year the fine structure of this source has been studied in greater detail, and the present communication summarizes the results obtained.

The investigation has been made on a frequency of 125 Mc./s. with an interferometer which has been described elsewhere[1,4,5]. It measures the variation of the correlation function ρ^2 with the aerial separation in terms of wave-lengths (N), where

$$\rho^2 = \frac{F^2 \cos \psi + F^2 \sin \psi}{F^2 \cos (o)}$$

and $F \sin \psi$ and $F \cos \psi$ are the sine and cosine terms of the Fourier transform of the angular distribution of intensity across the source[4]. The original measurements made by Hanbury Brown, Jennison and Das Gupta[1] with this interferometer were carried out at position angles of 58°, 113°, 179°, and the results indicated that the source was radially asymmetrical with a major axis near position angle 90°. In order to obtain a self-consistent set of results with these earlier measurements in a direction close to the major axis, the present investigation has been made at a number of spacings in the position angle 113°. The correlation function and the corresponding values of $N^\star$ 1
are given in the accompanying table.

The measurements of Smith[2] and Mills[3] were made in position angle 90°, and in order to compare the

Reprinted, with permission, from 'Nature' 172, 996–997 (1953).

three sets of results the present observations have been transformed into a position angle of 90°. This has been done by projecting the values of N from position angle 113° to 90°, and the values are given in the table.

The intensity distribution shown in Fig. 1 compares these results with those of Smith and Mills. The magnitude of the secondary maximum indicates that

INTERFEROMETER SPACINGS N AND CORRESPONDING CORRELATION COEFFICIENT ρ^2 FROM THE EXTRA-TERRESTRIAL RADIO SOURCE CYGNUS 1

N at 113°	N projected into 90°	$\rho^2 \pm 0{\cdot}04$
610	560	0·460
650	598	0·500
900	830	0·200
1,450	1,340	0·045
1,558	1,440	0·065
1,655	1,520	0·120
1,952	1,800	0·245
2,100	1,930	0·280
2,230	2,050	0·290

the source must have at least two distinct centres of emission. The distribution of intensity across the source cannot be determined uniquely since the phase of the transform is unknown ; however, the shape of the curve in Fig. 1 suggests that the phase of the transform is reversed in the secondary maximum, and this favours a symmetrical source with two centres of emission.

The simplest distribution which will yield the transform shown in Fig. 1 consists of two components of equal intensity, each of length 51″, separated by 1′ 28″. The additional information[1] supplied by the results obtained on bearings of 179° and 58° indicates that the source has a very small minor axis—less than 35″ in a position angle of approximately 180°—and it is apparent that the components forming the source must be distributed in a narrow strip as shown in Fig. 2.

The accurate position of the radio source in Cygnus has been determined by Smith[6]. It coincides with an extra-galactic object photographed at the 200-in. Hale telescope by Baade and Minkowski[7]. This object has a maximum diameter of 30″ in position

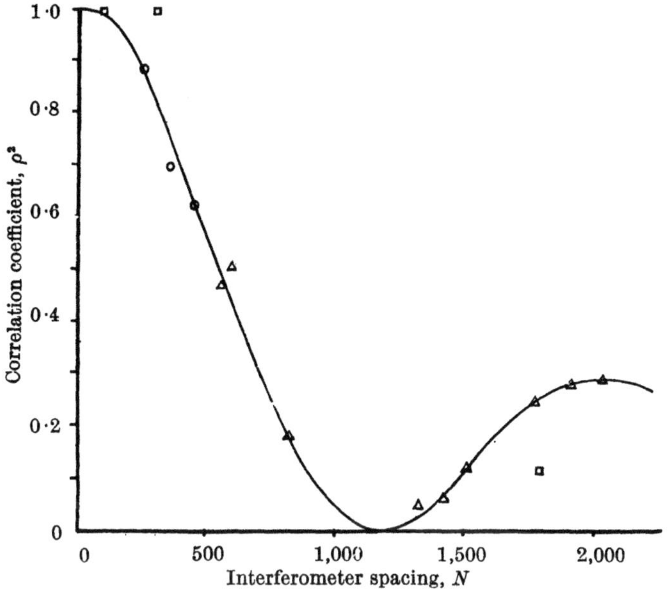

Fig. 1. Correlation coefficients ρ^2 measured at various interferometer spacings, N, for the radio source in Cygnus. Observations by Smith, ○ ; observations by Mills, □ ; observations by Jennison and Das Gupta, △. The continuous curve denotes the theoretical transform of the distribution shown in Fig. 2

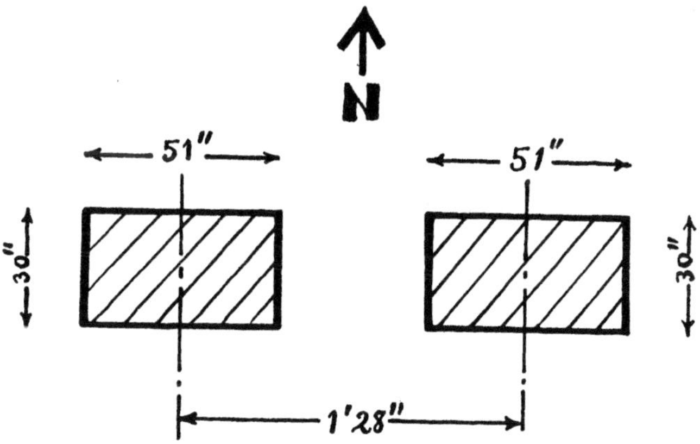

Fig. 2. Approximate intensity distribution of the extra-terrestrial radio source in Cygnus

angle 150° and shows a high excitation emission spectrum with marked signs of tidal distortion. It has been interpreted by Baade and Minkowski as two late-type spiral galaxies in collision. Although Fig. 2 represents the most compact distribution of the radio source which can be derived from the measurements shown in Fig. 1, it is nevertheless much larger than the visual object described above. The two components of the radio source straddle the visual object with little overlap between the regions of optical and radio emission. If this identification of the radio source is correct, there would appear to be no direct correlation between the radio emission and the visible light from the colliding galaxies.

The work was made possible by a grant from the Department of Scientific and Industrial Research. We wish to thank Prof. A. C. B. Lovell and Mr. R. Hanbury Brown for their interest in the investigation. We are also indebted to Drs. Baade and Minkowski for the information about their work on the colliding galaxies in advance of publication. One of us (M. K. D. G.) is indebted to the Government of India for a scholarship and to the Department of Scientific and Industrial Research for a maintenance grant, and the other (R. C. J.) to the Royal Society for the Mackinnon studentship.

R. C. JENNISON
M. K. DAS GUPTA

Jodrell Bank Experimental Station,
Lower Withington,
Cheshire.

Nov. 4.

[1] Hanbury Brown, R., Jennison, R. C., and Das Gupta, M. K., *Nature*, **170**, 1061 (1952).
[2] Smith, F. G., *Nature*, **170**, 1065 (1952).
[3] Mills, B. Y., *Nature*, **170**, 1063 (1952).
2 [4] Hanbury Brown, R., and Twiss, R. Q. (in publication).★
3 [5] Jennison, R. C., and Das Gupta, M. K. (in publication).★
[6] Smith, F. G., *Nature*, **168**, 555 (1951).
4 [7] Baade, W., and Minkowski, R. (in publication).★

EDITOR'S NOTES AND AMENDMENTS

1. *N* is the interferometer spacing measured in wavelengths.
2. *Phil. Mag.* Ser. 7, **45**, 663 (1954).
3. *Phil. Mag.* Ser. 8, **1**, 55 + 65 (1956).
4. Paper 29.

PART V

SPECTROSCOPY

The great advantage of spectroscopy over continuum work lies in the ability of a line profile to reveal accurate information concerning the temperature and density of the region in which the line originates. Furthermore, one can easily measure radial velocities and generally be certain of exactly which atomic or molecular species is responsible for the observed line. Thus, it was natural to attempt to extend this powerful astronomical technique into the radio domain. This section is overwhelmingly concerned with a single transition, the 21 cm hyperfine line of atomic hydrogen. The line was first predicted by Van de Hulst in 1944 (Paper 34); this work was improved upon in a largely independent manner by Shklovsky in 1948 (Paper 35). Despite these encouraging calculations, the difficulties of receiver technology were such that the line was not detected in the Milky Way until several years later, in 1951 (Papers 36 and 37). And yet, when this occurred, as so often happens in modern science, detection and subsequent confirmation were accomplished within a very short interval by three groups scattered around the globe – in the United States, Holland, and Australia.

It was realized from the beginning, especially by the Dutch group led by J. H. Oort, that this line would be an invaluable tool, not only for probing the conditions between the stars, but also for delineating galactic structure. For example, only nine months after the discovery of the 21 cm line, a lecture was given by Oort (1952) on the topic of galactic structure. It was a remarkable discourse on how radio techniques would contribute to our perception of the galaxy. Oort's single-mindedness of purpose was such that it almost seems that he knew exactly what was about to transpire over the rest of the decade, as if the observations were only necessary in order to fill in details here and there! It is ironic that 1951 was also the year in which optical spiral features were first recognized; Morgan *et al.* (1952) showed that the nearby regions of luminous stars and ionized hydrogen tend to lie in two "long, narrow belts" which were soon named the Orion and Perseus arms. The Australians followed this with evidence for possible spiral structure in the neutral hydrogen in a survey by Christiansen and Hindman (1952). The subsequent

Dutch and Australian surveys of opposite sides of the Milky Way culminated in a 1958 review by Oort, Kerr, and Westerhout, in which the entire jigsaw puzzle was assembled into the famous "Leiden–Sydney" map.

Another aspect of study of the 21-cm line which represents a basic finding in the early 1950s is that the line can be studied in *absorption* against a strong background source of radio continuum. This effect was discovered by Williams and Davies (1954) at Jodrell Bank and by Hagen, Lilley, and McClain (1955) at the U.S. Naval Research Laboratory. It was immediately used both to probe physical conditions in the interstellar medium and to estimate distances to galactic radio sources.

A few other theoretical studies of line emission deserve special mention. Saha (1946) speculated on detecting radio radiation which might occur if the energy levels of atomic nuclei were excited by a strong magnetic field. Following the laboratory work of Rabi, his hope (never realized) was that an RF-varying magnetic field might exist in sunspots so that one could do astrophysical experiments using nuclear magnetic resonance. A paper written by Wild (1952), immediately after the discovery of the 21 cm line, contains a thorough analysis of the atomic structure and spectroscopy of the hydrogen atom, notwithstanding its summary dismissal of high-n recombination lines (see the introduction to Paper 34). Paper 35 by Shklovsky is perhaps most notable for being the first to investigate the feasibility of detecting line radiation from interstellar *molecules.* Another landmark study along similar lines was that in 1955 by Townes (1957). These prognostications were, however, somewhat delayed in coming to fruition through successful observations. Hydroxyl was not detected until a decade later, followed by the deluge of complex interstellar molecules which began in the late 1960s.

SECTION P

PREDICTION OF THE 21 CM HYDROGEN LINE

Paper 34 by Van de Hulst is remarkable both for its scientific prescience and for the conditions under which it was produced. During the time of the Nazi occupation of Holland, Van de Hulst was a student under M. Minnaert at the University of Utrecht. When Minnaert was taken away to a detention camp, Van de Hulst chose to continue his studies at nearby Leiden. There he came into contact with Oort, who had recently been exposed to Reber's report of galactic radio waves (Paper 5) in an issue of the *Astrophysical Journal* which had been somehow smuggled into the country. Oort was immediately struck by the possibilities which these dust-penetrating radio waves held for the investigation of galactic structure. He therefore arranged for the topic of one of the inter-university meetings of the *Astronomenclub*, a Dutch society comprising amateurs and professionals alike, to be concerned with radio techniques and the observations of Jansky and Reber. The first principal speaker was C. J. Bakker, a radio engineer from the Philips Research Laboratory; Bakker summarized the applicable theory of radio noise and the measurements of Jansky, Reber, and Fränz. This paper was also eventually published in 1945 (immediately before Paper 34) once the war had ended.

The second speaker was Van de Hulst and he succeeded in far more than a summary of the prevalent theory. In Paper 34 we find many original contributions to the nascent field of radio astrophysics. After ruling out both stars and interstellar "smoke" (dust grains) as significant contributors to the observed radio waves, he derives the theory of free-free radio emission from the gas in a somewhat different manner and style than that of Henyey and Keenan in 1940 (Paper 8). The calculated spectrum (his Figure 2) agrees very well with the previous results as well as with modern theory.

The heart of the paper, and the section of most interest today, lies in the discussion of the possibility of discrete lines in the spectrum of interstellar hydrogen. Van de Hulst first considers the high-level transitions which are now called recombination lines. His expressions for line frequency, integrated line-to-continuum ratio ($\epsilon_{\alpha\text{-line}}\Delta\nu/\epsilon_{\text{cont}}$), and Doppler broadening are all in basic agreement with modern theory. He only goes wrong when considering the line broadening due to the Stark effect. He states that this broadening is "$\Delta\nu/\nu = (\text{wavelength}/100\text{ m})^{3/5}$", but a recent analysis of his original working notes has revealed that a transposition error occurred; the exponent, as derived from the work of Inglis and Teller, should have been 5/3.[1] In either case, the expected broadening at low frequencies is sufficiently great that the intensity ratio of line to continuum is extremely small and, therefore, the line is not observable. The different exponents, however, predict very different line-widths at high frequencies. For instance, in the case of the n: $110 \rightarrow 109$ transition at 6 cm, Van de Hulst's expression implies a line intensity fully ~100 times smaller than the "5/3-law" value of ~2–3% of the continuum signal. This latter value is, in fact, close to that observed today, but it required another theoretical prediction by Kardashev (1959) to stimulate Soviet experimenters to the first successful observations of radio recombination lines exactly two decades later.

Van de Hulst next turns to consider the detectability of the 21 cm transition arising from the hyperfine splitting of the hydrogen atom. He is not deterred by the fact that neither the frequency nor the intensity of the line had ever been measured in the laboratory. He derives a frequency of 1411 MHz and, using expected physical conditions for the interstellar medium, determines that the line may well be observable if the probability of spontaneous emission, A, is greater than 8×10^{-17} sec^{-1}. In fact, his notes reveal a bit of incredulity concerning this happenstance – a transition of the most abundant species at an ideal wavelength; at one point in his derivations he writes, "If now the whole sky isn't gleaming at this wavelength!" Since Van de Hulst's requirement on A is indeed more than satisfied (by a factor of ~35), the line is quite intense. Many years later it was soon found that one cannot direct an antenna towards any position near the galactic plane without receiving a very strong signal from the neutral hydrogen line. The cautious tone of the actual prediction in Paper 34, using a word like "speculative", arises from the lack of certainty in dealing with spectral emission of such low frequency and with such a long natural lifetime relative to those of the familiar optical lines.

The final *tour de force* of Paper 34 is the fact that it is the first to discuss the implications of radio data on various world models. The final section is based on Reber's spurious detection of the Andromeda Nebula (Paper 5) and the apparent flat radio spectra of galaxies. Using these, the author is able to adduce strong arguments against a static model of the Universe. Although his input data were faulty, the general thinking anticipates by a decade the eventual contributions which radio astronomy was to make to cosmology.

NOTE

[1] Current theory in fact yields an exponent of ~5.0.

ORIGIN OF THE RADIO WAVES FROM SPACE*

(Herkomst der radiogolven uit het wereldruim)

H. C. VAN DE HULST

1 SUMMARY.* Radio waves, received from any celestial object – they being the far infrared portion of its spectrum – deserve attention. Observations of small objects are prevented by diffraction. The Sun may be a measurable object for future instruments.

The radiation observed from our galaxy must be due to the interstellar gas, the stars
2 being ruled out by their small angular dimensions and the solid smoke particles* being ruled out by their low temperature.

The spectral emission of a homogeneous layer of ionized hydrogen is computed. The continuous spectrum arising from free-free transitions has the intensity of black-body radiation at wavelengths larger than 6 m and has a nearly constant intensity at wavelengths smaller than 2 m, corresponding to a large and to a small optical thickness, respectively. These intensities, shown in Figure 2, agree with those computed by Henyey
3 and Keenan* and tally fairly well with the observations. No better accordance is to be expected, owing to the unknown electron density and extension of the interstellar gas and to unsatisfactory data about the directional sensitivity of the antenna.

Discrete lines of hydrogen are proved to escape observation. The 21.2-cm line, due to transitions between hyperfine structure components of the hydrogen ground level, might be observable if the lifetime of the upper level does not exceed 4×10^8 years, which, however, is improbable.

Reber's observation of the Andromeda nebula suggests a rather high electron density. A cosmological remark concludes the article. The low background intensity due to remote nebulae contradicts the Hubble–Tolman static model.

ASTRONOMICAL SIGNIFICANCE

Although the existence of radio waves of extraterrestrial origin has already been known for a decade, astronomers have not yet paid much attention to them. This is partly due to the crude data thus far furnished by the observations; not much more has been established than the order of magnitude of the intensity and the rough directional dependence of the radiation. Thus, little can be expected from a detailed discussion of the observational facts.

Neither is the existence of these waves especially interesting from a purely theoretical viewpoint. The production of radio waves is not at all an essential feature of the physical conditions in the interstellar gas. The amount of

* [translated from the Dutch by W. T. Sullivan, III in 1973; first published in *Nederlandsch Tijdschrift voor Natuurkunde* 11, 210 (Dec. 1945).]

kinetic energy which is converted into radiation energy is negligible; it is like a small leak in the overall energy budget which begins with the ionization of interstellar atoms by starlight. From purely theoretical considerations of the production of these radio waves, we cannot expect any new insights into the physical state of the interstellar gas,[1] mainly characterized by its density, degree of ionization and distribution of electron velocities.

It is nevertheless the possibility of direct observation of these waves which makes this subject attractive. For twenty centuries astronomers have drawn all their knowledge from observations in the rather narrow frequency range around the visual frequencies. They have designed powerful instruments and for these observations have spared themselves no trouble. The Earth's atmosphere, however, leaves open another frequency range, near the radio frequencies, for astronomical observations. The first measurements have now been made in this range, but the observational technique is still in its infancy. In the previous article Bakker[2] has shown that these observations can be considerably improved with equipment already available.

The long wavelength does, however, entail one difficulty. Without telescopes of enormous aperture the radio waves will never furnish a detailed picture of the heavens. For the time being we must be quite content if a resolving power of about one degree can be realized. The Sun, the Milky Way, and the brightest extragalactic nebulae would then be measurable objects. In the Milky Way, the variation of intensity with longitude and latitude as well as other individual details, especially the distribution of H I and H II regions in the interstellar gas, could be investigated. There is the further possibility that the very distant extragalactic nebulae form a measurable diffuse background which would be of particular cosmological importance.

THE SPECTRUM OF THE MILKY WAY

We shall now attempt to derive theoretically the appearance of the spectrum of the Milky Way at these frequencies. We consider the entire width of the "radio window" in the Earth's atmosphere, i.e., from around 20 m to 1 cm wavelength (Figure 1). The observations (Figure 2) at wavelengths of 14 m, 16 m and 1.86 m produce intensities of a similar order of magnitude. Although a continuous spectrum is likely, we do not want to exclude the possibility of discrete spectral lines. That we are discussing the Milky Way implies no limitation of any kind. The Milky Way is what we see of our galaxy with its threefold population of stars, interstellar smoke and interstellar gas. The

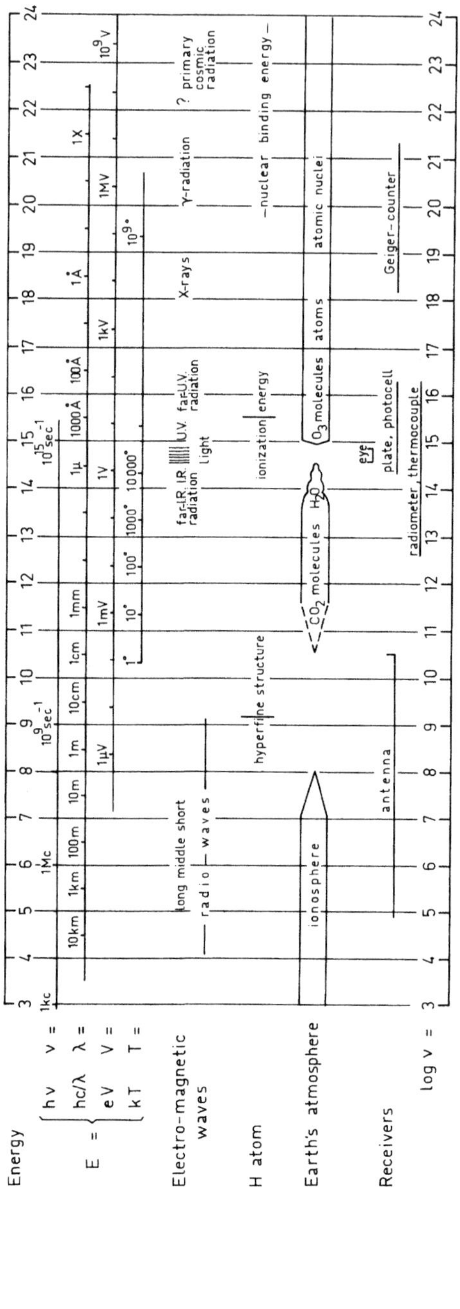

Fig. 1. A comparison of orders of magnitude. This article treats only the region of the radio window in the Earth's atmosphere.

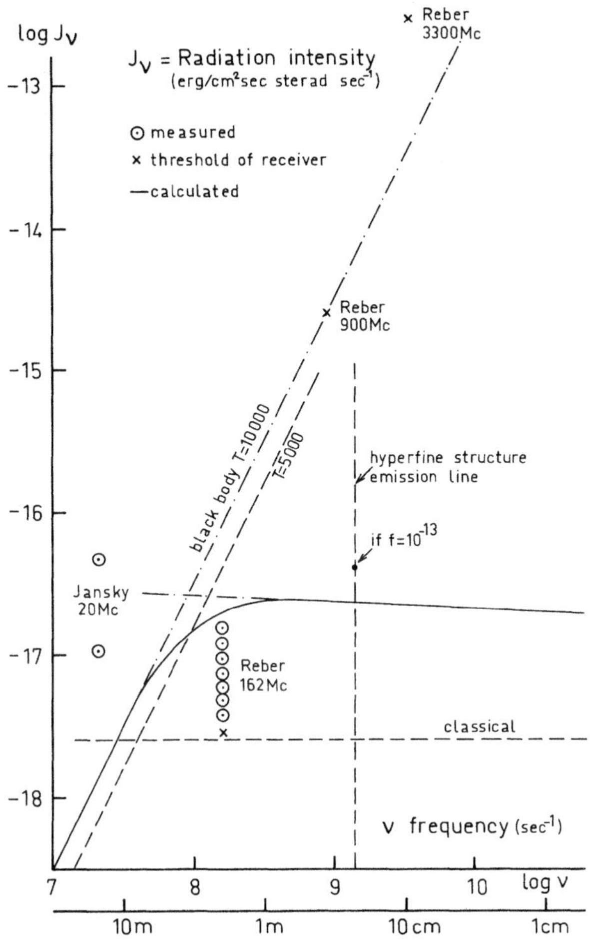

Fig. 2. Measured intensity of the Milky Way, compared with the calculated intensity of a homogeneous layer of ionized hydrogen gas ($T = 10\,000°$K, $n_e = n_p = 1$ cm^{-3}, $l = 16\,000$ pc). [Editor's note: The line - - - - - shows the expected spectrum using classical calculations, and the line — · — · — shows that for an optically thin layer of gas. The higher Jansky datum refers to 1932 measurements at a wavelength of 14.6 m (Paper 1); the lower is derived from 1937 measurements at 16.7 m. The range of values for Reber at 162 Mc/s comes from Figure 3 of Paper 4.]

greatest intensity of radio waves is observed in the direction of the constellation Sagittarius. Indeed we know from many other data that this is the direction towards the galactic center, where we look through the deepest and densest layers of our galaxy. As a rough model for calculations[3] we shall take the galaxy in this direction to be a homogeneous layer of depth 16 000 pc = 5×10^{22} cm, and with a composition the same as that which we know near the Sun.

First we try to determine from which population group of our galaxy the observed radio waves mainly originate. For that purpose we use the black-body radiation law, which is the classical Rayleigh–Jeans law for the wavelength range where $h\nu \ll kT$:

$$j_{\text{black}} = 2\nu^2 kT/c^2 \ . \tag{1}$$

For a comparison of orders of magnitude in this equation, consult Figure 1.

(a) *Stars.* These probably radiate approximately as black bodies with a mean $T \sim 5000$°K. The surface of the sun and of the stars thus has an intensity, indicated by the line $T = 5000°$ in Figure 2, which increases in proportion to ν^2. It does not seem impossible that in the future the intensity of the Sun might be measurable at decimeter wavelengths.[4] Reber's method of measurement was still not accurate enough for this. It is certain, however, that the stars together do not provide a sky background of sufficient intensity, since the stellar disks undoubtedly cover less than 10^{-10} of the sky area.

(b) *Interstellar Smoke.* These solid particles of the interstellar material have a temperature of ~3°K. Even if they covered the entire sky, their contribution could be neglected. This conclusion is not altered by the more accurate details which Whipple and Greenstein[5] included in their extensive calculation of this contribution, namely (1) the consideration that too small a particle cannot radiate at all in this frequency range since it has no eigenfrequencies, and (2) the possibility that the smoke particles near the galactic center might have a temperature as high as 30°K because of the greater energy density of stellar radiation.

(c) *Interstellar Gas.* The interstellar gas has a kinetic temperature of ~10 000°K and, in contrast to the stellar material, radiates from the entire area of the sky. The only question is whether the layers are optically thick enough to produce black-body radiation. And this question cannot be answered unless we go somewhat deeper into the mechanism of the emission.

RADIATION FROM A HOMOGENEOUS LAYER OF GAS

First we give a short summary of the well-known equations for the intensity of radiation emitted by a homogeneous layer of gas. If x is the path of the light and j_ν the intensity of the radiation at a frequency ν, then j_ν satisfies the differential equation

$$\frac{dj_\nu}{dx} = \epsilon_\nu J_\nu - \kappa_\nu j_\nu \,. \tag{2}$$

The second term is the product of the coefficient of absorption κ_ν and the local intensity j_ν. The first term is the product of the coefficient of emission ϵ_ν and the constant intensity $J_\nu = 2h\nu^3/c^2$. For a layer of thickness l, integration of this equation yields the intensity:

$$j_\nu = J_\nu \frac{\epsilon_\nu}{\kappa_\nu}(1 - e^{-\kappa_\nu l}). \tag{3}$$

For thin layers we can approximate this as $j_\nu = J_\nu \epsilon_\nu l$ and for thick layers as $j_\nu = J_\nu \epsilon_\nu / \kappa_\nu$.

The coefficient of emission represents the summed coefficients of all processes whereby radiation is emitted or scattered into this direction, and that for absorption represents all processes which absorb or scatter radiation from this direction. We have already demonstrated by a rough estimate that the only emission processes which can produce a sufficiently large contribution are the atomic processes in the interstellar gas. A similar estimate for "exotic" absorption processes, namely absorption by the smoke particles and scattering by free electrons, shows that they can be neglected since their contribution to the optical depth $\kappa_\nu l$ is $\ll 1$. Thus if there is any absorption at all, it must be self-absorption, i.e., absorption in the same atomic transitions in which the emission occurs.

Within a discrete spectral line ϵ_ν and κ_ν change quite sharply with frequency. Their values integrated over the width of the line, however, follow from quantum mechanical calculations of the transition probabilities or oscillator strengths. The following equations obtain:

$$\begin{aligned} \int \epsilon_\nu \, d\nu &= \int \epsilon'_\nu \, d\nu = \frac{\pi e^2}{mc} f_{nn'} N_{n'} \\ \int \kappa_\nu \, d\nu &= \frac{\pi e^2}{mc} f_{nn'} N_n \,. \end{aligned} \tag{4}$$

The coefficient of self-absorption is the difference between that for real absorption and that for stimulated emission. The letters e, m, c, and h are the usual physical constants, $f_{nn'}$ is the oscillator strength of a transition from a level n of unit statistical weight, and N_n is the population of that level. Analogous relations hold for a continuous spectrum except that ϵ and κ themselves are used and not their integrated values. The Boltzmann equation for thermodynamic equilibrium gives $N_{n'}/N_n = e^{h\nu/kT}$. It follows that, for a spectral line as well as a continuous spectrum, $\kappa/\epsilon = \mathrm{e}^{h\nu/kT} - 1$. Using this relation, the approximation given above for j_ν in the case of a thick layer becomes Planck's law for j_{black}.

THE CONTINUOUS SPECTRUM

The greatest part by far of the interstellar gas is hydrogen. Thus we must search in the spectra of ionized and un-ionized hydrogen for small quantum jumps of the order of a microvolt. The energy level diagram of hydrogen is well known. The states with energy $E > 0$, in which a free electron describes a hyperbolic orbit around a hydrogen ion, are not quantized. Those with $E < 0$, in which the atom is not ionized and the electron is bound in an elliptic orbit, are limited to $E_n = -hR/n^2$ where n is an integer and hR is the ionization energy.

If an electron jumps from a state of energy E to a lower one of energy E', then a light quantum $h\nu = E-E'$ is emitted. The possible transitions are: (I) free → free, (II) free → bound (i.e., capture of the electron by the proton), and (III) bound → bound. If for a moment we think of the bound states as well as the free states to be continuously distributed, then the possibilities (I), (II), and (III) can be treated together. For all three the following equation holds:

$$\epsilon_\nu = CN_e T^{-3/2} \nu^{-3} \int \mathrm{e}^{-E/kT} G\, dE. \tag{5}$$

This equation comes from the theory of Kramers, corrected by the quantum mechanical factor G.[6] The quantity N_e is the number of electrons per cm^3, which is also equal to the number of protons per cm^3. These quantities are also relevant to the bound-bound transitions since use is made of the equation of ionization equilibrium.[7] The integral must be extended over all possible energies E of the upper level. The constant is

$$C = \frac{2^4}{\pi 3\sqrt{3}} \left(\frac{h^2}{2\pi mk}\right)^{3/2} \frac{\pi e^2}{mc} \frac{R}{2h} = 10^{24.42}\ \text{erg}^{-1}\ \text{cm}^5\ \text{sec}^{-3}\ {}^\circ\text{K}^{3/2}\ .$$

We can estimate the ratios of the contributions (I), (II), and (III) by inserting the appropriate limits of integration in Equation (5). For (I) the integration extends from $h\nu$ to ∞ and the value of the integral is $\bar{G}_I kT$ since $h\nu \ll kT$. For (II) and (III) the values of G and the exponential factor are both practically equal to 1, and the value of the integral is equal to the range of integration dE. The integration for (III) must extend from the lowest level at which lines of this frequency are possible up to $E = 0$. The range of integration is roughly $(hR)^{1/3}(h\nu)^{2/3}$. The ratios of the contributions are thus

$$\mathrm{I} : \mathrm{II} : \mathrm{III} = \bar{G}_I kT/h\nu : 1 : \left(\frac{R}{\nu}\right)^{1/3} .$$

Since $\bar{G}_I \sim 8$, we conclude that the component arising from the free-free transitions is by far the largest. Both other components are negligible with respect to (I). The contribution of (II) is about a millionth of (I) and that of (III), if one thinks of the spectral lines as smeared out, is somewhat more than (II), but cannot touch (I).

We can now numerically compute the intensity of the continuous hydrogen spectrum which arises totally from free-free transitions. The coefficient of emission is

$$\epsilon_\nu = Ck\,\bar{G}\,N_e^2\,T^{-1/2}\nu^{-3}.^{\star} \qquad 4$$

An optically thin layer of gas of thickness l emits an intensity

$$j_\nu = 2hkc^{-2}C\,\bar{G}\,N_e^2\,T^{-1/2}\,l\,. \qquad (6)$$

It can be seen that ν^3 cancels out so that the classical value of the intensity is the same for all ν. The quantum mechanical correction factor changes only slightly [with frequency], from ~9 to ~6. Substitution of $N_e = 1\ \mathrm{cm}^{-3}$, $T = 10\,000\,^\circ\mathrm{K}$, and $l = 5 \times 10^{22}$ cm gives

$$j_\nu = \bar{G} \cdot 2.7 \times 10^{-18}\ \mathrm{erg\,cm}^{-2}\,, \qquad (7)$$

where now the units of time (sec) and frequency (sec^{-1}) have cancelled each other. Figure 2 shows our computation of this intensity as a function of frequency for the case where the interestellar gas is an optically thin layer. On the other hand if the gas is optically thick, the intensity then falls to the value of j_{black} on account of the self-absorption. From Figure 2 we see that this holds for wavelengths longer than 10 m. The Milky Way is optically thin for wavelengths shorter than 1 m. The transition region between 10 and 1 m is drawn in Figure 2 as given by Equation (3). The entire curve agrees very well with that calculated by Henyey and Keenan.

The agreement between the observed and computed intensities is satisfactory. It is obvious why Reber has not found anything at the two short wavelengths. His sensitivity would have been just good enough for black-body radiation, but the actual radiation is much weaker. Reber's highest intensity at 1.8 m agrees well with the computed intensity in the direction of the center of the Milky Way. We must not let our joy over this, however, cause us to overlook the large uncertainties. The observed intensities are first of all definitely uncertain by a factor of 2 or more because of difficulties with the directional characteristics [of the antennas employed]. There is no reasonable doubt concerning the theoretically derived Equation (6), but quite a bit of doubt about the actual numbers substituted into it. The thickness of the layer is probably a factor of 10 less, since only the HII regions are effective; the temperature T is rather well determined and has little influence; the electron density, however, may well be a factor of 2 higher, which makes the intensity a factor of 4 higher.

The comparison of Jansky's observations with the theory comes off somewhat differently. Since the gas is optically thick at these wavelengths, the intensity depends only on the temperature.

Henyey and Keenan fear that departures from thermodynamic equilibrium, in particular departures from the Maxwellian velocity distribution of the electrons, may be able to have a large unexpected influence here. The coefficient of self-absorption represents the difference between the real absorption and the stimulated emission; the ratio of the two processes is here $10^7 : (10^7 + 1)$. One can demonstrate by a trick, however, that this fear is unfounded. This trick is to choose the integration variables such that the difference integrand is not $(B_E - B_{E+h\nu}) \cdot G_{E,E+h\nu}$, but $B_E\, (G_{E,E+h\nu} - G_{E-h\nu,E})$. The gradual variation of the Gaunt factor G then guarantees approximately the same result, even when the population factor B_E is not precisely a Boltzmann factor.

We must thus conclude that the measured points of Jansky in Figure 2 lie too high by a factor of 10 and more. It seems quite possible to place most of the blame on the insufficiently known directional characteristics [of his antenna]. We cannot reasonably gain more than a factor of 2 by assuming a higher temperature.

ARE THERE ALSO DISCRETE SPECTRAL LINES?

We have established that on average the bound-bound transitions contribute an imperceptible amount to the continuous spectrum. The energy liberated in these transitions, however, is emitted in discrete spectral lines, and it is conceivable that within the rather narrow lines the intensity could be appreciably higher than in the continuum.

The strongest lines are the *α-lines*, emitted in transitions from a level $n+1$ to the next lower level n. They are a direct continuation of the Lyman α series ($2 \to 1$), the Balmer α series ($3 \to 2$), etc. Near 1.8 m wavelength lie the α-lines $341 \to 340$, $342 \to 341$, etc.; their separation is 1600 kc sec^{-1}. The ratio of the coefficient of emission within the α-lines to that calculated above for the continuum is

$$\frac{\epsilon_{\alpha\text{-line}}}{\epsilon_{\text{cont}}} = \frac{\nu}{\Delta\nu}\,\frac{h\nu}{kT}\,\frac{G_\alpha}{\overline{G}}\,.^{\star} \qquad (8)$$ 5

Because the area of the line is a constant, the coefficient of emission within the line is inversely proportional to the width $\Delta\nu$ of the emission line.

The broadening [of the line] has various causes. Thermal motion of the hydrogen atoms accounts for at least $\Delta\nu = \nu/10^4$. Galactic rotation and large-scale motions of the clouds of interstellar gas contribute the same order of magnitude. Still more important, however, is the Stark-effect broadening which is caused by the electric fields of nearby electrons and ions. We are inclined to overlook these since the gas is very rarefied; on the other hand, the fact is that the highly-excited atoms have very large dimensions. We roughly estimate[8] that the Stark-effect broadening is $\Delta\nu/\nu = (\text{wavelength}/100\ \text{m})^{3/5}$.★ Thus the Stark-effect broadening dominates over the Doppler 6
broadening for all α-lines in the radio window.

If we substitute [into Equation (8)] the value of $\Delta\nu/\nu$ just given, the proper $h\nu/kT$, and $G_\alpha/\overline{G} = 0.78/8 = 0.1$, then it is evident that all of the α-lines are *unobservable*. Due to the severe line broadening, the lines are not narrow enough to compensate for the ratio $h\nu/kT$. The same is true for lines arising in other transitions between the high energy levels of hydrogen.

All of these lines, just as the free-free continuum, would only be formed in H II regions. But there still remains a completely different possibility. The ground state of hydrogen is split by *hyperfine structure* into two levels with a separation of 0.047 cm^{-1}.[9] The spins of the electron and proton are pointed in the same direction in one state and are opposite in the other state. A quantum of wavelength 21.2 cm★ is emitted due to a spontaneous flip of the spin. 7
Such a transition is, of course, forbidden. On the other hand, the ground state, on account of the extreme dilution of radiation in interstellar space, is preferred above all other states (including the free states) by a factor 10^{14} compared with that in thermodynamic equilibrium. In the H I regions, where the ionized state is completely absent, practically all of the atoms are in the

ground state. We presume that these atoms are about equally divided between the two sublevels.

It would be especially interesting if this line were observable. With no self-absorption the intensity within the line is

$$j_{\text{line}} = \frac{1}{4\pi} lNAh\nu/\Delta\nu\,. \tag{9}$$

In this equation A is the transition probability for spontaneous emission, which can be theoretically calculated, but which is at present [experimentally] still unknown. We substitute $l = 5 \times 10^{22}$ cm, $N = 1$ cm^{-3} and $\nu/\Delta\nu = 10^4$. The line will stand out well above the continuum if $j_{\text{line}} > j_{\text{cont}} = 2 \times 10^{-17}$ erg cm^{-2}. This will be satisfied if $A > 0.8 \times 10^{-16}$ sec^{-1}, i.e., if the lifetime of the hydrogen atom in the upper hyperfine-structure level of the ground state is less than 4×10^8 years! That would not seem to be a heavy demand. For the usual definition of oscillator strength f, this requirement is $f > 5 \times 10^{-14}$; for Shortley's definition[10] of magnetic dipole strength S_m, the requirement is $S_m > 0.03$. Since S_m is usually of order 1, this possibility does not appear hopeless, even when we consider that the sensitivity of today's receiver installations must be improved by still another factor of 100. There are reasons to suppose, however, that the actual S_m may be quite a bit smaller. Until a rigid calculation is made, the existence of this line remains speculative.

THE ANDROMEDA NEBULA

We are accustomed to speaking about our own galaxy, but let us now take a look at the situation in the other galaxies.

From the nearest galaxy, the Andromeda Nebula, Reber has received radio
8 radiation of barely observable intensity.* A conversion of his measurement to absolute intensity is even more difficult than for the Milky Way. The Andromeda Nebula is a point-like object; as a result Reber's record of the nebula should show a peak agreeing [in shape] with the directional characteristics of his receiver. The fact that the recorded peak has a width at half intensity of $0^{\text{h}}.6 = 7^\circ$, while theoretically it ought to be about 11°, points to observational errors as yet incalculable. Nevertheless it is remarkable that the radiation from the Andromeda Nebula is observable at all. The surface brightness of such a small area must be a good bit higher than that of the Milky Way. The layer thickness l is appreciably less on account of our direction of

view, and the temperature must be about the same. Thus we must find everything in the factor N_e^2 which we can easily enlarge by a factor 100 by assuming a ten times higher electron density. We must then ask, however, if there are a sufficient number of very hot stars in the Andromeda Nebula to ionize the interstellar hydrogen over extended regions.

A COSMOLOGICAL REMARK

What do we observe of the remaining spiral nebulae? They are individually too small, but together they should be able to produce a detectable background brightness. An even stronger statement can be made: if (i) the universe is perfectly homogeneous, (ii) the spiral nebulae are all opaque, and (iii) there exists no absorption between the nebulae, nor any other causes by which the distant nebulae are attenuated, then the entire sky must be covered with nebulae of a surface brightness equal to that of a nearby nebulae. This argument has already been applied by Herschel to the stars which surround us. The evident fact that the night sky is not everywhere as bright as the solar disk is a proof of the finite size of our galaxy. The "universe of stars" is not homogeneous.

Conclusions are not so easy to draw for the "universe of nebulae". We are now entering the domain of cosmology. Instead of distance we shall use the time T that the light from a nebula has required to reach us. Hubble's observations with the 100-inch telescope on Mt Wilson reach to $T = 6 \times 10^8$ years. Out to that distance about 1% of the sky area has already been covered and yet no departures from homogeneity have been discovered. The extrapolation to a completely covered sky thus does not seem senseless.

The other causes by which the light from distant nebulae is weakened are also important. We see the distant nebulae in a very early stage of their development. This makes the intensity uncertain, but we would sooner expect a higher rather than a lower brightness. The most important factor is the *red-shift.* All of the nebular light is shifted to longer wavelengths by an amount $\delta\lambda$ that, out to the deepest observation of $T = 2 \times 10^8$ years, is practically always proportional to the distance. I thank Prof. Oort for this essential point in the argument. With normal photography the more distant the nebula, the less of the ultraviolet region of the spectrum which is photographed. This results in a considerable apparent attenuation of the distant nebulae. In the radio region the radiation of [originally] shorter wavelength, which one receives from the distant nebulae, is practically as intense as that from the nearby nebulae. The radio region should therefore be very suitable for tracing the influence of the remaining effects on the background brightness of the sky.

We have in particular investigated whether a decision is possible between the two models put forward for comparison by Hubble and Tolman[11] to explain the numbers of nebulae photographed at Mt Wilson. The first model is an expanding universe where the redshift is due to the Doppler effect. In the second model the redshift is ascribed to an unknown cause and the universe is static.

The complete calculation cannot be reproduced here. It is noteworthy that the result of both models, although derived in entirely different manners, is formally the same:

$$h = \int h_1(T) f^{-\alpha}(T)\, dT/T_0 \,, \tag{10}$$

where h = mean surface brightness of the sky,
$h_1(T)$ = surface brightness of a nearby spiral nebula,
$f(T)$ = redshift factor = $1 + \delta\lambda/\lambda$,
T_0 = free [travel] time of a light ray between the nebulae = 0.9×10^{11} years,
α = index of the spectral intensity distribution approximated as $j_\nu = \text{const}\ \nu^{-\alpha}$.

The integral must extend from 0 to the finite age T_l in the expanding model and to ∞ in the static model. In both cases we can use the extrapolation $f(T) = (1 + pT/T_\nu)^{1/p}$. We derive from the analysis of the observed redshifts given by Heckmann:[12]

$$T_\nu = 1.83 \times 10^9 \text{ years}$$

9 expanding universe: $p = -1.0 \pm 0.8$ and finite age $T_l = T_0/p$ *
static universe: $p = +1.0 \pm 0.8$ and infinite age.

The integration can now be carried out and yields for all cases:

$$h = h_1 \frac{T_\nu}{T_0} \frac{1}{\alpha - p} \quad \text{if } \alpha > p;\ \text{divergent if } \alpha < p\,.$$

Taking $T_\nu/T_0 = 0.01$, the results for h/h_1 are summarized in Table I.

TABLE I
Values of the ratio of the brightness of the sky background to that of a nearby nebula

Model	Frequency region	
	Photographic	Radio
	$\alpha = 2.5 \pm 0.5$	$\alpha = 0.1$
Expanding $p = -1.0 \pm 0.8$	0.002 to 0.005	0.005 to 0.03
Static $p = +1.0 \pm 0.8$	0.003 to 0.05	divergent

The rough calculation above indeed confirms the suggestion of Prof. Oort. In the photographic region the background for both models is about one per cent of the brightness of a nearby nebula. The results in the radio region, however, are entirely different. The expanding model again gives about 1% and the static model a divergent result. This means that we must take into account the shielding, which results [instead] in a uniformly bright sky. We compare these calculations with the observations which definitely show that also in the radio region both the Milky Way and the Andromeda Nebula stand out as bright objects against a dark background. Thus we may conclude that the static model is incorrect. We must bear in mind that this argument has only eliminated a certain combination of hypotheses; it is not possible to decide *which* of the hypotheses is wrong. Nevertheless, it is satisfying that these observations rule out a model which also has many theoretical objections.

In closing it is my desire to express my thanks to Prof. Oort for his attention and advice on several subjects and to Prof. Rosenfeld for a discussion concerning the hyperfine structure. I also thank the Kapteyn Astronomical Foundation which made possible my stay at the Leiden Observatory during which a part of this study was carried out.

Utrecht, Zonnenburg Observatory June 1944

NOTES

[1] See the review in Section 2 of the Astronomical Colloquium of the Nederlandsche Astronomen-Club No. 2, *Ned. Tijd. Natuurkunde* **10**, 243 (1943) and the literature cited therein.

[2] *Ned. Tijd. Natuurkunde* **11**, 201 (1945).

[3] Chosen in agreement with Henyey, L. G. and Keenan, P. C., 1940, *Ap. J.* **91**, 625.

[4] In the meantime, Southworth (1945, *J. Franklin Inst.* **239**, 285) has succeeded in measuring the solar radiation in the wavelength range 1 to 10 cm.

[5] Whipple, F. L. and Greenstein, J. L., 1937, *Proc. Nat. Acad. Sci.* **23**, 177.

[6] Full derivation by Menzel, D. H., 1937, *Ap. J.* **85**, 330.

[7] Taken from Menzel, D. H. and Pekeris, C. L., 1935, *Monthly Notices Roy. Astron. Soc.* **96**, 77.

[8] Cf. Inglis, D. R. and Teller, E., 1939, *Ap. J.* **90**, 439.

[9] Kopfermann, H., 1940, *Kernmomente*, p. 15, with substitution of $\mu_p = 2.785$.

[10] Shortley, G. H., 1940, *Phys. Rev.* **57**, 225.

[11] Hubble, E. and Tolman, R. C., 1935, *Ap. J.* **82**, 302.

[12] Heckmann, O., 1942, *Theorien der Kosmologie* (Berlin), Section 17d.

EDITOR'S NOTES AND AMENDMENTS

1. This summary originally appeared in English.
2. The modern term for these *solid smoke particles* is "dust grains".
3. Original had "Greenstein" instead of "Keenan".
4. Original had $T^{-1/3}$ instead of $T^{-1/2}$ in this equation.
5. Original had G instead of $\bar{G}$ in this equation.
6. The exponent here was mistakenly calculated; see the introduction.
7. The actual value for the wavelength is 21.1 cm, corresponding to a frequency of 1420.4 MHz.
8. This detection was in fact spurious (see Editor's Note 5 for Paper 5 and also Note 5 for Paper 6).
9. Original had T_e instead of T_0 in this equation.

SECTION Q

PREDICTION OF MOLECULAR LINE EMISSION

Paper 35 by Shklovsky is not well known in the West, but it certainly deserves a place in this collection. Although Shklovsky's calculations concerning the 21 cm line of neutral hydrogen were independent of those by Van de Hulst in Paper 34, the initial idea was, in fact, sparked by a brief mention of the latter's results in a review by Reber and Greenstein (1947). Despite lack of access to the original Dutch article, Shklovsky was not deterred and started his investigation from scratch. His paper is the first to calculate the probability for the hyperfine transition and he turns out to be off by only a factor 4. The author argues for collisional excitation of the line at a temperature of 5000 K. This extremely high adopted temperature is influenced by the work of Eddington prevalent at that time and it leads Shklovsky to derive a quite small optical depth in the line. Nevertheless, he clearly demonstrates the feasibility of detection of the line with available equipment.

The second portion of the paper concerning possible radio line radiation from deuterium and from interstellar molecules is entirely original. In fact, it was not until six years later that anyone other than Shklovsky (who also wrote other papers (1952b, 1953b) on this topic) published any such work on molecular radio lines in the astrophysical context. In Paper 35 the deuterium line, which actually falls at a wavelength of 92 cm, is only given an outside chance. On the other hand, the author is enthusiastic about the odds for detection of the Λ-doublet transitions arising in the ground rotational states of the molecules OH and CH. His calculations for the wavelength of the OH "line"[1] go awry from the correct value of 18.0 cm due to the fact that the approximation which he uses (Hund's "Case b") for the amount of splitting is too simplified for the case of the OH molecule. But this is a detail which should not be allowed to overshadow the genius of Shklovsky in anticipating, by two decades, the rich field of galactic microwave spectroscopy.

NOTE

[1] Hyperfine splitting, which, in fact, causes four transitions to arise in the Λ-doublet, is not considered in this article.

MONOCHROMATIC RADIO EMISSION FROM THE GALAXY AND THE POSSIBILITY OF ITS OBSERVATION*

(МОНОХРОМАТИЧЕСКОЕ РАДИОИЭЛУЧЕНИЕ ГАЛАКТИКИ И ВОЭМОЖНОСТЬ ЕГО НАБЛЮДЕНИЯ)

I. S. SHKLOVSKY
P.K. Sternberg State Astronomical Institute

SUMMARY. When transitions take place between the components of the ground-state hyperfine levels of interstellar hydrogen atoms, monochromatic radio waves are emitted at a frequency of 1421.3 megacycles (λ = 21 cm). The calculated probability of such a transition is 0.72×10^{-15} sec^{-1}. The half-width of this emission line due to the Doppler effect is $\Delta\nu_D = 4 \times 10^4$ sec^{-1}; the optical thickness of the interstellar hydrogen in this line is small. The intensity at λ = 21 cm over a bandwidth of $2\Delta\nu_D$ is ten times greater than the intensity of the general continuum of galactic noise.

The possibility of experimentally detecting this emission has been investigated. It has been shown that, using modern radio equipment of high sensitivity, this emission can be detected provided that the gain of the antenna is greater than 65, a value easily attainable.

It has been further shown that it is impossible to expect monochromatic radio emission from other interstellar atoms.

The molecules CH, CH^+ and OH are capable of emitting detectable radio lines (for OH, λ = 9.54 cm) which arise from transitions between the components of a Λ-doublet in the ground rotation state.

The investigation of the intensity distribution over the sky of the monochromatic radio emission yields extremely important data concerning the physical state of the interstellar gas.

Van de Hulst [1] was the first to point out the probability of the existence of monochromatic radio emission from the galaxy. This emission arises when a transition takes place between two very close discrete levels of atoms or molecules. Since the overwhelming majority of atoms and molecules in interstellar space will be in the ground state, it is necessary to consider transitions between the sublevels of these states. We have examined all possible transitions of both interstellar atoms and molecules. The main results are given below.

When transitions take place between the fine structure states which comprise multiplets in atoms, they emit at frequencies far higher than the limits of the radio spectrum. It is clear that the same applies to the rotational

* *Extensive modification of the translation by C. W. Wouters from Astr. Zh.* **26,** 10–14 (1949).

spectrum of the interstellar molecules, the hydrides. However, transitions between hyperfine states produce emission lines which lie as a rule in the decimeter and meter bands.

Let us examine the most interesting case, that of hydrogen. Here the splitting of the ground state can be obtained from the formula of Fermi [2]. However, quite recently this value has been found experimentally with very great precision by Nafe, Nelson and Rabi [3] using the magnetic-resonance method. According to these authors, the above transition between the hyperfine states occurs at

$$\nu_{\mathrm{H}} = (1421.3 \pm 0.2) \times 10^{6}\ \mathrm{sec}^{-1}, \tag{1}$$

corresponding to a wavelength of $\lambda_{\mathrm{H}} = 21$ cm.

When a transition takes place between the hyperfine states, magnetic-dipole radiation will arise. For this radiation the probability of the transition, according to Condon and Shortley [4], is given by

$$A_m = 2.696 \times 10^{-11} \frac{\nu^3 S_m}{2J+1}, \tag{2}$$

where ν is expressed in reciprocal centimeters, $g = 2J + 1$ = the statistical weight of the original state, and S_m is the "strength" of the line:

$$S_m = \frac{\{J^2 - (L - S)^2\}\{(S + L + 1)^2 - J^2\}}{4J}.$$

J, L and S are the inner, azimuthal and spin quantum numbers. In the case of transitions between the hyperfine states of hydrogen, these quantum numbers must be replaced by F, J and I,* respectively, where I is the spin of the 1
nucleus and $F = J + I$. The application of the simple formula for S_m, which is valid only for Russell–Saunders coupling, is fully permissible in this case (see Reference [5]).

Numerically, $I = ½$, $J = ½$, $F = 1$, $S_m = ¾$; consequently $A_m = 0.72 \times 10^{-15}$ sec^{-1}.

The life of an atom in the excited state will be $t_1 = 1.39 \times 10^{15}$ sec or 4.4×10^{7} years! On the other hand the interval t_2 between two successive collisions of an atom in the zones of un-ionized hydrogen will be of the order of several hundred years. Since $t_1 \gg t_2$, the return transition to the ground state will be realized in the overwhelming majority of cases with help from a collision of the second kind. This means that the ratio between the density of the atoms of hydrogen in the upper and lower levels is given by

2
$$\frac{N_2}{N_1} = \frac{g_2}{g_1} \exp\left(-\frac{h\nu_H}{kT}\right) \approx \frac{g_2}{g_1} = \frac{3}{1},^{\star} \tag{3}$$

where T is the kinetic temperature of the interstellar hydrogen.

In order to calculate the intensity of the emission originating from these transitions between the hyperfine states of hydrogen atoms, it is necessary to determine the equation of transfer

$$\frac{dI_\nu}{dS} = -k'_\nu I + \epsilon_\nu;\; k'_\nu = k_\nu\left(1 - \exp\left(-\frac{h\nu}{kT}\right)\right) = k_\nu \frac{h\nu}{kT}. \tag{4}$$

It is evident that absorption will only occur in the presence of the Doppler effect. We have

$$k_\nu = \frac{2\pi^{3/2} e^2}{mc} N_1 f_{12} \frac{1}{\Delta\nu_D} \exp\left(-\left(\frac{\Delta\nu}{\Delta\nu_D}\right)^2\right) \text{cm}^{-1}, \tag{5}$$

and

$$\epsilon_\nu\, d\nu = \frac{1}{4\pi} N_2 A_m\, h\nu \frac{1}{\sqrt{\pi}} \frac{1}{\Delta\nu_D} \exp\left(-\left(\frac{\Delta\nu}{\Delta\nu_D}\right)^2\right) d\nu \frac{\text{erg}}{\text{cm}^2 \text{ sec ster}}. \tag{6}$$

The oscillator strength is

$$f_{12} = \frac{g_1}{g_2} \frac{mc^3}{8\pi^2 e^2 \nu^2} A_m = 1.62 \times 10^{-13}.$$

The Doppler half-width of the line is

$$\Delta\nu_D = \frac{\nu_H}{c} \sqrt{\frac{2kT}{\mu_H}}.$$

Assuming $T = 5000°$, we obtain

$$\frac{k_\nu}{f_{12}} = 2.47 \times 10^{-7}\ \text{cm}^{-1},\ \Delta\nu_D = 4.2 \times 10^4\ \text{sec}^{-1};\ \frac{h\nu_H}{kT} = 1.27 \times 10^{-5}.$$

The optical thickness in the center of the line is $\tau_{\nu_H} = k'_\nu N_1 l$. Assuming $N_1 = 1\ \text{cm}^{-3}$ and $l = 3 \times 10^{22}$ cm (10^4 pc), we obtain $\tau_{\nu_H} = 0.04$.

Consequently, it is possible as a first approximation to consider the galaxy as transparent to its own emission of monochromatic radio waves. By solving Equation (4) we find

$$I_\nu = \frac{\epsilon_\nu}{k'_\nu}(1 - e^{-\tau_\nu}) \approx \epsilon_\nu l. \tag{7}$$

On the other hand, it is known that there exists in the galaxy a continuous radio emission spectrum produced by hyperbolic transitions of interstellar electrons. From experimental and theoretical data we know that the intensity of this radio continuum component from the galaxy does not depend on wavelength in the meter and decimeter bands [6]. At a wavelength of 62 cm, Reber discovered radio emission from the constellation Sagittarius of intensity

$$6 \times 10^{-22} \frac{\text{Watt}}{\text{cm}^2\ (\text{circ deg})\ \text{megacycles}} \qquad \text{(see Reference [6])}.$$

The intensity of the radio emission from other parts of the galaxy was lower, but the sensitivity of the apparatus was such that it was still possible to detect.

The full intensity of the line of interstellar hydrogen at $\lambda_H = 21$ cm is

$$\begin{aligned} I &= \int_0^\infty I_\nu \, d\nu = \frac{1}{4\pi} N_2 A_m h\nu_H l \\ &= 4.8 \times 10^{-11} \text{ erg cm}^{-2} \text{ sec}^{-1} \text{ ster}^{-1} \\ &= 1.15 \times 10^{-21} \text{ Watt cm}^{-2} \ (\text{circ deg})^{-1} . \end{aligned} \tag{8}$$

The width of the line is $2\Delta\nu_H = 0.1$ megacyle. *Consequently the intensity of this line over its width will be nearly 20 times higher than the intensity at this frequency of the continuous emission.*[1]

Let us now examine the general conditions for the detection of monochromatic radio emission. The power accepted by an antenna connected to a matched receiver is equal to [7]

$$P = \frac{\lambda^2}{16} G_a I \Omega ,$$

where Ω is the solid angle from which the antenna receives radio emission and G_a is the so called "gain" of the antenna, measured relative to that of a half-wave dipole. It is evident that G_a is determined by the dimensions of the antenna. On the other hand, the level of the fluctuation noise entering the receiver, taking also into account the noise in the receiver itself, will be $P_n = NkT_a \Delta\nu$ where N is the "noise factor", T_a the antenna temperature, and $\Delta\nu$ the bandwidth of the receiver. One can assume that monochromatic radio emission from the galaxy will be detectable if

$$\frac{P}{P_n} = \frac{\lambda^2 G_a I \Omega}{16 NkT_a \Delta\nu} \geqslant 0.02 . \tag{9}$$

It is clear that the condition $\Delta\nu \geqslant 2\Delta\nu_D$ must be fulfilled; when $\Delta\nu = \Delta\nu_D$ we have the optimum conditions for detection of the radio emission. Assuming that $N = 10$, $T_a = 300°K$, and $\Omega = 25$ square degrees = 0.0076 ster, we find that $G_a \geqslant 65$.

This gives us a lower limit for the dimensions of the antenna. Such a gain is easily obtained in the decimeter band, e.g., for a mirror of 5 m radius (approximately corresponding to the dimensions of Reber's equipment) the gain will exceed 3000.

It is a pity that it is hardly likely that one will be able to detect monochromatic radio emission from the other interstellar atoms. The nuclear spins of He^4, C^{12} and O^{16} are all equal to zero. The carbon isotope C^{13} does have a spin, but its abundance in interstellar space is very small; on the Earth the abundance of C^{13} is approximately 1% of that of C^{12} whereas in interstellar space the relative abundance of carbon to hydrogen is 1 : 1000. The spin of N^{14} is equal to 1, its ground state being $^4S_{3/2}$. However, in view of the comparatively small quantity of nitrogen (H/N $\sim 10^3$), the intensity will be very small. The same must be said for deuterium, which gives a magnetic dipole transition at $\lambda = 84.5$ cm between the hyperfine levels of its ground state. The net magnetic dipole moment of the electrons in the neon atom, which occurs quite abundantly, is equal to zero. Other elements in interstellar space are much less abundant.

The molecules, however, can produce an unexpected effect. The hydrides
3 CH, CH^+ and NH have been discovered in interstellar space.★ According to
4 Dunham [8]★ the density of CH is 2×10^{-6} cm^{-3} and that of CH^+ is 10^{-6}
cm^{-3}. Emission in the radio spectrum arises during transitions between the hyperfine levels of the ground rotational state. This structure (so-called "Λ-doubling") arises when the projection of the electronic orbital moment upon the axis of the molecules, Λ, differs from zero. The ground state of the molecules CH and CH^+ is Π. The same is true for the molecule OH. Although this molecule has not yet been detected in interstellar space (the corresponding resonance band must begin at $\lambda = 3066$ A), its existence can hardly be doubted. None of the well-known selection rules forbids transitions between components of the Λ-doublet. The frequency of this dipole radio emission for the OH molecule can be calculated according to the formula of Van Vleck [9]. The ground state of OH corresponds to the so-called Hund's "Case b", i.e., the rotational frequency $B_0 \ll$ the width of the doublet A_0 [10]. In the case of OH, we have $B_0 = 18.48$ cm^{-1}, $A_0 = 137.9$ cm^{-1} and

$$\nu_{OH} = \frac{4B_0^2}{\nu_{el}} K(K+1), \tag{10}$$

where $\nu_{el} = 32\,600\ \text{cm}^{-1}$ is the frequency of the electronic transition and $K = 1$. Numerically, $\nu_{\text{OH}} = 0.105\ \text{cm}^{-1}$ * and $\lambda_{\text{OH}} = 9.52$ cm. The probability of the transition is $A = 0.65 \times 10^{-9}\ \text{sec}^{-1}$. Assuming that the abundance of OH molecules in interstellar space is the same as for CH, we can show that the intensity of the line at $\lambda_{\text{OH}} = 9.52$ cm is comparable to the intensity of the hydrogen line at $\lambda_{\text{H}} = 21$ cm. The half-width of this line is twice as small as that for the 21-cm line.

For the CH molecule we have $B_0 = 14.189\ \text{cm}^{-1}$ and $A_0 = 28.4\ \text{cm}^{-1}$, i.e., B_0 and A_0 are comparable in magnitude. Consequently, we have here a transition from Hund's "Case b" to "Case a" and it is impossible to calculate the frequencies exactly. We can only say that the line will lie in the decimeter range and possess an intensity comparable to the intensity of the lines at $\lambda_{\text{OH}} = 9.52$ cm and $\lambda_{\text{H}} = 21$ cm. The same also holds for the molecule CH^+. There is no Σ ground state or Λ doubling in the NH molecule.

In summary, we must say that the resources of contemporary radio techniques fully allow one to detect and measure monochromatic radio emission from the galaxy.

In general the sources of this radio emission are the zones of un-ionized hydrogen (in contrast with the continuous emission of radio waves from the galaxy [11]). This emission penetrates through the atmosphere of the Earth and is not noticeably absorbed. Observations can be conducted at any time of day or night and depend only little on weather.

It is of fundamental importance that the interstellar hydrogen can in this case be directly studied in its ground state. In direct contrast, all methods now existing in astrophysics allow one to determine the population of only the excited states of the hydrogen atom. The calculation of the population of the ground state, as a rule, cannot be carried out with sufficient accuracy.

Study of the intensity distribution of monochromatic radio emission from the Milky Way would directly give the distribution of interstellar gas in various regions of the galaxy; we may well find interesting structural peculiarities. Finally, an examination of the half-widths of the lines will give the distribution of the kinetic temperature of the interstellar gas averaged along the line of sight.

Soviet radiophysicists and astronomers should endeavor to solve this absorbing and important problem.

NOTE

[1] Prof. Khaikin has been so kind as to inform us that the bandwidth of modern receivers can easily be made as small as 0.1 megacycles.

LITERATURE

1. Van de Hulst, 1945, *Ned. Tijd. Natuurkunde* **11**, 201, 210.
2. Frisch, 1933, *Atomic Spectra.*
3. Nafe, Nelson, and Rabi 1947, *Phys. Rev.* **71**, 914.
4. Pasternack, 1940, *Ap. J.* **92**, 129.
5. Bethe and Bacher, 1936, *Rev. Mod. Phys.* **8**, 82.
6. Townes, 1947, *Ap. J.* **105**, 235.
7. Ginzburg, 1947, *Usp. Fiz. Nauk.* **32**, 26.
8. Dunham, 1937, *Pub. Astron. Soc. Pac.* **49**, 26.
9. Van Vleck, 1929, *Phys. Rev.* **33**, 502.
10. Kondrat'ev, 1946, *Struktura Atomov i Molekul.*
11. Shklovsky, 1946, *Astr. Zh.* **23**, 333.

EDITOR'S NOTES AND AMENDMENTS

1. Because (J, L, S) cannot be simply replaced by (F, J, I) as stated, the derived value for the transition probability A_m is in fact $1/4$ of the correct value.
2. Original had '$1/3$', but '$3/1$' is correct.
3. It is not clear to which observations of NH Shklovsky is here referring; the only generally accepted detections of molecules at that time were CH, CH^+ and CN.
4. The indicated reference in fact gives no information on CH or CH^+ densities; Shklovsky must be referring to later work by Dunham, e.g., *Pub. Amer. Astron. Soc.* **10**, 123 (1940).
5. From the values given it would seem that Equation (10) yields $\nu_{OH} = 0.084\ cm^{-1}$.

SECTION R

DISCOVERY OF THE 21 CM HYDROGEN LINE

Despite the previous theoretical predictions, progress was slow in experimental verification of the existence of the 21 cm line in galactic regions. Finally, in 1951, two groups converged on the discovery from quite different paths. H. I. Ewen was working on his thesis research under E. M. Purcell in the Physics Department of Harvard University. These men were physicists hoping to employ the line as a probe of the conditions in interstellar regions. On the other hand, across the Atlantic, Oort was pushing for detection of the line in order to study the spatial structure of our galaxy. To this end he had brought into existence the Netherlands Foundation for Radio Astronomy as the vehicle by which to move Holland full-scale into this new arena.

It is hardly coincidence that the line was detected by both groups within seven weeks of each other. By a quirk of fate, Van de Hulst himself was spending a sabbatical year at Harvard at that time. He was able to closely follow the developments in each camp and, in fact, acted as a vital catalyst to the success of each group. To Holland he sent information on the technique of frequency switching which C. A. Muller, the new and relatively inexperienced Dutch engineer, quickly adopted in order to greatly reduce receiver instabilities. On the other hand, Ewen and Purcell started off switching to a comparison band only 10 kHz removed from the signal band. When they had no success with this, they extensively modified their electronics in order to make a greater offset possible, in agreement with Van de Hulst's expectation of a broad line width from any interstellar HI. The net result of these exchanges was a trade-off of technology for astrophysics which benefited both groups.

After the initial detection this spirit of cooperation was further fostered when Purcell held off from instant publication and, instead, immediately notified both Van de Hulst and F. J. Kerr, also temporarily away from his home institution (CSIRO in Sydney), as well as their colleagues abroad. While the effect of this communication on the Dutch effort was primarily to increase the already great enthusiasm, the Australians had no such project in progress. Nevertheless, Christiansen and Hindman seized on the opportunity, slapped together a suitable receiver, and within a couple of months had also detected the line (see the telegram from Pawsey immediately following Paper

37). Having their confirmation, Ewen and Purcell then insisted that the initial reports from all three groups should simultaneously appear, as they did in the 1 September 1951 issue of *Nature*.

A surprising amount of information is gathered in Paper 36 from drift scans taken at a single declination, especially when one considers that $\delta = -5°$ was mainly chosen so as to be convenient for a large horn stuck out a window of the Lyman Laboratory at Harvard! The neutral hydrogen is shown to be concentrated towards the galactic plane in the constellations of Ophiucus, Scutum and Aquila ($l^I \sim 355°$). It is argued that the spin temperature T_s of the transition is probably determined by gas collisions with hydrogen atoms and that T_s is ~35 K if the line is optically thick as expected. We can now discern that there are two reasons why Ewen and Purcell, in fact, measured too weak a signal. Firstly, their ~12° resolution was not adequate to resolve the main hydrogen emission in the galactic plane. Secondly, their frequency-switching interval of 75 kHz (15 km s^{-1}), while adequate to detect the line, was still too small for the extremely broad profiles which are observed in the galactic plane. The full hydrogen profile at $l^I = 355°$ is ~200 km s^{-1} broad and exhibits complex structure; thus, in effect, Ewen and Purcell were only detecting the tip of the iceberg. The measured values for intensity and line width are consequently too small and the frequency shifts arising from galactic rotation are missed.

Paper 37 describes the initial Dutch observations. Muller had been hired only a few months previous in order to put life into the radio astronomy project, earlier hampered by lack of know-how and even a disastrous fire. The antenna was a 7.5 m Würzburg paraboloid which had been abandoned by the German army at the close of the war.[1] The receiver was of a similar sensitivity to Ewen's and employed frequency-switching over an interval of 110 kHz (23 km s^{-1}), still not adequate to properly measure the line profiles. The effect on the reported results is nevertheless not as severe as for the Harvard experiment for two reasons: (1) Muller and Oort do not concern themselves with astrophysical results derived from the line intensities, and (2) they are able to measure the frequency shifts of the peak line intensity by examining *several* different longitudes in the plane. Applying the principles of galactic structure and dynamics which Oort himself pioneered in the 1920s, the authors use the observed linewidths and distribution in latitude to argue that most of the hydrogen lies in an extremely thin plane and is situated nearby (within 400 pc). They then demonstrate the effects of differential rotation of our galaxy in the direction towards Cygnus. The closing paragraph on future plans led directly into the first look at the large-scale spiral structure of the

Milky Way. Van de Hulst, Muller, and Oort (1954) derived a map of HI density over much of the galactic plane at radii exterior to the sun, discovered trailing spiral-like features, and derived detailed physical conditions in the cool gas pervading the Milky Way.

NOTE

[1] These Würzburg dishes played an important role in the experiments of many early radio astronomy groups, e.g., the interferometer used by F. G. Smith in Paper 28 consisted of two such dishes.

OBSERVATION OF A LINE IN THE GALACTIC RADIO SPECTRUM

Radiation from Galactic Hydrogen at 1,420 Mc./sec.

THE ground-state of the hydrogen atom is a hyperfine doublet the splitting of which, determined by the method of atomic beams, is 1,420·405 Mc./sec.[1]. Transitions occur between the upper ($F = 1$) and lower ($F = 0$) components by magnetic dipole radiation or absorption. The possibility of detecting this transition in the spectrum of galactic radiation, first suggested by H. C. van de Hulst[2], has remained one of the challenging problems of radio-astronomy. In interstellar regions not too near hot stars, hydrogen atoms are relatively abundant, there being, according to the usual estimate, about one atom per cm.[3]. Most of these atoms should be in the ground-state. The detectability of the hyperfine transition hinges on the question whether the temperature which characterizes the distribution of population over the hyperfine doublet—which for want of a better name we shall call the hydrogen 'spin temperature'—is lower than, equal to, or greater than the temperature which characterizes the background radiation field in this part of the galactic radio spectrum. If the spin temperature is lower than the temperature of the radiation field, the hyperfine line ought to appear in absorption; if it is higher, one would expect a 'bright' line; while if the temperatures are the same no line could be detected. The total intensity within the line, per unit band-width, should depend only on the difference between these temperatures, providing the source is thick enough to be opaque.

We can now report success in observing this line. A micro-wave radiometer, built especially for the purpose, consists mainly of a double superheterodyne receiver with pass band of 17 kc., the band being shifted back and forth through 75 kc. thirty times per second. The conventional phase-sensitive detector and narrow (0·016 c./s.) filter then enable the

Reprinted, with permission, from 'Nature' 168, 356–358 (1951).

radiometer to record the apparent radio temperature *difference* between two spectral bands 75 kc. apart. These bands are slowly swept in frequency through the region of interest. The overall noise figure of the receiver, measured by the glow-discharge method[3], is 11 db., and the mean output fluctuation at the recorder corresponds to a temperature change of 3·5°. The antenna is a pyramidal horn of about 12° half-power beam-width. It is rigidly mounted at declination −5° ; scanning is effected by the earth's rotation.

The line was first detected on March 25, 1951. It appeared in emission with a width of about 80 kc., and was most intense in the direction 18 hr. right ascension. Many subsequent observations have established the following facts. At declination −5° the line is detectable, by our equipment, over a period of about six hours, during which the apparent temperature at the centre of the line rises to a maximum of 25° above background and then subsides into the background. The source appears to be an extended one approximately centred about the galactic plane. The frequency of the centre of the line, which was measured with an accuracy of ±5 kc., was displaced some 150 kc. above the laboratory value, and this shift varied during an observing period. Both the shift and its variation are reasonably well accounted for by the earth's orbital motion and the motion of the solar system toward Hercules. The period of reception shifts two hours per month, in solar time, as it ought to.

Some conclusions can already be drawn from these results. Extrapolation of radio temperature data for somewhat lower frequencies[4] suggests that the background radiation temperature near the 21-cm. line is not more than 10° K. Then the hydrogen spin temperature is not more than 35° K., if the source is 'thick'. But we can calculate the opacity of the source on the assumption of a spin temperature of 35° K. and 1 atom/cm.^{3}, using only the observed line-width and the matrix element of the transition in question, and we obtained 900 light-years for the absorption-length. As this is much smaller than galactic dimensions, we conclude that the temperature observed corresponds indeed to the spin temperature at the source. To the extent that 'self-absorption' contributes to the observed line-width, the true absorption length at the frequency of the centre may be *less* than that computed. Further evidence for relatively high opacity is the absence of large frequency-shifts, which would be expected to arise from galactic rotation were the opacity-thickness comparable to the size of the galaxy. This conclusion is contrary to the prediction of Shklovsky[5], who has recently discussed the possibility of detecting galactic line radiation.

We have made rough theoretical estimates of the efficacy of various processes through which energy is exchanged between the hydrogen hyperfine levels and the other thermal reservoirs in the interstellar matter plus radiation complex. Of these we find exchange with the radiation field (involving spontaneous emission) and exchange with gas-kinetic energy of the hydrogen atoms (via H–H collisions) much the most important, with the latter process probably dominant. This is consistent with the observation, and if correct implies that the gas-kinetic temperature of the hydrogen exceeds, but not greatly, the spin temperature. The estimated spin relaxation time for these processes is of the order of 10^5 years.

This work was assisted by a grant from the Rumford Fund of the American Academy of Arts and Sciences.

H. I. Ewen
E. M. Purcell

Lyman Laboratory,
Harvard University,
Cambridge,
Mass.
June 14.

[1] Kusch, P., and Prodell, A. G., *Phys. Rev.*, **79**, 1009 (1950).
[2] van de Hulst, H. C., *Nederl. Tij. Natuurkunde*, 's-Gravenhage-Martinus Nyhoff, **11**, 201 (1945).
[3] Mumford, W. W., *Bell Syst. Tech. J.*, **28**, 608 (1949).
[4] Herbstreit, J. W., and Johler, J. R., *Nature*, **161**, 515 (1948).
[5] Shklovsky, I. S., *Astronomicheskii Zhurnal*, **26**, 10 (1949) (in Russian).

The Interstellar Hydrogen Line at 1,420 Mc./sec., and an Estimate of Galactic Rotation

FOLLOWING a suggestion made by Dr. H. C. van de Hulst in 1944 [1], attempts have been made to measure the radiation at 1,420 Mc./sec. (λ21 cm.) emitted by atomic interstellar hydrogen. The first experimental evidence for the presence of this interstellar emission line was obtained by Ewen and Purcell on March 25, 1951 (see the preceding communication). In the Netherlands, the first successful measurements were made on May 11.

The receiver consists of a double superheterodyne instrument, with a crystal-controlled first local oscillator, which is switched between two frequencies 110 kc. apart thirty times per second by a reactance modulator. By varying the frequency of the second local oscillator, the two frequencies can be moved together through a 4-Mc. wide pass band at 1,420 Mc. The pass band of the second i.f. channel is 25 kc. wide. Behind the usual phase-sensitive detector a narrow-band filter with a time constant of 12 sec. is used. So the difference between two frequency bands 110 kc. apart is measured. The noise factor of the receiver has not yet been measured. The losses in the coaxial antenna cable are rather high, and an effective noise factor of the whole receiving system of 25 has been deduced from other measurements with the sun as a source of noise. Important parts of the receiver have been constructed at the Philips Laboratory in Eindhoven under the supervision of Dr. F. L. Stumpers.

The receiver has been mounted behind a movable paraboloid of 7·5 m. aperture and 1·7 m. focal-length at the radio station at Kootwijk, which was kindly put at our disposal for these measurements by the Radio Department of the Post and Telegraph Service. The beam-width at half-power is 2·8°.

The results of the measurements made on a few of the first tracings are shown in the accompanying graphs. While these tracings were being made, the instrument was left in a fixed position relative to the earth, the motion across the sky being provided by the earth's rotation. The frequency of the second local oscillator is switched every 3 min. between positions in which either of the two pass bands coincides with the spectral line. The first position gives a positive, the other a negative, deflexion on the recording meter. The curves show the intensity as function of the right-ascension; the interval between successive vertical lines is 20 min. of time. The point at which each sweep crossed the galactic equator has been marked with an arrow, accompanied by a number giving the corresponding longitude. The galactic co-ordinates at the beginning and end of each tracing have also been indicated. The curves shown may be slightly distorted because the radial components of the orbital motion of the earth, of the sun's motion relative to the nearby interstellar clouds and of differential galactic rotation are different for the various directions. However, we do not believe that these effects will have seriously affected the general shapes of the records, except for latitudes less than $1\frac{1}{2}°$, where in some directions the line is greatly widened by galactic rotation, and the measured intensities will be too small.

It is evident from the wide spread in galactic latitude that the clouds observed must be relatively close to the earth. From the width of the emission line observed in the region of the centre of the Galaxy, where the rotation of the galactic system does not affect the line, it is found that the random velocities average about 5 km./sec. in one co-ordinate, which agrees approximately with what had been found from absorption lines in the visual region. With such small velocities it is unlikely that the gas extends to more than an average distance of about 50 parsecs from the galactic plane. With an average latitude of the order of 8°, the gas seen in the general direction of the centre cannot then be farther away than 300 or 400 parsecs. In Cygnus the distance may be twice this amount.

These small distances might be taken as an indication either that the more central parts of the Galaxy are devoid of atomic hydrogen, or else that within a distance of between 500 and 1,000 parsecs

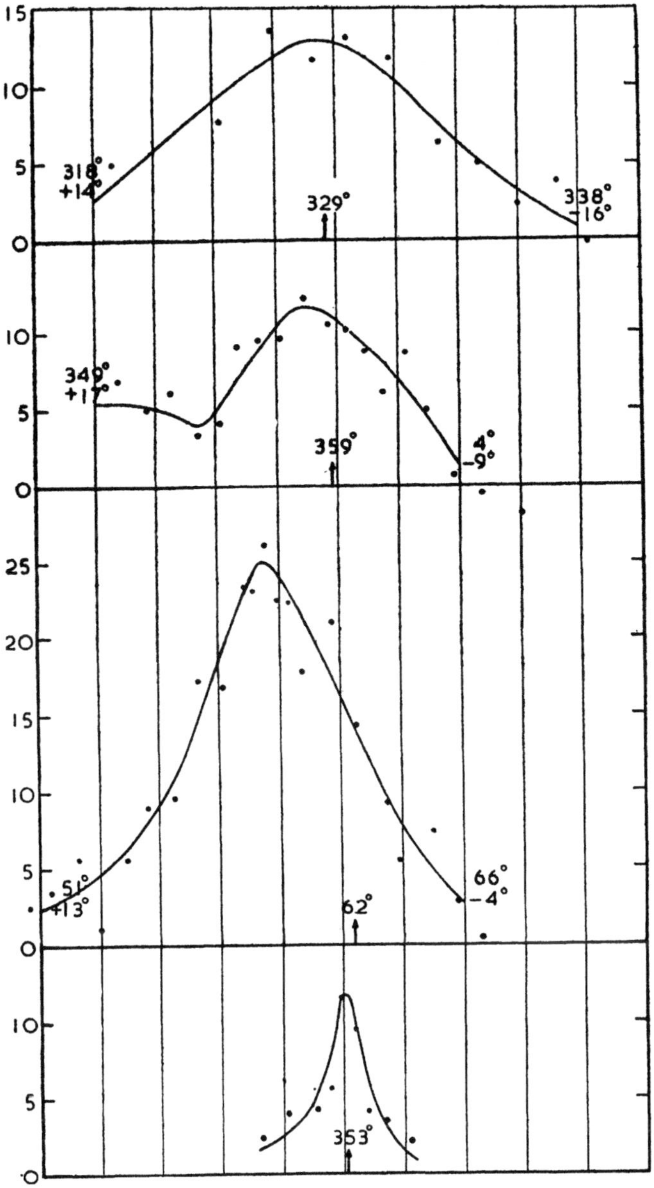
15
10
5
0
318°
+14°
329°
338°
-16°
10
5
0
349°
+17°
359°
4°
-9°
25
20
15
10
5
0
51°
+13°
62°
66°
-4°
10
5
0
353°

in the galactic plane the gas becomes optically thick in the wave-length of this line. Although we should like to reserve a definite judgment until more complete measures have been obtained, we believe the latter alternative is the more probable on the basis of the results so far available.

In order to test the presence of radiating hydrogen in the inner regions of the Galaxy, measures have been made across the Milky Way at $l = 355°$, 30° from the centre, where the change of frequency due to differential galactic rotation should make it possible to observe the gas up to large distances. Provisional measures indicate that the emission line in this direction has an effective width of about 330 kc./sec., or 70 km./sec., and that the rotational velocity of the galactic system at a distance from the centre equal to half the sun's distance from the centre is approximately 190 km./sec. This agrees well enough with the rotational velocity of 205 km./sec. computed from a schematic model of the galactic system[2]. (In both cases a value of 270 km./sec. has been assumed for the rotation near the sun.)

The graph at the bottom shows the results of a registration across the Milky Way at 355° longitude and at a frequency 250 kc./sec. lower than the normal frequency of the line, corresponding to a differential radial velocity of +55 km./sec. The gas clouds observed at this frequency are presumably situated at an average distance of about 8 kiloparsecs. It may be noted that the spread in latitude is quite small, and may roughly correspond to the beam-width of the instrument, as would be expected for these large distances.

The tracings reveal many irregularities in distribution. The extension to high positive latitudes in the second graph may possibly be connected with the well-known dense clouds in Ophiuchus. The maxima in the Cygnus region lie systematically about 2½° south of the galactic circle at longitudes between 40° and 45°, while between 56° and 62° they occur roughly 3° north of the plane. Though no weight should be attached to the exact differences in height of the various sweeps, the intensities in Cygnus appear to be definitely higher than in other regions (except in that opposite the centre). This may be ascribed to the effects of galactic rotation. The fact that the intensity is also lower in the direction towards the centre, where these effects are presumably negligible, is possibly due to the superposition of continuous background radiation, which has probably a considerable intensity in this direction.

The measurements reported in this communication are of a provisional nature. Improvements in the receiver are possible, and are now being made in preparation for more systematic measurements.

The investigation has been made possible by grants from the Netherlands Organization for Pure Research (Z.W.O.).

C. A. Muller
J. H. Oort
Netherlands Foundation for Radio Astronomy,
Kootwijk Radio Station.
Observatory,
Leyden.
June 26.

[1] Van de Hulst, H. C., *Nederl. Tij. Natuurkunde*, **11**, 201 (1945).
[2] Oort, J. H., *Bull. Astro. Inst. Netherlands*, **9**, 193 (1941).

The following cable dated July 12 has been received from Sydney, N.S.W.:

Referring Prof. Purcell's letter of June 14 announcing the discovery of hyperfine structure of the hydrogen line in galactic radio spectrum, confirmation of this has been obtained by Christiansen and Hindman, of the Radio Physics Laboratory, Commonwealth Scientific and Industrial Research Organization, using narrow-beam aerial. Intensity and line-width are of same order as reported, and observations near declination 20° S. show similar extent about galactic equator. J. L. Pawsey

REFERENCES

1896

* Wilsing, J. and Scheiner, J., 'On an attempt to detect electrodynamic solar radiation and on the change in contact resistance when illuminating two conductors by electric radiation' [in German], *Ann. Phys. Chem.* 59, 782 (Paper 13).

1900

Lodge, O. J., *Signalling Across Space Without Wires*, The Electrician Publ. Co., London, 3rd ed., p. 33.

1902

* Deslandres, H. and Décombe, L., 'On the search for Hertzian radiation emanating from the sun' [in French], *Comptes Rendus Acad. Sci.* **134**, 527 (Paper 15).

* Nordmann, C. (a), 'A search for Hertzian waves emanating from the sun' [in French], *Comptes Rendus Acad. Sci.* **134**, 273 (Paper 14).

Nordmann, C. (b), 'Research on the role of Hertzian waves in physical astronomy' [in French], *Rev. Gén. Sci.* **13**, 379.

1932

* Jansky, K. G., 'Directional studies of atmospherics at high frequencies', *Proc. IRE* **20**, 1920 (Paper 1).

1933

* Jansky, K. G. (a), 'Electrical disturbances apparently of extraterrestrial origin', *Proc. IRE* **21**, 1387 (Paper 2).

Jansky, K. G. (b), 'Radio waves from outside the solar system', *Nature* **132**, 66.

Jansky, K. G. (c), 'Electrical phenomena that apparently are of interstellar origin', *Pop. Astron.* **41**, 548.

1935

* Jansky, K. G., 'A note on the source of interstellar interference', *Proc. IRE* **23**, 1158 (Paper 3).

Langer, R. M., 'Radio noises from the galaxy', *Phys. Rev.* **49**, 209 (abstract).

1937

Friis, H. T. and Feldman, C. B., 'A multiple unit steerable antenna for shortwave reception', *Proc. IRE* **25**, 841.

* An asterisk indicates a paper reprinted in this volume.

Jansky, K. G., 'Minimum noise levels obtained on short-wave radio receiving systems', *Proc. IRE* **25**, 1517.

Whipple, F. L. and Greenstein, J. L., 'On the origin of interstellar radio disturbances', *Proc. Nat. Acad. Sci.* **23**, 177.

1940

* Henyey, L. G. and Keenan, P. C., 'Interstellar radiation from free electrons and hydrogen atoms', *Astrophys. J.* **91**, 625 (Paper 8).

* Reber, G. (a), 'Cosmic static', *Proc. IRE* **28**, 68 (Paper 4).

* Reber, G. (b), 'Cosmic static', *Astrophys. J.* **91**, 621 (Paper 5).

1942

Duyvendak, J. J. L., 'Further data bearing on the identification of the Crab Nebula with the supernova of 1054 A.D. Part I. The ancient oriental chronicles', *Publ. Astron. Soc. Pacific* **54**, 91.

Fränz, K., 'Measurement of receiver sensitivity for short waves' [in German], *Hochfreq. Electroakustik* **59**, 143.

Mayall, N. U. and Oort, J. H., 'Further data bearing on the identification of the Crab Nebula with the supernova of 1054 A.D. Part II. The astronomical aspects', *Publ. Astron. Soc. Pacific* **54**, 95.

Reber, G. 'Cosmic static', *Proc. IRE* **30**, 367.

1944

* Reber, G. 'Cosmic static', *Astrophys. J.* **100**, 279 (Paper 6).

1945

Appleton, E. V., 'Departure of long-wave solar radiation from blackbody intensity', *Nature* **156**, 534.

Bakker, C. J., 'The reception of radio waves from space' [in Dutch], *Ned. Tijd. Natuurkunde* **11**, 201.

* Hulst, van de, H. C., 'The origin of radio waves from space' [in Dutch], *Ned. Tijd. Natuurkunde* **11**, 210 (Paper 34).

* Southworth, G. C., 'Microwave radiation form the sun', *J. Franklin Institute* **239**, 285 (Paper 17); (erratum: **241**, following p. 166 (1946)).

1946

Appleton, E. V. and Hey, J. S. (a), 'Circular polarization of solar radio noise', *Nature* **158**, 339.

Appleton, E. V. and Hey, J. S. (b), 'Solar radio noise', *Phil. Mag.*, Ser. 7, **37**, 73.

* Dicke, R. H., 'The measurement of thermal radiation at microwave frequencies', *Rev. Sci. Instr.* **17**, 268 (Paper 11).

* Dicke, R. H. and Beringer, R., 'Microwave radiation from the sun and moon', *Astrophys. J.* **103**, 375 (Paper 22).

Dicke, R. H., Beringer, R., Kyhl, R. L., and Vane, A. B., 'Atmospheric absorption measurements with a microwave radiometer', *Phys. Rev.* **70**, 340.

Ginzburg, V. L., 'On solar radiation in the radio spectrum' [in Russian], *Dokl. Akad. Nauk SSSR* **52**, 487.

* Hey, J. S., 'Solar radiations in the 4–6 metre radio wave-length band', *Nature* **157**, 47 (Paper 16).
* Hey, J. S., Parsons, S. J., and Phillips, J. W., 'Fluctuations in cosmic radiation at radio-frequencies', *Nature* **158**, 234 (Paper 23).
Hey, J. S., Phillips, J. W. and Parsons, S. J., 'Cosmic radiations at 5 metres wavelength', *Nature* **157**, 296.
Hey, J. S. and Stewart, G. S., 'Derivation of meteor stream radiants by radio reflection methods', *Nature* **158**, 481.
Martyn, D. F. (a), 'Polarization of solar radio-frequency emissions', *Nature* **158**, 308.
* Martyn, D. F. (b), 'Temperature radiation from the quiet sun in the radio spectrum', *Nature* **158**, 632 (Paper 20).
Moxon, L. A., 'Variation of cosmic radiation with frequency', *Nature* **158**, 758.
* Pawsey, J. L., 'Observation of million degree thermal radiation from the sun at a wave-length of 1.5 metres', *Nature* **158**, 633 (Paper 21).
* Ryle, M. and Vonberg, D. D., 'Solar radiation on 175 Mc/s', *Nature* **158**, 339 (Paper 18).
Saha, M. N., 'Origin of radio waves from the sun and the stars', *Nature* **158**, 717.
Sander, K. F., 'Measurements of galactic noise at 60 Mc/s', *J. Inst. Elec. Eng.* **93**, 1487.
Shklovsky, I. S., 'On the radiation of radio waves by the galaxy and by upper layers of the solar atmosphere' [in Russian], *Astron. Zh.* **23**, 333.
Unsöld, A., 'The cosmic short-wave radiation' [in German], *Naturwiss.* **33**, 37.

1947
Elder, F. R., Gurewitsch, A. M., Langmuir, R. V., and Pollock, H. C., 'Radiation from electrons in a synchrotron', *Phys. Rev.* **71**, 829.
* McCready, L. L., Pawsey, J. L., and Payne-Scott, R., Solar radiation at radio frequencies and its relation to sunspots', *Proc. Roy. Soc.* **A190**, 357 (Paper 19).
Reber, G. and Greenstein, J. L., 'Radio-frequency investigations of astronomical interest', *Observatory* **67**, 15.
Schott, E., '175-MHz radiation from the sun' [in German], *Phys. Blatt.* **3**, 159.

1948
Bolton, J. G., 'Discrete sources of galactic radio-frequency noise', *Nature* **162**, 141.
* Bolton, J. G. and Stanley, G. J. (a), 'Variable source of radio-frequency radiation in the constellation of Cygnus', *Nature* **161**, 312 (Paper 24).
Bolton, J. G. and Stanley, G. J. (b), 'Observations on the variable source of cosmic radio frequency radiation in the constellation of Cygnus', *Aust. J. Sci. Res.* **A1**, 58.
Herbstreit, J. W. and Johler, J. R., 'Frequency variation of the intensity of cosmic radio noise', *Nature* **161**, 515.
Hey, J. S., Phillips, J. W., and Parsons, S. J., 'An investigation of galactic radiation in the radio spectrum', *Proc. Roy. Soc.* **A192**, 425.
Martyn, D. F., 'Solar radiation in the radio spectrum', *Proc. Roy. Soc.* **A193**, 44.
Northcott, R. J. and Williamson, R. E., 'Galactic noise and the plane of the galaxy', *J. Roy. Astron. Soc. Canada* **42**, 269.
* Reber, G. (a), 'Cosmic static', *Proc. IRE* **36**, 1215 (Paper 7).
Reber, G. (b), 'Cosmic radio noise', *Radio News* (*Radio-Electronic Engineering* edition) **11**, No. 1 (July), 3.

* Ryle, M. and Smith, F. G., 'A new intense source of radio-frequency radiation in the constellation of Cassiopeia', *Nature* **162**, 462 (Paper 25).

Ryle, M. and Vonberg, D. D., 'An investigation of radio frequency radiation from the Sun', *Proc. Roy. Soc.* **A193**, 98.

1949

* Bolton, J. G., Stanley, G. J. and Slee, O. B. 'Positions of three discrete sources of galactic radio-frequency radiation', *Nature* **164**, 101 (Paper 26).

Piddington, J. H. and Minnett, H. C., 'Microwave thermal radiation from the moon', *Aust. J. Sci. Res.* **A2**, 63.

* Shklovsky, I. S., 'Monochromatic radio emission from the galaxy and the possibility of its observation' [in Russian], *Astron. Zh.* **26**, 10 (Paper 35).

Struve, O., 'Progress in radio astronomy – I and II', *Sky and Telescope* **9**, 27 + 55 (Dec. 1949 and Jan. 1950).

Unsöld, A., 'On the origin of the radio-frequency and shortwave radiation in the Milky Way' (in German), *Z. Astrophys.* **26**, 176.

1950

* Alfvén, H. and Herlofson, N., 'Cosmic radiation and radio stars', *Phys. Rev.* 78, 616 (Paper 27).

Allen, C. W. and Gum, C. S., 'Survey of galactic radio noise at 200 Mc/s', *Aust. J. Sci. Res.* **A3**, 224.

Bolton, J. G. and Westfold, K. C., 'Galactic radiation at radio frequencies I. 100-Mc/s survey', *Aust. J. Sci. Res.* **A3**, 19.

Hanbury Brown, R. and Hazard, C., 'Radio-frequency radiation from the great nebula in Andromeda (M31)', *Nature* **166**, 901.

* Kiepenheuer, K. O., 'Cosmic rays as the source of general galactic radio emission', *Phys. Rev.* **79**, 738 (Paper 9).

Little, C. G. and Lovell, A. C. B., 'Origin of the fluctuations in the intensity of radio waves from galactic sources', *Nature* **165**, 423.

Ryle, M., 'Radio astronomy', *Rept. Prog. Phys.* **13**, 184.

Ryle, M., Smith, F. G. and Elsmore, B., 'A preliminary survey of the radio stars in the northern hemisphere', *Monthly Notices Roy. Astron. Soc.* **110**, 508, (erratum: **111**, 641 (1951)).

Smith, F. G., 'Origin of the fluctuations in the intensity of radio waves from galactic sources', *Nature* **165**, 422.

Stanier, H. M., 'Distribution of radiation from the undisturbed Sun at a wavelength of 60 cm', *Nature* **165**, 354.

Stanley, G. J. and Slee, O. B., 'Galactic radiation at radio frequencies II. The discrete sources', *Aust. J. Sci. Res.* **A3**, 234.

Wild, J. P. and McCready, L. L., 'Observations of the spectrum of high intensity solar radiation at metre wavelengths. I. The apparatus and spectral types', *Aust. J. Sci. Res.* **A3**, 387.

1951

Dewhirst, D. W., (report of a discussion during the 12 October 1951 meeting of the Royal Astronomical Society), *Observatory* **71**, 209.

* Ewen, H. I. and Purcell, E. M., 'Radiation from galactic hydrogen at 1420 Mc/s', *Nature* **168**, 356 (Paper 36).

* Ginzburg, V. L, 'Cosmic rays as the source of galactic radio emission' [in Russian], *Dokl. Akad. Nauk SSSR* **76**, 377 (Paper 10).

Hanbury Brown, R. and Hazard, C., 'Radio emission from the Andromeda Nebula', *Monthly Notices Roy. Astron. Soc.* **111**, 357.

Machin, K. E., Distribution of radiation across the solar disk at a frequency of 81.5 Mc/s', *Nature* **167**, 889.

Mills, B. Y. and Thomas, A. B., 'Observations of the source of radio-frequency radiation in the constellation of Cygnus', *Aust. J. Sci. Res.* **A4**, 158.

* Muller, C. A. and Oort, J. H., 'The interstellar hydrogen line at 1420 Mc/s and an estimate of galactic rotation', *Nature* **168**, 357 (Paper 37).

Ryle, M. and Elsmore, B., 'A search for long-period variations in the intensity of radio stars', *Nature* **168**, 555.

* Smith, F. G. (a), 'An accurate determination of the positions of four radio stars', *Nature* **168**, 555 (Paper 28).

Smith, F. G. (b), 'An attempt to measure the annual parallax or proper motion of four radio stars', *Nature* **168**, 962.

Westerhout, G. and Oort, J. H., 'A comparison of the intensity distribution of radio-frequency radiation with a model of the galactic system', *Bull Astron. Inst. Netherlands* **11**, 323.

1952

Christiansen, W. N. and Hindman, J. V., 'A preliminary survey of 1420 Mc/s line emission from galactic hydrogen', *Aust. J. Sci. Res.* **A5**, 437.

Getmantsev, G. G., 'Cosmic electrons as a source of the radio emission from the Galaxy' [in Russian], *Dokl. Akad. Nauk SSSR* 83, 557.

*Hanbury Brown, R., Jennison, R. C., and Das Gupta, M. K., 'Apparent angular sizes of discrete radio sources', *Nature* **170**, 1061 (Paper 30).

Hutchinson, G. W., 'On the possible relation of galactic radio noise to cosmic rays', *Phil. Mag.* Ser. 7 **43**, 847.

Mills, B. Y. (a), 'The distribution of the discrete sources of cosmic radio radiation', *Aust. J. Sci. Res.* **A5**, 266; (erratum: *Austral. J. Phys.* **6**, 125 (1953)).

* Mills, B.Y. (b), 'Apparent angular sizes of discrete radio sources', *Nature* **170**, 1063 (Paper 31).

Mills, B.Y. (c), 'The positions of 6 discrete sources of cosmic radio radiation', *Aust. J. Sci. Res.* **A5**, 456.

Morgan, W. W., Sharpless, S., and Osterbrock, D., 'Some features of galactic structure in the neighborhood of the sun', *Astron. J.* 57, 3.

Oort, J. H., 'Problems of galactic structure', *Astrophys. J.* **116**, 233.

* Ryle, M., 'A new radio interferometer and its application to the observation of weak radio stars', *Proc. Roy. Soc.* **A211**, 351 (Paper 12).

Shklovsky, I. S. (a), 'On the nature of the radio emission from the Galaxy' [in Russian], *Astr. Zh.* **29**, 418.

Shlovsky, I. S. (b), 'Radio spectroscopy of the Galaxy' [in Russian], *Astr. Zh.* **29**, 144.

* Smith, F. G. (a), 'Apparent angular sizes of discrete radio sources', *Nature* **170**, 1065 (Paper 32).

Smith, F. G. (b), 'The determination of the position of a radio star', *Monthly Notices Roy. Astron. Soc.* **112**, 497.

Wild, J. P., 'The radio-frequency line spectrum of atomic hydrogen and its applications in astronomy', *Astrophys. J.* **115**, 206.

1953

Christiansen, W. N., 'A high-resolution aerial for radio astronomy', *Nature* **171**, 831.

* Jennison, R. C. and Das Gupta, M. K., 'Fine structure of the extra-terrestrial radio source Cygnus I', *Nature* **172**, 996 (Paper 33).

Mills, B. Y. and Little, A. G., 'A high resolution aerial system of a new type', *Aust. J. Phys.* **6**, 272.

O'Brien, P. A., 'Distribution of radiation across the solar disk at metre wavelengths', *Monthly Notices Roy. Astron. Soc.* **113**, 597.

Shklovsky, I. S. (a), 'The problem of cosmic radio waves' [in Russian], *Astron. Zh.* **30**, 15.

Shklovsky, I. S. (b), 'The possibility of observing the monochromatic radio radiation of interstellar molecules' [in Russian], *Dokl. Akad. Nauk SSSR* **92**, 25.

Shklovsky, I. S. (c), 'On the nature of the luminescense of the Crab Nebula' [in Russian], *Dokl. Akad. Nauk SSSR* **90**, 983.

1954

* Baade, W. and Minkowski, R. (a), 'Identification of the radio sources in Cassiopeia, Cygnus A. and Puppis A', *Astrophys. J.* **119**, 206 (Paper 29).

Baade, W. and Minkowski, R. (b), 'On the identification of radio sources', *Astrophys. J.* **119**, 215.

Dombrovsky, V. A., 'On the nature of the radiation from the Crab Nebula' [in Russian], *Dokl. Akad. Nauk SSSR* **94**, 1021.

Hanbury Brown, R., 'The remnants of supernovae as radio sources in the galaxy', *Observatory* **74**, 185.

Hanbury Brown, R. and Twiss, R. Q., 'A new type of interferometer for use in radio astronomy', *Phil. Mag.* Ser. 7 **45**, 663.

Hulst, van de, H. C., Muller, C. A., and Oort, J. H., 'The spiral structure of the outer part of the galactic system derived from the hydrogen emission at 21-cm wavelength', *Bull. Astron. Inst. Netherlands* **12**, 117.

Kwee, K. K., Muller, A. C. and Westerhout, G., 'The rotation of the inner parts of the galactic system', *Bull. Astron. Inst. Netherlands* **12**, 211.

Minkowski, R. and Aller, L. H., 'The spectrum of the radio source in Cassiopeia', *Astrophys. J.* **119**, 232.

Minkowski, R. and Greenstein, J. L., 'The power radiated by some discrete sources of radio noise', *Astrophys. J.* **119**, 238.

Shklovsky, I. S., 'Identification of the strongest discrete sources of radio radiation in the galaxy with the remnants of supernovae which exploded in the last 200 years' [in Russian], *Dokl. Akad. Nauk SSSR* **94**, 417.

Vashakidze, M. A., 'On the degree of polarization of the light near extragalactic nebulae and the Crab Nebula' [in Russian], *Astron. Tsirk.*, No 147, 11.

Williams, D. R. W. and Davies, R. D., 'A method for the measurement of the distance of radio stars, *Nature* **173**, 1182.

1955

Christiansen, W. N. and Warburton, J. A., 'The distribution of radio brightness over the solar disk at a wavelength of 21-cm. III. The quiet sun. Two-dimensional observations', *Aust. J. Phys.* **8**, 474.

Hagen, J. P., Lilley, A. E., and McClain, E. F., 'Absorption of 21 cm radiation by interstellar hydrogen', *Astrophys. J.* **122**, 361.

Hanbury Brown, R. and Lovell, A. C. B., 'Large radio telescopes and their use in radio astronomy', in A. Beer (ed.), *Vistas in Astronomy*, Vol. 1, Pergamon, London, p. 542.

Pawsey, J. L., 'A catalogue of reliably known discrete sources of cosmic radio waves', *Astrophys. J.* **121**, 1.

Pawsey, J. L. and Bracewell, R. N., *Radio Astronomy*, Clarendon Press, Oxford.

Ryle, M., Radio stars and their cosmological significance'. *Observatory* **75**, 137.

Ryle, M. and Scheuer, P. A. G., 'The spatial distribution and the nature of radio stars', *Proc. Roy. Soc.* **A230**, 448.

Shakeshaft, J. R., Ryle, M., Baldwin, J. E., Elsmore, B., and Thomson, J. H., 'A survey of radio sources between declinations –38° and +83°', *Mem. Roy. Astron. Soc.* **67**, 106.

1956

Dröge, F. and Priester, W., 'Distribution of galactic radiation at 200 Mhz' (in German), *Z. Astrophys.* **40**, 236.

Piddington, J. H. and Trent, G. H., 'A survey of cosmic radio emission at 600 Mc/s', *Aust J. Phys.* **9**, 481.

Southworth, C. G., 'Early history of radio astronomy', *Sci. Monthly* **82**, 55.

1957

Blythe, J. H., 'A new type of pencil beam aerial for radio astronomy', *Monthly Notices Roy. Astron. Soc.* **117**, 644.

Mayer, C. H., McCullough, T.P., and Sloanaker, R. M., 'Evidence for polarized radio radiation from the Crab Nebula', *Astrophys. J.* **126**, 468.

Mills, B. Y. and Slee, O. B., 'A preliminary survey of radio sources in a limited region of the sky at a wavelength of 3.5 m', *Aust. J. Phys.* **10**, 162.

Ryle, M., 'The Mullard Radio Astronomy Observatory, Cambridge', *Nature* **180**, 110.

Townes, C. H., 'Microwave, and radio frequency resonance lines of interest to radio astronomy', in H. C. van de Hulst (ed.), *Radio Astronomy*, I.A.U. Symp. No. 4, Cambridge Univ. Press, p. 92 (paper presented August, 1955).

1958

Oort, J. H., Kerr, F. J., and Westerhout, G., 'The galactic system as a spiral nebula', *Monthly Notices Roy. Astron. Soc.* **118**, 379.

Reber, G., 'Early radio astronomy at Wheaton, Illinois', *Proc. IRE* **46**, 15.

Ryle, M., 'The nature of cosmic radio sources', *Proc. Roy. Soc.* **A248**, 289.

Shane, C. D., 'Radio astronomy in 1890: a proposed experiment', *Pub. Astron. Soc. Pacific* **70**, 303 (reprint of a letter sent from A. E. Kennelly to E. S. Holden).

1959

Kardashev, N. S., 'On the possibility of detection of allowed lines of atomic hydrogen in the radio frequency spectrum', *Sov. Astron.- A. J.* **3**, 813 = *Astron. Zh.* **36**, 838 [in Russian].

1960

Ryle, M. and Hewish, A., 'The synthesis of large radio telescopes', *Monthly Notices Roy. Astron. Soc.* **120**, 220.

1961

Earl, J. A., 'Cloud-chamber observations of primary cosmic-ray electrons', *Phys. Rev. L.* **6**, 125.

Meyer, P. and Vogt, R., 'Electrons in the primary cosmic radiation', *Phys. Rev. L.* **6**, 193.

1973

Hey, J. S., *The Evolution of Radio Astronomy*, Elek Science, London.

Hanbury Brown, R., *The Intensity Interferometer*, Taylor and Francis, London.

1976

Edge, D. O. and Mulkay, M., *Astronomy Transformed: The Emergence of Radio Astronomy in Britain*, Wiley-Interscience, New York.

1978

Sullivan, W. T., III, 'A new look at Karl Jansky's original data', *Sky and Telescope* **56**, 101.

1979

Lang, K. R. and Gingerich, O., *A Source Book in Astronomy and Astrophysics, 1900–1975*, Harvard Univ. Press, Cambridge.

NAME INDEX

Italic figures in parentheses denote paper numbers

SUBJECT INDEX

Italic figures in parentheses denote paper numbers

STUDIES IN THE HISTORY OF MODERN SCIENCE

Editors:

1. F. Gregory, *Scientific Materialism in Nineteenth Century Germany*. 1977, XXIII+279 pp.
2. E. Manier, *The Young Darwin and His Cultural Circle*. 1978, XII+242 pp.
3. G. Canguilhem, *On the Normal and the Pathological*. 1978, XXIV+208 pp.
4. H. C. Kennedy, *Peano*. 1980, XII+230 pp.
5. L. Badash, J. O. Hirschfelder, and H. P. Broida (eds.), *Reminiscences of Los Alamos, 1943–1945*. 1980, XXI+188 pp.
6. L. L. Bucciarelli and N. Dworsky, *Sophie Germain: An Essay in the History of the Theory of Elasticity*. 1980, XI+147 pp.
7. D. J. Struik, *The Land of Stevin and Huygens*. 1981, XIV+162 pp.
8. I. Langham, *The Building of British Social Anthropology*. 1981, XXXII+392 pp.
9. Z. Bechler (ed.), *Contemporary Newtonian Research*. 1982.

GPSR Compliance

The European Union's (EU) General Product Safety Regulation (GPSR) is a set of rules that requires consumer products to be safe and our obligations to ensure this.

If you have any concerns about our products, you can contact us on ProductSafety@springernature.com

In case Publisher is established outside the EU, the EU authorized representative is:

Springer Nature Customer Service Center GmbH
Europaplatz 3
69115 Heidelberg, Germany

Batch number: 10815974

Printed by Printforce, the Netherlands